Realities and Impacts of Climate Change

Realities and Impacts of Climate Change

Edited by **Daisy Mathews**

R CALLISTO
REFERENCE

New York

Published by Callisto Reference,
106 Park Avenue, Suite 200,
New York, NY 10016, USA
www.callistoreference.com

Realities and Impacts of Climate Change
Edited by Daisy Mathews

© 2015 Callisto Reference

International Standard Book Number: 978-1-63239-527-6 (Hardback)

Printed in the United States of America.

Contents

Preface

This book has been an outcome of determined endeavour from a group of educationists in the field. The primary objective was to involve a broad spectrum of professionals from diverse cultural background involved in the field for developing new researches. The book not only targets students but also scholars pursuing higher research for further enhancement of the theoretical and practical applications of the subject.

The earth is the only planet where life exists. The evolution of human life was only possible due to its favorable environmental conditions i.e. availability of water, oxygen and suitable temperature. Climate change is a significant and lasting change in the statistical distribution of weather patterns over periods ranging from decades to millions of years. Hence, the field of climate control requires special attention to control the rise in atmospheric temperature by controlling the emission of greenhouse gases. There is a need to conserve natural resources and look for some alternative to these resources. The book provides basic information and facts about climate change with special focus on its impact on vegetation, adapting rain fed agriculture, climate change assessment due to soil moisture change and adaptation to future climate change for crop farms. Additionally, latest technologies applied and adopted to reduce the impact of climate change are also discussed.

It was an honour to edit such a profound book and also a challenging task to compile and examine all the relevant data for accuracy and originality. I wish to acknowledge the efforts of the contributors for submitting such brilliant and diverse chapters in the field and for endlessly working for the completion of the book. Last, but not the least; I thank my family for being a constant source of support in all my research endeavours.

Editor

Climate Change Realities and Its Evidences

A Study About Realities of Climate Change: Glacier Melting and Growing Crises

Bharat Raj Singh and Onkar Singh

Additional information is available at the end of the chapter

1. Introduction

Climate change has ceased to be a scientific curiosity since long, and is no longer just one of many environmental and regulatory concerns. As the Secretary General of United Nations has said, it is the major, overriding environmental issue of our time, and the single greatest challenge faced by environmental regulators. It is a growing crisis with economic, health and safety, food production, security, and other dimensions.

Climate change is expected to hit developing countries the hardest. Its effects; higher temperatures, changes in precipitation patterns, rising sea levels, and more frequent weather-related disasters-pose risks for agriculture, food, and water supplies. The fight against poverty, hunger and disease, and the lives and livelihoods of billions of people in developing countries are at stake. Tackling this immense challenge must involve both mitigation-to avoid the unmanageable and adaptation- to manage the unavoidable while maintaining a focus on its social dimensions.

1.1. What is weather and climate?

The weather, as we experience it, is the fluctuating state of the atmosphere around us, characterised by the temperature, wind, precipitation, clouds and other weather elements. This weather is the result of rapidly developing and decaying weather systems such as mid-latitude low and high pressure systems with their associated frontal zones, showers and tropical cyclones. Weather has only limited predictability. Mesoscale convective systems are predictable over a period of hours only; synoptic scale cyclones may be predictable over a period of several days to a week. Beyond a week or two individual weather systems are unpredictable.

Climate- It refers to the average weather in terms of the mean and its variability over a certain time-span and a certain area. Classical climatology provides a classification and description of the various climate regimes found on the Earth. It varies from place to place, depending on latitude, distance to the sea, vegetation, presence or absence of mountains or other geographical factors. Climate also varies with time; from season to season, year to year, decade to decade or on much longer time-scales, such as the Ice Ages. Statistically significant variations of the mean state of the climate or of its variability, typically persisting for decades or longer, are referred to as "climate change".

Climate variations and change, caused by external factors, may be partly predictable, particularly on the larger, continental and global, spatial scales. Because human activities, such as the emission of greenhouse gases or change in land-use, do result in external forces, it is believed that the large-scale aspects of human-induced climate change are also partly predictable. However the ability to actually do so is limited because we cannot accurately predict population change, economic change, technological development, and other relevant characteristics of future human activity. Therefore, one has to rely on carefully constructed scenarios of human behaviour and determine climate projections on the basis of such scenarios.

Climate variables-The traditional knowledge of weather and climate focuses on those variables that affect daily life directly i.e.; average, maximum and minimum temperature, wind near the surface of the Earth, precipitation in its various forms, humidity, cloud type and amount, and solar radiation. These are the variables observed hourly by a large number of weather stations around the globe.

However, this is only part of the reality that determines weather and climate. The growth, movement and decay of weather systems also depend on the vertical structure of the atmosphere, the influence of the underlying land and sea and many other factors not directly experienced by human beings. Climate is determined by the atmospheric circulation and by its interactions with the large-scale ocean currents and the land with its features such as albedo, vegetation and soil moisture. The climate of the Earth as a whole depends on factors that influence the radiative balance, such as for example, the atmospheric composition, solar radiation or volcanic eruptions. To understand the climate of our planet Earth and its variations and to predict the changes of the climate brought about by human activities, one cannot ignore any of these many factors and components that determine the climate. We must understand the climate system, the complicated system consisting of various components, including the dynamics and composition of the atmosphere, the ocean, the ice and snow cover, the land surface and its features, the many mutual interactions between them, and the large variety of physical, chemical and biological processes taking place in and among these components. "Climate" in a wider sense refers to the state of the climate system as a whole, including a statistical description of its variations.

1.2. What is greenhouse effect?

A natural system known as the "greenhouse effect" regulates temperature on the Earth. Just as glass in a greenhouse keeps heat in, our atmosphere traps the sun's heat near earth's surface, primarily through heat-trapping properties of certain "greenhouse gases". Earth is heated by sunlight and most of the sun's energy passes through the atmosphere, to warm the earth's surface, oceans and atmosphere. However, in order to keep the atmosphere's energy budget in balance, the warmed earth also emits heat energy back to space as infrared radiation. As this energy radiates upward, most is absorbed by clouds and molecules of greenhouse gases in the lower atmosphere. These re-radiate the energy in all directions, some back towards the surface and some upward, where other molecules higher up can absorb the energy again. This process of absorption and re-emission is repeated until; finally, the energy does escape from the atmosphere to space. However, because much of the energy has been recycled downward, surface temperatures become much warmer than if the greenhouse gases were absent from the atmosphere. This natural process is known as the greenhouse effect. Without greenhouse gases, Earth's average temperature would be -19°C instead of +14°C, or 33°C colder. Over the past 10,000 years, the amount of greenhouse gases in our atmosphere has been relatively stable. Then a few centuries ago, their concentrations began to increase due to the increasing demand for energy caused by industrialization and rising populations, and due to changing land use and human settlement patterns.

1.3. What are greenhouse gases?

Water vapour is the most common constituent of greenhouse gases. But others are equally important and some occur naturally while some come from human activity. Carbon Dioxide or CO_2 is the significant greenhouse gas released by human activities, mostly through the burning of fossil fuels. It is the main contributor to climate change.

Methane is produced when vegetation is burned, digested or rotted with no oxygen present. Garbage dumps, rice paddies, and grazing cows and other livestock release lots of methane.

Nitrous oxide can be found naturally in the environment but human activities are increasing the amounts. Nitrous oxide is released when chemical fertilizers and manure are used in agriculture.

Halocarbons are a family of chemicals that include CFCs (which also damage the ozone layer), and other human-made chemicals that contain chlorine and fluorine. Since greenhouse gases make up such a small percentage of the atmosphere, why do changes in their concentrations have such a big effect on climate?

Most greenhouse gases are extremely effective at absorbing heat escaping from the earth and keeping it trapped. In other words, it takes only small amounts of these gases to significantly change the properties of the atmosphere. 99% of the dry atmosphere consists of nitrogen and oxygen, which are relatively transparent to the sunlight and infrared energy, and have little effect on the flow of the sunlight and heat energy through the air. By comparison, the atmospheric greenhouse gases that cause the earth's natural greenhouse effect total less than 1% of the atmosphere. But that tiny amount increases the earth's average surface temperature from -19°C to +14°C - a difference of about 3°C. A little bit of greenhouse gas goes a long way. Because the concentration of greenhouse gases in the atmosphere is so low, human emissions can have a significant effect. For example, human emissions of carbon dioxide (CO_2)

currently amount to roughly 28 billion tonnes per year. Over the next century human emissions will increase the concentration of carbon dioxide in the atmosphere from about 0.03% today to almost certainly 0.06% (a doubling), and possibly to 0.09% (a tripling).

1.4. What causes climate change?

Earth's climate changes naturally and such changes in the intensity of sunlight reaching the earth cause cycles of warming and cooling that have been a regular feature of the Earth's climatic history. Some of these solar cycles - like the four glacial-interglacial swings during the past 400,000 years - extend over very long time scales and can have large amplitudes of 5 to 6°C. For the past 10,000 years, the earth has been in the warm interglacial phase of such a cycle. Other solar cycles are much shorter, with the shortest being the 11 year sunspot cycle. Other natural causes of climate change include variations in ocean currents (which can alter the distribution of heat and precipitation) and large eruptions of volcanoes (which can sporadically increase the concentration of atmospheric particles, blocking out more sunlight). Still, for thousands of years, the Earth's atmosphere has changed very little. Temperature and the balance of heat-trapping greenhouse gases have remained just right for humans, animals and plants to survive. But today we're having problems keeping this balance, because we burn fossil fuels to heat our homes, run our cars, produce electricity, and manufacture all sorts of products, we're adding more greenhouse gases to the atmosphere. By increasing the amount of these gases, the warming capability of the natural greenhouse effect is enhanced. It's the human-induced enhanced greenhouse effect that causes environmental concern, because it has the potential to warm the planet at a rate that has never been experienced in human history.

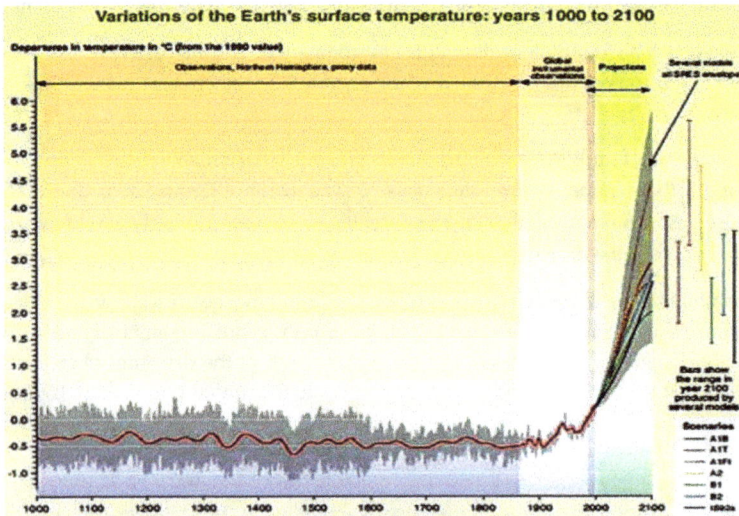

Figure 1. Variations in average surface temperature from year 1000 to year 1860; year 1860 to 2000 annually averaged and year 2000-2100 projected average temperature

From year 1000 to year 1860 variations in average surface temperature of the Northern Hemisphere are shown in **Fig. 1**, but corresponding data from the Southern Hemisphere was not available and hence it was reconstructed from proxy data (tree rings, corals, ice cores, and historical records). The line shows the 50-year average, the grey region the 95% confidence limit in the annual data. From years 1860 to 2000 are shown variations in observations of globally and annually averaged surface temperature from the instrumental record; the line shows the decadal average. From years 2000 to 2100 projections of globally averaged surface temperature are shown for the six illustrative SRES scenarios and IS92a using a model with average climate sensitivity. The grey region marked "several models all SRES envelope" shows the range of results from the full range of 35 SRES scenarios in addition to those from a range of models with different climate sensitivities. The temperature scale is departure from the 1990 value.

1.5. What happens due to climate change?

Shifting weather patterns, threaten food production through increased unpredictability of precipitation, rising sea levels contaminate coastal freshwater reserves and increase the risk of catastrophic flooding, and a warming atmosphere aids the pole-ward spread of pests and diseases once limited to the tropics. Ice-loss from glaciers and ice sheets has continued, leading, for example, to the second straight year with an ice-free passage through Canada's Arctic islands, and accelerating rates of ice-loss from ice sheets in Greenland and Antarctica. Combined with thermal expansion—warm water occupies more volume than cold—the melting of ice sheets and glaciers around the world is contributing and an ultimate extent of sea-level rise that could far outstrip those anticipated in the most recent global scientific assessment.

2. Alarming evidences due to climate change

There is alarming evidence that important tipping points, leading to irreversible changes in major ecosystems and the planetary climate system, may already have been reached or passed. Ecosystems as diverse as the Amazon rainforest and the Arctic tundra, for example, may be approaching thresholds of dramatic change through warming and drying. Mountain glaciers are in alarming retreat and the downstream effects of reduced water supply in the driest months will have repercussions that transcend generations. Climate feedback systems and environmental cumulative effects are building across Earth systems demonstrating behaviours we cannot anticipate.

2.1. Deforestation and climate change

Forests are vital for life, home to millions of species, they protect soil from erosion, produce oxygen, store carbon dioxide, and help control climate. Forests are also vital for us to live as they provide us with food, shelter and medicines as well as many other useful things. They also purify the air we breathe and water that we need to survive. Deforestation by humans is causing reduction in all of these necessary functions, and hence damaging the atmosphere even further.

Forests play a huge role in the carbon cycle on our planet. When forests are cut down, not only does carbon absorption cease, but also the carbon stored in the trees is released into the atmosphere as CO_2 if the wood is burned or even if it is left to rot after the deforestation process. Smaller crops e.g. plants and agricultural crops also draw in carbon dioxide and release oxygen, however forests store up to 100 times more carbon than agricultural fields of the same area. Deforestation is an important factor in global climate change. Climate change is because of a buildup of carbon dioxide in out atmosphere and if we carry on cutting down the main tool we have to diminish this CO_2 build up, we can expect the climate of our planet to change dramatically over the next decades.

It is estimated that more than 1.5 billion tons of carbon dioxide is released to the atmosphere due to deforestation, mainly the cutting and burning of forests, every year. Over 30 million acres of forests and woodlands are lost every year due to deforestation; causing a massive loss of income to poor people living in remote areas who depend on the forest to survive.

3. Risk due to climate change

Climate vulnerability and risk management is a part of dialogue and work with developing countries. Key sectors affected by climate change include health, water supply and sanitation, energy, transport, industry, mining, construction, trade, tourism, agriculture, forestry, fisheries, environmental protection, and disaster management as detailed ahead.

3.1. Potential health impact due to climate change

Change in world climate would influence the functioning of many ecosystems and their member species. Likewise, there would be impacts on human health. Some of these health impacts would be beneficial. For example, milder winters would reduce the seasonal winter-time peak in deaths that occurs in temperate countries, while in currently hot regions a further increase in temperatures might reduce the viability of disease-transmitting mosquito populations. Overall, however, scientists consider that most of the health impacts of climate change would be adverse.

Climatic changes over recent decades have probably already affected some health outcomes. Indeed, the World Health Organisation estimated, in its "World Health Report 2002", that climate change was estimated to be responsible in 2000 for approximately 2.4% of worldwide diarrhoea, and 6% of malaria in some middle-income countries. However, small changes, against a noisy background of ongoing changes in other causal factors, are hard to identify. Once spotted causal attribution is strengthened; if there are similar observations in different population settings.

The first detectable changes in human health may well be alterations in the geographic range (latitude and altitude) and seasonality of certain infectious diseases – including vector-borne infections such as malaria and dengue fever, and food-borne infections (e.g. salmonellosis) which peak in the warmer months. Warmer average temperatures combined with increased climatic variability would alter the pattern of exposure to thermal extremes and resultant health impacts, in both summer and winter. By contrast, the public health consequences of the disturbance of natural and managed food-producing ecosystems, rising sea-levels and population displacement for reasons of physical hazard, land loss, economic disruption and civil strife, may not become evident for up to several decades.

3.2. Glacier melting

3.2.1. Greenland ice sheet may melt completely with 1.6 degrees of global warming[1]

The Greenland ice sheet is likely to be more vulnerable to global warming than previously thought. The temperature threshold for melting the ice sheet completely is in the range of 0.8 to 3.2 degrees Celsius of global warming, with a best estimate of 1.6 degrees above pre-industrial levels, shows a new study by scientists from the Potsdam Institute for Climate Impact Research (PIK) and the Universidad Complutense de Madrid. Today, already 0.8 degrees of global warming has been observed. Substantial melting of land ice could contribute to long-term sea-level rise of several meters and therefore it potentially affects the lives of many millions of people.

The time it takes before most of the ice in Greenland is lost strongly depends on the level of warming. "The more we exceed the threshold, the faster it melts," says Alexander Robinson, lead-author of the study now published in Nature Climate Change. In a business-as-usual scenario of greenhouse-gas emissions, in the long run humanity might be aiming at 8 degrees Celsius of global warming. This would result in one fifth of the ice sheet melting within 500 years and a complete loss in 2000 years, according to the study. "This is not what one would call a rapid collapse," says Robinson. "However, compared to what has happened in our planet's history, it is fast. And we might already be approaching the critical threshold."

In contrast, if global warming would be limited to 2 degrees Celsius, complete melting would happen on a timescale of 50.000 years. Still, even within this temperature range often considered a global guardrail, the Greenland ice sheet is not secure. Previous research suggested a threshold in global temperature increase for melting the Greenland ice sheet of a best estimate of 3.1 degrees, with a range of 1.9 to 5.1 degrees. The new study's best estimate indicates about half as much.

"Our study shows that under certain conditions the melting of the Greenland ice sheet becomes irreversible. This supports the notion that the ice sheet is a tipping element in the Earth system," says team-leader Andrey Ganopolski of PIK. "If the global temperature significantly overshoots the threshold for a long time, the ice will continue melting and not

---

[1] Science Daily (Mar. 11, 2012)

regrow -- even if the climate would, after many thousand years, return to its preindustrial state." This is related to feedbacks between the climate and the ice sheet: The ice sheet is over 3000 meters thick and thus elevated into cooler altitudes as shown in **Fig. 2(a)**. When it melts its surface comes down to lower altitudes with higher temperatures, which accelerates the melting. Also, the ice reflects a large part of solar radiation back into space. When the area covered by ice decreases, more radiation is absorbed and this adds to regional warming.

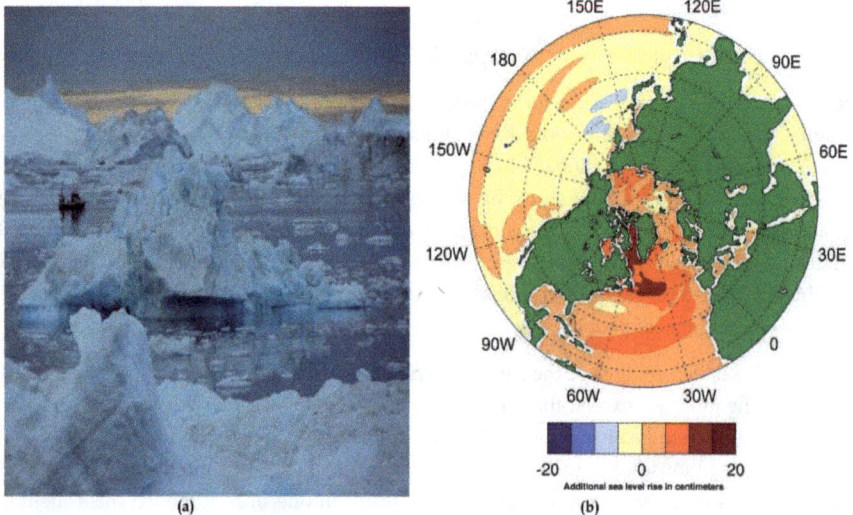

(a)   (b)

Figure 2. (a) The Greenland ice sheet is likely to be more vulnerable to global warming than previously thought. The temperature threshold for melting the ice sheet completely is in the range of 0.8 to 3.2 degrees Celsius of global warming, with a best estimate of 1.6 degrees above pre-industrial levels, shows a new study. Today, already 0.8 degrees global warming has been observed. (Credit: © Martin Schwan / Fotolia);
(b)This visualization, based on new computer modeling, shows that sea level rise may be an additional 10 centimeters (4 inches) higher by populated areas in northeastern North America than previously thought. Extreme northeastern North America and Greenland may experience even higher sea level rise. (Credit: Graphic courtesy Geophysical Research Letters, modified by UCAR)

The scientists achieved insights by using a novel computer simulation of the Greenland ice sheet and the regional climate. This model performs calculations of these physical systems including the most important processes, for instance climate feedbacks associated with changes in snowfall and melt under global warming. The simulation proved able to correctly calculate both the observed ice-sheet of today and its evolution over previous glacial cycles, thus increasing the confidence that it can properly assess the future. All this makes the new estimate of Greenland temperature threshold more reliable than previous ones as shown in **Fig. 2(b).**

3.2.2. Arctic sea ice shrinks to smallest extent ever recorded

Sea ice in the Arctic has shrunk to its smallest extent ever recorded, smashing the previous record minimum and prompting warnings of accelerated climate change. Satellite images show that the rapid summer melt has reduced the area of frozen sea to less than 3.5 million square kilometres this week from 27 August 2012 – less than half the area typically occupied four decades ago. Arctic sea ice cover has been shrinking since the 1970s when it averaged around 8m sq km a year, but such a dramatic collapse in ice cover in one year is highly unusual.

A record low in 2007 of 4.17 million sq km was broken on Monday, 27 August 2012; further melting has since amounted to more than 500,000 sq km. The record, which is based on a five-day average, is expected to be officially declared in the next few days by the National Snow and Ice Data Centre (NSIDC) in Colorado. The NSIDC's data shows the sea ice extent is bumping along the bottom, with a new low of 3.421m sq km on Tuesday, which rose very slightly to 3.429m sq km on Wednesday and 3.45m sq km on Thursday as seen in **Fig. 3**.

Arctic sea ice
Extent, millions of sq km where there is at least 15% sea ice

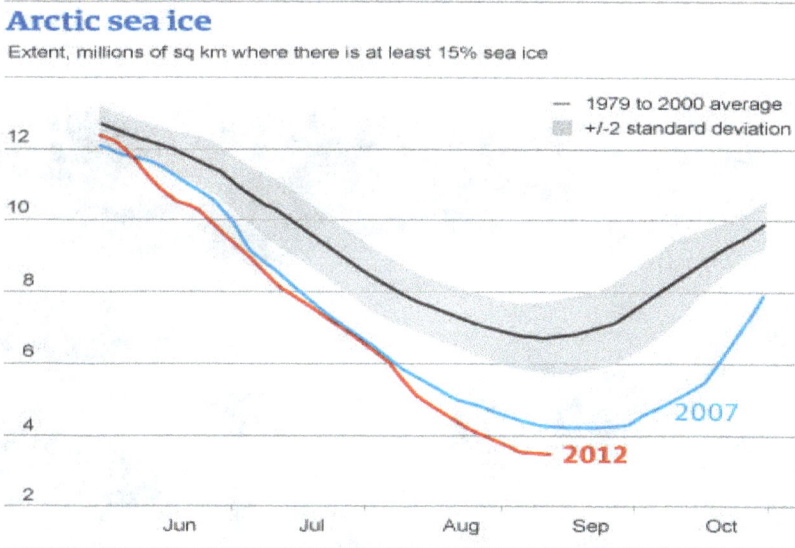

Figure 3. The shrinking of the ice cap was interpreted by environment groups as a signal of long-term global warming caused by man-made greenhouse gas emissions. A study published in July in the journal Environmental Research Letters, that compared model projections with observations, estimated that the radical decline in Arctic sea ice has been between 70-95% due to human activities.

Scientists have predicted on 31st August 2012 that the Arctic Ocean could be ice-free in summer months within 20 years, leading to possibly major climate impacts. "I am surprised. This is an indication that the Arctic sea ice cover is fundamentally changing. The trends all show less ice and thinner ice," said Julienne Stroeve, a research scientist with the NSIDC.

"We are on the edge of one of the most significant moments in environmental history as sea ice heads towards a new record low. The loss of sea ice will be devastating, raising global temperatures that will impact on our ability to grow food and causing extreme weather around the world," said John Sauven, director of Greenpeace UK.

Sea ice experts said that they were surprised by the collapse because weather conditions were not conducive to a major melt this year. The ice is now believed to be much thinner than it used to be and easier to melt.

Arctic sea ice follows an annual cycle of melting through the warm summer months and refreezing in the winter. The sea ice plays a critical role in regulating climate, acting as a giant mirror that reflects much of the Sun's energy, helping to cool the Earth.

David Nussbaum, chief executive of WWF-UK, said: "The disappearance of Arctic ice is the most visible warning sign of the need to tackle climate change and ensure we have a world fit to pass on to the next generation. The sheer scale of ice loss is shocking and unprecedented. This alarm call from the Arctic needs to reverberate across Whitehall and boardrooms. We can all take action to cut carbon emissions and move towards a 100% renewable economy."

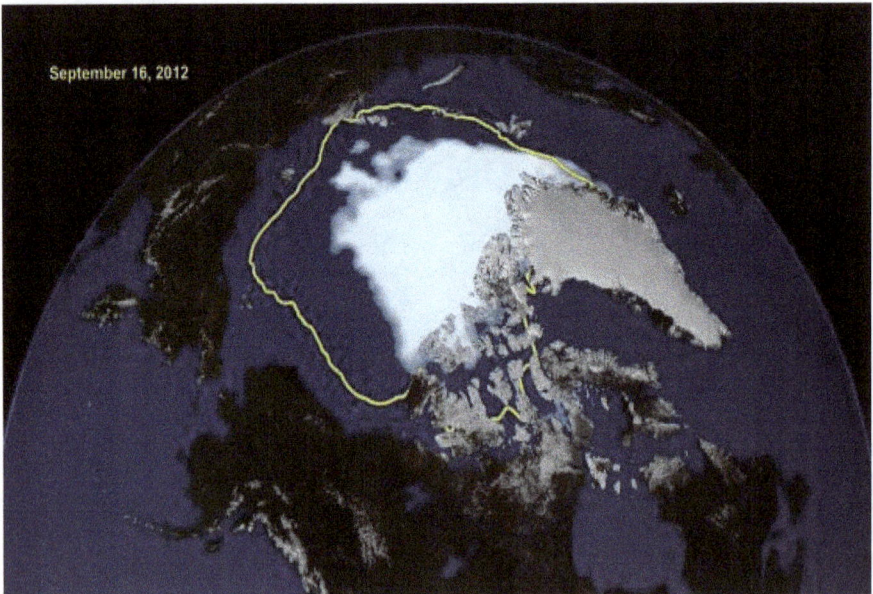

Figure 4. Arctic sea ice extent on 16 September 2012, in white, compared with the, Satellite data reveal how the new record low Arctic sea ice extent, from Sept. 16, 2012, compares to the average minimum extent over the past 30 years (in yellow) with reference to 1979, NASA/Goddard Scientific Visualization Studio

Ed Davey, the UK climate and energy secretary, said: "These findings highlight the urgency for the international community to act. We understand that Arctic sea-ice decline has accelerated over recent years as global warming continues to increase Arctic temperatures at a faster rate than the global average.

"This Government is working hard to tackle climate change and we are working closely with our international partners not to exceed 2 degrees above pre industrial levels. I am calling for the EU to increase its emission target from 20% to 30% and will be taking an active lead at the UNFCCC Climate change talks in Doha later this year, where I will push for further progress towards a new global deal on climate change and for more mitigation action now. The fact is that we cannot afford to wait".

Canadian scientists said that the record melt this year could lead to a cold winter in the UK and Europe, as the heat in the Arctic water will be released into the atmosphere this autumn, potentially affecting the all-important jet stream. While the science is still developing in this area, the Met Office said in May that the reduction in Arctic sea ice was contributing in part to the colder, drier winters the UK has been experiencing in recent years as shown in **Fig. 4**.

3.2.3. Loss of Arctic sea ice '70% man-made'

Study finds only 30% of radical loss of summer sea ice is due to natural variability in Atlantic – and it will probably get worse. Since the 1970s, there has been a 40% decrease in the extent of summer sea ice. Photograph: Alaska Stock/Corbis. The radical decline in sea ice around the Arctic is at least 70% due to human-induced climate change, according to a new study, and may even be up to 95% down to humans – rather higher than scientists had previously thought. The loss of ice around the Arctic has adverse effects on wildlife and also opens up new northern sea routes and opportunities to drill for oil and gas under the newly accessible sea bed as shown in **Fig. 5**. The reduction has been accelerating since the 1990s and many scientists believe the Arctic may become ice-free in the summers later this century, possibly as early as the late 2020s.

"Since the 1970s, there's been a 40% decrease in the summer sea ice extent," said Jonny Day, a climate scientist at the National Centre for Atmospheric Science at the University of Reading, who led the latest study.

"We were trying to determine how much of this was due to natural variability and therefore imply what aspect is due to man-made climate change as well."

To test the ideas, Day carried out several computer-based simulations of how the climate around the Arctic might have fluctuated since 1979 without the input of greenhouse gases from human activity.

He found that a climate system called the Atlantic multi-decadal oscillation (AMO) was a dominant source of variability in ice extent. The AMO is a cycle of warming and cooling in the North Atlantic that repeats every 65 to 80 years – it has been in a warming phase since the mid-1970s.

Figure 5. Science correspondent The Guardian, Thursday 26 July 2012 by Alok Jha,

Comparing the models with actual observations, Day was able to work out what contribution the natural systems had made to what researchers have observed from satellite data.

"We could only attribute as much as 30% [of the Arctic ice loss] to the AMO," he said. "Which implies that the rest is due to something else, and this is most likely going to be man-made global change?"

Previous studies had indicated that around half of the loss was due to man-made climate change and that the other half was due to natural variability. Looking across all his simulations, Day found that the 30% figure was an upper limit – the AMO could have contributed as little as 5% to the overall loss of Arctic ice in recent decades.

The research is published online in the journal Environmental Research Letters. Day said that there are a number of feedback effects that could see the Arctic ice loss continue in the coming years, as the Earth warms up. "[There is] something called the ice-albedo feedback, which means that when you have less ice, it means there's more open water and therefore the ocean absorbs more radiation and will continue to warm," he said.

"It's unclear what will happen – it definitely seems like it's going in that direction."

3.3. Sea level rise due to global warming

3.3.1. Sea level rise poses threat to New York City[2]

Global warming is expected to cause the sea level along the northeastern U.S. coast to rise almost twice as fast as global sea levels during this century, putting New York City at

---

[2] Science Daily (Mar. 16, 2009)

greater risk for damage from hurricanes and winter storm surge, according to a new study led by a Florida State University researcher as shown in **Fig. 6**.

Figure 6. New York Skyline. Global warming is expected to cause the sea level along the northeastern U.S. coast to rise almost twice as fast as global sea levels during this century, putting New York City at greater risk for damage from hurricanes and winter storm surge. (Credit: iStockphoto/Klaas Lingbeek-Van Kranen)

Jianjun Yin, a climate modeler at the Center for Ocean-Atmospheric Prediction Studies (COAPS) at Florida State, said there is a better than 90 percent chance that the sea level rise along this heavily populated coast will exceed the mean global sea level rise by the year 2100. The rising waters in this region -- perhaps by as much as 18 inches or more -- can be attributed to thermal expansion and the slowing of the North Atlantic Ocean circulation because of warmer ocean surface temperatures.

Yin and colleagues Michael Schlesinger of the University of Illinois at Urbana-Champaign and Ronald Stouffer of Geophysical Fluid Dynamics Laboratory at Princeton University are the first to reach that conclusion after analyzing data from 10 state-of-the-art climate models, which have been used for the Intergovernmental Panel on Climate Change (IPCC) Fourth Assessment Report. Yin's study is published in the journal Nature Geoscience.

"The northeast coast of the United States is among the most vulnerable regions to future changes in sea level and ocean circulation, especially when considering its population density and the potential socioeconomic consequences of such changes," Yin said. "The most populous states and cities of the United States and centers of economy, politics, culture and education are located along that coast."

The researchers found that the rapid sea-level rise occurred in all climate models whether they depicted low, medium or high rates of greenhouse-gas emissions. In a medium greenhouse-gas emission scenario, the New York City coastal area would see an additional rise of about 8.3 inches above the mean sea level rise that is expected around the globe because of human-induced climate change.

Thermal expansion and the melting of land ice, such as the Greenland ice sheet, are expected to cause the global sea-level rise. The researchers projected the global sea-level rise of 10.2 inches based on thermal expansion alone. The contribution from the land ice melting was not assessed in this study due to uncertainty.

Considering that much of the metropolitan region of New York City is less than 16 feet above the mean sea level, with some parts of lower Manhattan only about 5 feet above the mean sea level, a rise of 8.3 inches in addition to the global mean rise would pose a threat to this region, especially if a hurricane or winter storm surge occurs, Yin said.

Potential flooding is just one example of coastal hazards associated with sea-level rise, Yin said, but there are other concerns as well. The submersion of low-lying land, erosion of beaches, conversion of wetlands to open water and increase in the salinity of estuaries all can affect ecosystems and damage existing coastal development.

Although low-lying Florida and Western Europe are often considered the most vulnerable to sea level changes, the northeast U.S. coast is particularly vulnerable because the Atlantic meridional overturning circulation (AMOC) is susceptible to global warming. The AMOC is the giant circulation in the Atlantic with warm and salty seawater flowing northward in the upper ocean and cold seawater flowing southward at depth. Global warming could cause an ocean surface warming and freshening in the high-latitude North Atlantic, preventing the sinking of the surface water, which would slow the AMOC.

3.3.2. Significant sea-level rise in a two-degree warmer World[3]

Sea levels around the world can be expected to rise by several metres in coming centuries, if global warming carries on. Even if global warming is limited to 2 degrees Celsius, global-mean sea level could continue to rise, reaching between 1.5 and 4 metres above present-day levels by the year 2300, with the best estimate being at 2.7 metres, according to a study just published in **Nature Climate Change.** However, emissions reductions that allow warming to drop below 1.5 degrees Celsius could limit the rise strongly.

The study is the first to give a comprehensive projection for this long perspective, based on observed sea-level rise over the past millennium, as well as on scenarios for future greenhouse-gas emissions.

"Sea-level rise is a hard to quantify, yet critical risk of climate change," says Michiel Schaeffer of Climate Analytics and Wageningen University, lead author of the study. "Due

---

[3] Science Daily (June 24, 2012)

to the long time it takes for the world's ice and water masses to react to global warming, our emissions today determine sea levels for centuries to come."

Limiting global warming could considerably reduce sea-level rise

While the findings suggest that even at relatively low levels of global warming the world will have to face significant sea-level rise, the study also demonstrates the benefits of reducing greenhouse-gas emissions. Limiting global warming to below 1.5 degrees Celsius and subsequent temperature reductions could halve sea-level rise by 2300, compared to a 2-degree scenario. If temperatures are allowed to rise by 3 degrees, the expected sea-level rise could range between 2 and 5 metres, with the best estimate being at 3.5 metres.

The potential impacts are significant. "As an example, for New York City it has been shown that one metre of sea level rise could raise the frequency of severe flooding from once per century to once every three years," says Stefan Rahmstorf of the Potsdam Institute for Climate Impact Research, co-author of the study. Also, low lying deltaic countries like Bangladesh and many small island states are likely to be severely affected.

Sea-level rise rate defines the time for adaptation

The scientists further assessed the rate of sea-level rise. The warmer the climate gets, the faster the sea level climbs. "Coastal communities have less time to adapt if sea-levels rise faster," Rahmstorf says.

"In our projections, a constant level of 2-degree warming will sustain rates of sea-level rise twice as high as observed today, until well after 2300" as shown in **Fig. 7**, adds Schaeffer, "but much deeper emission reductions seem able to achieve a strong slow-down, or even a stabilization of sea level over that time frame".

Building on data from the past

Previous multi-century projections of sea-level rise reviewed by the Intergovernmental Panel on Climate Change (IPCC) were limited to the rise caused by thermal expansion of the ocean water as it heats up, which the IPCC found could reach up to a metre by 2300. However, this estimate did not include the potentially larger effect of melting ice, and research exploring this effect has considerably advanced in the last few years. The new study is using a complementary approach, called semi-empirical, that is based on using the connection between observed temperature and sea level during past centuries in order to estimate sea-level rise for scenarios of future global warming.

"Of course it remains open how far the close link between temperature and global sea level found for the past will carry on into the future," says Rahmstorf. "Despite the uncertainty we still have about future sea level, from a risk perspective our approach provides at least plausible, and relevant, estimates."

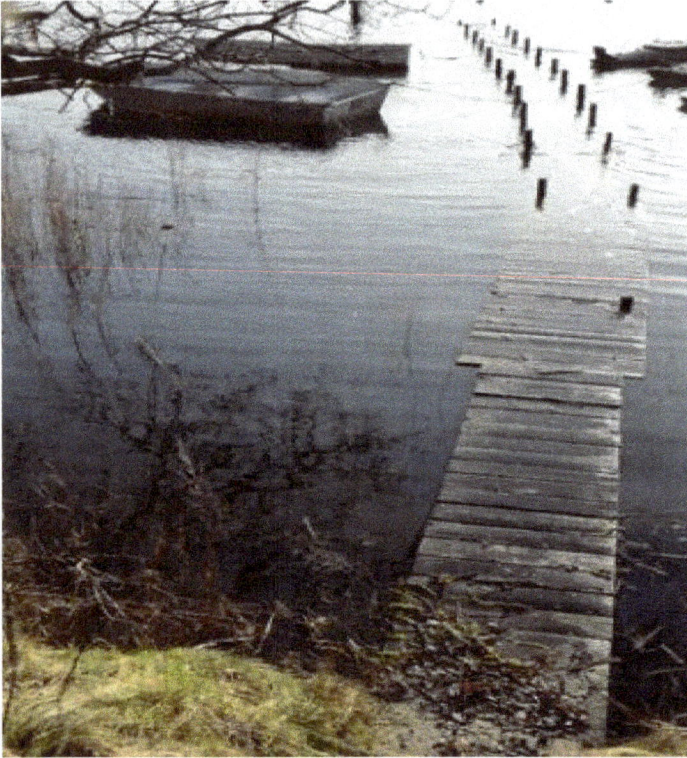

Figure 7. Flooded dock. Sea levels around the world can be expected to rise by several metres in coming centuries, if global warming carries on. Even if global warming is limited to 2 degrees Celsius, global-mean sea level could continue to rise, reaching between 1.5 and 4 metres above present-day levels by the year 2300, with the best estimate being at 2.7 metres. (Credit: © knuderik / Fotolia)

3.4. Hurricanes and global warming[4]

Debate over climate change frequently conflates issues of science and politics. "There's a push on climatologists to say something about extremes, because they are so important. But that can be very dangerous if we really don't know the answer" (Henson 2005). In this article we focus on a particular type of extreme event—the tropical cyclone—in the context of global warming (tropical cyclones are better known in the United States as hurricanes, i.e., tropical cyclones that form in the waters of the Atlantic and eastern Pacific oceans with maximum 1-min-averaged surface winds that exceeds 32 m s^{-1}). We follow distinctions between event risk and outcome risk presented by Sarewitz et al. (2003). "Event risk" refers to the occurrence of a particular phenomenon, and in the context of hurricanes we focus on trends and projections of storm frequencies and intensities. "Vulnerability" refers to "the

---

[4] American Metrology Society; November 2005

inherent characteristics of a system that create the potential for harm," but are independent from event risk. In the context of the economic impacts of tropical cyclones vulnerability has been characterized in terms of trends in population and wealth that set the stage for storms to cause damage. "Outcome risk" integrates considerations of vulnerability with event risk to characterize an event that causes losses. An example of outcome risk is the occurrence of a $100 billion hurricane in the United States. To calculate such a probability requires consideration of both vulnerability and event risk. This article discusses hurricanes and global warming from both of these perspectives.

3.4.1. Event risk

At the end of the 2004 Atlantic hurricane season, many scientists, reporters, and policymakers looked for simple answers to explain the extent of the devastation, which totaled more than $40 billion according to the National Hurricane Center. Some prominent scientists proposed that the intense 2004 hurricane season and its considerable impacts, particularly in Florida, could be linked to global warming resulting from the emissions of greenhouse gases into the atmosphere (e.g., Harvard Medical School 2004; NCAR 2004). But the current state of climate science does not support so close a linkage (Trenberth 2005).

Tropical cyclones can be thought of to a first approximation as a natural heat engine or Carnot cycle (Emanuel 1987). From this perspective global warming can theoretically influence the maximum potential intensity of tropical cyclones through alterations of the surface energy flux and/or the upper-level cold exhaust (Emanuel 1987; Lighthill et al. 1994; Henderson-Sellers et al. 1998). But no theoretical basis yet exists for projecting changes in tropical cyclone frequency, though empirical studies do provide some guidance as to the necessary thermodynamical and dynamical ingredients for tropical cyclogenesis (Gray 1968, 1979).

Since 1995 there has been an increase in the number of storms, and in particular the number of major hurricanes (categorys 3, 4, and 5) in the Atlantic. But the changes of the past decade in these metrics are not so large as to clearly indicate that anything is going on other than the multidecadal variability that has been well documented since at least 1900 (Gray et al. 1997; Landsea et al. 1999; Goldenberg et al. 2001). Consequently, in the absence of large or unprecedented trends, any effect of greenhouse gases on the frequency of storms or major hurricanes is necessarily very difficult to detect in the context of this documented variability. Perspectives on hurricanes are no doubt shaped by recent history, with relatively few major hurricanes observed in the 1970s, 1980s, and early 1990s, compared with considerable activity during the 1940s, 1950s, and early 1960s. The period from 1944 to 1950 was particularly active for Florida. During that period 11 hurricanes hit the state, at least one per year, resulting in the equivalent of billions of dollars in damage in each of those years (Pielke and Landsea 1998).

Globally there has been no increase in tropical cyclone frequency over at least the past several decades (Webster et al. 2005; Lander and Guard 1998; Elsner and Kocher 2000). In addition to a lack of theory for future changes in storm frequencies, the few global modeling results are contradictory (Henderson Sellers et al. 1998; Houghton et al. 2001). Because historical and observational data on hurricanes and tropical cyclones are relatively robust, it is clear that storm frequency has not tracked recent tropical climate trends. Research on possible future changes in hurricane frequency due to global warming is ambiguous, with most studies suggesting that future changes will be regionally dependent, and showing a lack of consistency in projecting an increase or decrease in the total global number of storms (Henderson-Sellers et al. 1998; Royer et al. 1998; Sugi et al. 2002). These studies give such contradictory results as to suggest that the state of understanding of tropical cyclogenesis provides too poor a foundation to base any projections about the future. While there is always some degree of uncertainty about the future and model-based results are often fickle, the state of current understanding is such that we should expect hurricane frequencies in the future to have a great deal of year-to-year and decade-to-decade variation as has been observed over the past decades and longer.

The issue of trends in tropical cyclone intensity is more complicated, simply because there are many possible metrics of intensity (e.g., maximum potential intensity, average intensity, average storm lifetime, maximum storm lifetime, average wind speed, maximum sustained wind speed, maximum wind gust, accumulated cyclone energy, power dissipation, and so on), and not all such metrics have been closely studied from the standpoint of historical trends, due to data limitations among other reasons. Statistical analysis of historical tropical cyclone intensity shows a robust relationship to the thermodynamic potential intensity (Emanuel 2000), suggesting that increasing potential intensity should lead to an increase in the actual intensity of storms. The increasing potential intensity associated with global warming as predicted by global climate models (Emanuel 1987) is consistent with the increase in modeled storm intensities in a warmer climate, as might be expected (Knutson and Tuleya 2004).But while observations of tropical and subtropical sea surface temperature have shown an overall increase of about 0.2°C over the past ~50 years, there is only weak evidence of a systematic increase in potential intensity (Bister and Emanuel 2002; Free et al. 2004).

Emanual (2005) reports a very substantial upward trend in power dissipation (i.e., the sum over the lifetime of the storm of the maximum wind speed cubed) in the North Atlantic and western North Pacific, with a near doubling over the past 50 years (Webster et al. 2005). The precise causation for this trend is not yet clear. Moreover, in the North Atlantic, much of the recent upward trend in Atlantic storm frequency and intensity can be attributed to large multidecadal fluctuations. Emanuel (2005) has just been published as of this writing and is certain to motivate a healthy and robust debate in the community. Other studies that have addressed tropical cyclone intensity variations (Landsea et al. 1999; Chan and Liu 2004) show no significant secular trends during the decades of reliable records.

Because the global earth system is highly complicated, until a relationship between actual storm intensity and tropical climate change is clearly demonstrated and accepted by the broader community, it would be premature to conclude with certainty that such a link exists or is significant (from the standpoints of either event or outcome risk) in the context of variability. Additionally, any such relationship between trends in sea surface temperature and various measures of tropical cyclone intensity would not necessarily mean that the storms of 2004 or 2005 or their associated damages could be attributed directly or indirectly to increasing greenhouse gas emissions.

Looking to the future, global modeling studies suggest the potential for relatively small changes in tropical cyclone intensities related to global warming. Early theoretical work suggested an increase of about 10% in wind speed for a 2°C increase in tropical sea surface temperature (Emanuel 1987). A 2004 study from the Geophysical Fluid Dynamics Laboratory in Princeton, New Jersey, that utilized a mesoscale model downscaled from coupled global climate model runs indicated the possibility of a 5% increase in the wind speeds of hurricanes by 2080 (Knutson and Tuleya 2004; cf. Houghton et al. 2001). Michaels et al. (2005) suggest that even this 5% increase may be overstated, and that a more realistic projection is on the order of only half of that amount. Even if one accepts that the Knutson and Tuleya results are in the right ballpark, these would imply that changes to hurricane wind speeds on the order of 0.5–1.0 m s^{-1} may be occurring today. This value is exceedingly small in the context of, for example, the more than doubling in numbers of major hurricanes between quiet and active decadal periods in the Atlantic (Goldenberg et al. 2001). Moreover, such a change in intensities would not be observable with today's combination of aircraft reconnaissance and satellite-based intensity estimates, which only resolves wind speeds of individual tropical cyclones to at best 2.5 m s^{-1} increments.

3.4.2. Vulnerability and outcome risk

Understanding of trends and projections in tropical cyclone frequencies and intensities takes on a different perspective when considered in the context of rapidly growing societal vulnerability to storm impacts (Pielke and Pielke 1997; Pulwarty and Riebsame 1997). There is overwhelming evidence that the most significant factor underlying trends and projections associated with hurricane impacts on society is societal vulnerability to those impacts, and not the trends or variation in the storms themselves (Pielke and Landsea 1998). Growing population and wealth in exposed coastal locations guarantee increased economic damage in coming years, regardless of the details of future patterns of intensity or frequency (Pielke et al. 2000). Tropical cyclones will also result in death and suffering, in less developed countries in particular, as seen in Haiti during Hurricane Jeanne (cf. Pielke et al. 2003).

Over the long term the effects of changes in society dwarf the effects of any projected changes in tropical cyclones according to research based on assumptions of the Intergovernmental Panel on Climate Change (IPCC), the scientific organization convened to report on the science of climate change. By 2050, for every additional dollar in damage that the IPCC expects to result from the effects of global warming on tropical cyclones, we should expect between $22 and $60 of increase in damage due to population growth and wealth (Pielke et al. 2000). The primary factors that govern the magnitude and patterns of future damages and causalities are how society develops and prepares for storms rather than any presently conceivable future changes in the frequency and intensity of the storms (see **Fig. 8**).

Figure 8. Flooding at costal area

Consider that if per capita wealth and population grow at a combined 5% per year, this implies a doubling in the real costs of hurricanes about every 15 years. In such a context, any climate trend would have to be quite large to be discernible in the impacts record.

With no trend identified in various metrics of hurricane damage over the twentieth century (Pielke and Landsea 1998), it is exceedingly unlikely that scientists will identify large changes in historical storm behavior that have significant societal implications. In addition, looking to the future, until scientists conclude:

a. that there will be changes to storms that are significantly larger than observed in the past,
b. that such changes are correlated to measures of societal impact, and
c. that the effects of such changes are significant in the context of inexorable growth in population and property at risk, then it is reasonable to conclude that the significance of any connection of human-caused climate change to hurricane impacts necessarily has been and will continue to be exceedingly small.

Thus a great irony here is that invoking the modulation of future hurricanes to justify energy policies to mitigate climate change may prove counterproductive. Not only does this provide a great opening for criticism of the underlying scientific reasoning, it leads to advocacy of policies that simply will not be effective with respect to addressing future hurricane impacts. There are much, much better ways; to deal with the threat of hurricanes than with energy policies (e.g., Pielke and Pielke 1997). There are also much, much better ways to justify climate mitigation policies than with hurricanes (e.g., Rayner 2004), with energy policies (e.g., Pielke and Pielke 1997) and to justify climate mitigation policies than with hurricanes (e.g., Rayner 2004).

3.4.3. Worrying beyond hurricane Sandy

With the last hurricane to directly hit New York City dating back to the 1800's, residents have so far lacked the impetus to demand concrete strategies for dealing with the potential devastation to housing, the subway system and the electrical infrastructure from a major modern-day storm. Now Hurricane Sandy threatens major flooding from a storm surge that could reach up to 14 feet above the average sea level here.

Some scientists suggested on Monday 29th October' 2012 that once New Yorkers have moved to higher ground and weathered the hurricane, they should begin to take more decisive steps to adapt to more of the same. As it was reported last month, pressure has been growing for aggressive action as shown in **Fig. 9**.

Figure 9. Awaiting an onslaught: wave activity at Rockaway Beach on Monday morning, when Hurricane Sandy was 425 miles southeast of New York City (Oct 29' 2012).

It is small comfort to sodden and stranded New Yorkers that Hurricane Sandy's flooding of the city's infrastructure, from power lines to subways to low-lying communities, was predicted in grimly precise detail by scientists in the latest state and city climate studies. Deeper and more frequent flooding from Rockaway to Lower Manhattan and the city's transit tunnels has been a repeated warning that largely went unnoticed by the public and most politicians.

But now, with the floods from Sandy and Tropical Storm Irene last year on his watch, Gov. Andrew Cuomo is pointedly stressing what he considers the inevitability of more such disasters. "Climate change is reality," the governor said on Wednesday, 31st Oct' 2012, estimating Sandy's economic damage up to $6 billion. "Given the frequency of these extreme weather situations that we've had-and I believe that it's an increasing frequency - for us to sit here today and say this is a once-in-a-generation and it's not going to happen again, I think would be shortsighted." Mr. Cuomo admits that he does not have all the answers nor enough government money for all the proposed solutions. And we can all hope that he is wrong in his forecast. But the urgency of his warning is rooted in a basic fact of nature underpinning the government studies: New York's coastal waters, which rose an inch per decade in the last century, are heading toward rates of 6 inches per decade as the oceans warm and expand. That would be a disastrous rise of 2 feet across the next 40 years, for anyone planning ahead. And there aren't many in government planning ahead as the postrecession political debate grinds along the question of how to slash government improvements, not expand them.

Just last September, Klaus Jacob, an adviser to the city on climate change, warned of the certainty of flooded Manhattan highways and tunnels and of stranded subway riders and subway commuters if the next storm surge topped Irene's. "I'm disappointed that the political process hasn't recognized that we're playing Russian roulette," said Mr. Jacob, a research scientist at Columbia University's Earth Institute and an author of a 2011 state study that predicted tens of billions in economic losses from worsening floods.

This is why the problem underlined by Mr. Cuomo deserves heightened public debate. Mayor Michael Bloomberg agreed. "What's clear is that the storms that we've experienced in the last year or so around this country and around the world are much more severe than before," the mayor said. "Whether that's global warming or what, I don't know. But we'll have to address those issues."

Mr. Cuomo proposed consideration of the sort of storm surge barriers in use in Europe. Gates like those guarding London's riverfront could be closed in disastrous weather at three main points of ocean inflow — the Verrazano-Narrows Bridge, the upper East River and the mouth of the Arthur Kill between Staten Island and New Jersey. This idea and others involve billions of dollars, which could be a bargain if Sandy and Irene truly are harbingers of more frequent disasters eating deeper into the city's heart.

(**Source: The New York edition;** Worrying Beyond Hurricane Sandy, October 31, 2012**.**)

4. How to prevent climate change?

The potential for runaway greenhouse warming is real and has never been more prominent as now. The most dangerous climate changes may still be avoided if we transform our hydrocarbon based energy systems and if we initiate rational and adequately financed adaptation programmes to forestall disasters and migrations at unprecedented scales. The tools are available, but they must be applied immediately and aggressively.

4.1. What can I do to help prevent climate change?

In the United States, approximately 6.6 tons (almost 15,000 pounds carbon equivalent) of greenhouse gases are emitted per person every year. And emissions per person have increased about 3.4% between 1990 and 1997. Most of these emissions, about 82%, are from burning fossil fuels to generate electricity and power our cars. The remaining emissions are from methane from wastes in our landfills, raising livestock, natural gas pipelines, and coal, as well as from industrial chemicals and other sources. (Source: US EPA).

With this said, also keep in mind that emissions vary based on the country and state in which you live. At the present time, the United States emits more greenhouse gasses per person than any other country. Emissions also vary by state as they are based on the many factors such as the types of fuel used to generate electricity, the total population of a state, and the amount of (and distance traveled by) commuters.

As an individual there are three areas where we can make the most impact in reducing carbon emissions:

- the electricity we use in our homes,
- the waste we produce, and
- the transportation we choose to use.

According to the U.S. EPA, you can affect the emissions of about 4,800 pounds of carbon equivalent, or nearly 32% of the total emissions per person by the choices we make in these three areas. The other 68% of emissions are affected more by the types of industries in the U.S. the types of offices we use, how our food is grown and other factors (source: U.S. EPA). Below are tips on how to reduce carbon emissions and help stop climate change.

4.2. How can I do to help prevent climate change?

When people talk about Climate Change, they are talking about the temperature and weather on the Planet Earth changing. In the last one hundred years the temperature on our planet has gone up by a little bit, about 1 degree. The reason it's getting warmer is because too much heat is getting trapped in earth's atmosphere and making the planet too hot. Every time we use energy, we send more heat (Carbon Dioxide) into the atmosphere. Carbon Dioxide, or Mr. Carbon as we call him, is created when humans drive cars and make electricity for things like lights and computers. But, if we are all more careful and don't waste energy or use more than we need, we can help to cool off the earth!

Here are some great ideas to get you started:

- **Use less energy at home-** In winter, wear a sweater and turn the thermostat down. In summer, turn off the lights and use the natural light of the sun!
- **Take shorter showers-** Heating shower water uses energy. Even just a few minutes can add up to a big difference over time!
- **Ride your bike!** -If you live close enough to school, hop on your bike or walk. Replace car rides whenever possible.

- **Carry a reusable water bottle-** Picture how many disposable water bottles pile up after a week, a month, or a year. Skip all that waste by getting a cool reusable one!
- **Power down-** Even when they're "off," many appliances like computers continue to use energy. Ask your family to unplug these items when they're not in use, or to turn off the power strip they're attached to.
- **Eat Your Veggies-** Livestock like cows create carbon on farms. Even just one day a week of vegetarian meals can make a big difference!

4.3. Ten basic tips to help stop climate change

Don't have a lot of times, but want to take action? Here are ten, simple, everyday things each of us can do to help stop climate change. Pick one, some, or all. Every little effort helps and adds up to a whole lot of good.

a. **Change a light-** Replacing a regular light bulb with a compact fluorescent one saves 150 pounds of carbon dioxide each year.*

b. **Drive less-**Walk, bike, carpool; take mass transit, and/or trip chain. All of these things can help reduce gas consumption and one pound of carbon dioxide for each mile you do not drive.

c. **Recycle more and buy recycled-** Save up to 2,400 pounds of carbon dioxide each year just by recycling half of your household waste. By recycling and buying products with recycled content you also save energy, resources and landfill space!

d. **Check your tyres-** Properly inflated tyres mean good gas mileage. For each gallon of gas saved, 20 pounds of carbon dioxide are also never produced.

e. **Use less hot water-** It takes a lot of energy to heat water. Reducing the amount used means big savings in not only your energy bills, but also in carbon dioxide emissions. Using cold water for your wash saves 500 pounds of carbon dioxide a year, and using a low flow showerhead reduces 350 pounds of carbon dioxide. Make the most of your hot water by insulating your tank and keeping the temperature at or below 120.

f. **Avoid products with a lot of packaging-** Preventing waste from being created in the first place means that there is less energy wasted and fewer resources consumed. When you purchase products with the least amount of packaging, not only do you save money, but you also help the environment! Reducing your garbage by 10% reduces carbon dioxide emissions by 1,200 pounds.

g. **Adjust your thermostat-** Keeping your thermostat at 68 degrees in winter and 78 degrees in summer not only helps with your energy bills, but it can reduce carbon dioxide emissions as well. No matter where you set your dial, two degrees cooler in the winter or warmer in the summer can mean a reduction of 2,000 pounds of carbon dioxide a year.

h. **Plant a tree-** A single tree can absorb one ton of carbon dioxide over its lifetime.

i. **Turn off electronic devices when not in use-** Simply turning off your TV, VCR, computer and other electronic devices can save each household thousands of pounds of carbon dioxide each year.

j. **Stay informed-** Use the Earth 911 Web site to help stay informed about environmental issues, and share your knowledge with others. Together, we can and do Make Every Day Earth Day!

5. Conclusion

From the studies and reports, it is evident that the potential for runaway greenhouse warming due to release of carbon dioxide and other gases in the atmosphere which is the cause of potential increase of the global temperature, and subsequent melting of ice cap, rise in sea level, and it triggers the disasters. The following major issues are noticed:

- Emissions from human activities are increasing the frequency of extreme weather events.
- Due to climate change there are likely to be many more heatwaves, droughts and changes in rainfall patterns.
- By the mid-2020s, sea level rise around Manhattan and Long Island could be up to 10 inches, if the rapid melting of polar sea ice continues at same pace. By 2050, sea-rise could reach 2.5ft and more than 4.5ft by 2080 under the same conditions.
- Global warming threatens the planet in a new and unexpected way – by triggering earthquakes, tsunamis, avalanches and volcanic eruptions.
- Irene-like storms of the future would put a third of New York City streets under water and flood many of the tunnels leading into Manhattan in under an hour because of climate change.

Climate changes may still be avoided if we transform our hydrocarbon based energy systems and if we initiate rational and adequately financed adaptation programmes to forestall disasters and migrations at unprecedented scales.

Author details

Bharat Raj Singh
School of Management Sciences, Technical Campus, Lucknow, Uttar Pradesh, India

Onkar Singh
Harcourt Butler Technological Institute, Kanpur, Uttar Pradesh, India

Acknowledgement

Authors indebted to extend their thanks to the Management of School of Management Sciences, Technical Campus, Lucknow, UP, India and Harcourt Butler Technological Institute, Kanpur, UP, India for providing the support of Library.

6. References

Bister, M., and K. A. Emanuel, 2002: Low frequency variability of tropical cyclone potential intensity. 1: Interannual to interdecadal variability. J. Geophys. Res., 107, 4801, doi: 10.1029/2001JD000776.

Chan, J. C. L., and S. L. Liu, 2004: Global warming and western North Pacific typhoon activity from an observational perspective. J. Climate, 17, 4590–4602.

Eilperin, J., 2005: Hurricane scientist leaves U.N. team. Washington Post, 23 January, a13.

Elsner, J. B., and B. Kocher, 2000: Global tropical cyclone activity: A link to the North Atlantic Oscillation. Geophys. Res. Lett., 27, 129–132.

Emanuel, K., 1987: The dependence of hurricane intensity on climate. Nature, 326, 483–485.

Emanuel, K., 2000: A statistical analysis of tropical cyclone intensity. Mon. Wea. Rev., 128, 1139–1152.

Emanuel, K., 2005: Increasing destructiveness of tropical cyclones over the past 30 years. Nature, 436, 686–688.

Epstein, P., and J. McCarthy, 2004: Assessing climate stability. Bull. Amer. Meteor. Soc., 85, 1863–1870.

Free, M., M. Bister, and K. Emanuel, 2004: Potential intensity of tropical cyclones: Comparison of results from radiosonde and reanalysis data. J. Climate, 17, 1722–1727.

Goldenberg, S. B., C. W. Landsea, A. M. Mestas-Nuñez, and W. M. Gray, 2001: The recent increase in Atlantic hurricane activity: Causes and implications. Science, 293, 474–479.

Gray, W. M., 1968: Global view of the origin of tropical disturbances and storms. Mon. Wea. Rev., 96, 669–700.

Gray, W. M., 1979: Hurricanes: Their formation, structure and likely role in the tropical circulation. Meteorology over Tropical Oceans, D. B. Shaw, Ed., Royal Meteorological Society, 155–218.

Gray, W. M., J. D. Sheaffer, and C. W. Landsea, 1997: Climate trends associated with multidecadal variability of Atlantic hurricane activity. Hurricanes: Climate and Socioeconomic Impacts, H. F. Diaz and R. S. Pulwarty, Eds., Springer-Verlag, 15–53.

Harvard Medical School cited 2004: Experts to warn global warming likely to continue spurring more outbreaks of intense hurricane activity. [Available online at www.med.harvard.edu/chge/hurricanespress.html; full transcript of the press conference can be found online at www.ucar.edu/news/record/transcripts/hurricanes102104.shtml.]

Henderson-Sellers, A., and Coauthors, 1998: Tropical cyclones and global climate change: A post-IPCC assessment. Bull. Amer. Meteor. Soc., 79, 9–38.

Henson, B., 2005: Going to extremes. UCAR Quarterly, Winter 2004–05. [Also available online at www.ucar.edu/communications/quarterly/winter04/extremes.html.]

Houghton, J. T., Y. Ding, D. J. Griggs, M. Noguer, P. J. van der Linden, and D. Xiaosu, Eds., 2001: Climate Change 2001: The Scientific Basis: Contributions of Working Group I to the Third Assessment Report of the Intergovernmental Panel on Climate Change. Cambridge University Press, 881 pp.

Knutson, T. R., and R. E. Tuleya, 2004: Impact of CO_2-induced warming on simulated hurricane intensity and precipitation: Sensitivity to the choice of climate model and convective parameterization. J. Climate, 17, 3477–3495.

Lander, M. A., and C. P. Guard, 1998: A look a global tropical cyclone activity during 1995: Contrasting high Atlantic activity with low activity in other basins. Mon. Wea. Rev., 126, 1163–1173.

Landsea, C. W., R. A. Pielke Jr., A. M. Mestas-Nuñez, and J. A. Knaff, 1999: Atlantic basin hurricanes: Indices of climatic changes. Climatic Change, 42, 89–129.

Lighthill, J., and Coauthors, 1994: Global climate change and tropical cyclones. Bull. Amer. Meteor. Soc., 75, 2147–2157.

Michaels, P. J., P. C. Knappenberger, and C. W. Landsea, 2005: Comments on "Impacts of CO2-induced warming on simulated hurricane intensity and precipitation: Sensitivity to the choice of climate model and convective scheme. J. Climate, in press.

NCAR, 2004: Hurricanes and climate change: Is there a connection? NCAR Staff Notes Monthly, October. [Available online at www.ucar.edu/communications/staffnotes/0410/hurricane.html.]

Pielke, R. A., Jr., and R. A. Pielke Sr., 1997: Hurricanes: Their nature and impacts on society. John Wiley and Sons, 279 pp.

Pielke, R. A., and C. W. Landsea, 1998: Normalized U.S. hurricane damage, 1925–1995. Wea. Forecasting, 13, 621–631.

Pielke, R. A., R. Klein, and D. Sarewitz, 2000: Turning the big knob: Energy policy as a means to reduce weather impacts. Energy Environ., 11, 255–276.

Pielke, R. A., J. Rubiera, C. Landsea, M. L. Fernandez, and R. Klein, 2003: Hurricane vulnerability in Latin America and the Caribbean: Normalized damage and loss potentials. Nat. Hazards Rev., 4, 101–114.

Pulwarty, R. S., and W. E. Riebsame, 1997: The political ecology of vulnerability to hurricane-related hazards. Hurricanes: Climate and Socioeconomic Impacts, H. F. Diaz and R. S. Pulwarty, Eds., 292 pp.

Rayner, S., 2004: The international challenge of climate change: UK leadership in the G8 and EU. Memo. to the Environmental Audit Committee House of Commons, 16 pp. [Available online at www.cspo.org/ourlibrary/documents/EACmemo.pdf.]

Royer, J.-F., F. Chauvin, B. Timbal, P. Araspin, and D. Grimal, 1998: A GCM study of impact of greenhouse gas increase on the frequency of occurrence of tropical cyclones. Climate Dyn., 38, 307–343.

Sarewitz, D., R. A. Pielke Jr., and M. Keykyah, 2003: Vulnerability and risk: Some thoughts from a political and policy perspective. Risk Anal., 23, 805–810.

Sugi, M., A. Noda, and N. Sato, 2002: Influence of the global warming on tropical cyclone climatology: An experiment with the JMA global model. J. Meteor. Soc. Japan, 80, 249–272.

Trenberth, K., 2005: Uncertainty in hurricanes and global warming. Science, 308, 1753–1754.

Walsh, K., 2004: Tropical cyclones and climate change: Unresolved issues. Climate Res., 27, 78–83.

Webster, P. J., G. J. Holland, J. A. Curry, and H.-R. Chang, 2005. Changes in Tropical Cyclone Number, Duration, and Intensity in a Warming Environment, Science, 309, 1844–1846.

Chapter 2

A Methodology to Interpret Climate Change Due to Influences of the Orbital Parameter on Changes of Earth's Rotation Rate and Obliquity

Xinhua Liu

Additional information is available at the end of the chapter

1. Introduction

Astroclimatology is the branch of paleoclimatology in which climate change over geological history is interpreted in terms of changes in astronomical elements or in the Earth's orbital parameters. There are two mature theories in astroclimatology. Milankovitch theory considers the Pleistocene glacial cycles to be the result of tiny changes in three orbital elements (obliquity, eccentricity, and the equation of the equinoxes) (Berger, A., et al., 1984). Another theory proposes that oscillations between glacial and interglacial periods result from large fluctuations of obliquity (Williams, G. E., 1981). It is clear that changes in orbital parameters can have an important effect on climate. Reconstructions of paleoclimate and its simulation are vital for us to understand the evolution of climate and to forecast future climate change.

The influences of the orbital parameters have multiple time scales from several hours to geological time scales. Zheng (1994) and Qian et al. (1995) investigated the influence on the atmosphere and ocean of changes in the Earth's rotation rate on a time scale of less than 100 years, using observations combined with a dynamic method. The results suggest that the rotation rate of the Earth has a close relationship with Southern Oscillation, El Nino, magnitude and location of the Subtropical High, sea surface temperature, and precipitation, respectively. These researches have deepened our knowledge of the relationship between rotation rate and the atmosphere and ocean. Long-term changes in rotation rate and changes in other orbital parameters can also affect the atmospheric circulation and have generally been investigated through numerical simulations of paleoclimate. The aim of paleoclimate research is to use our knowledge of paleoclimate change to understand the mechanism of climate change, and thus to improve our ability to forecast climate change. Models used in the simulation of paleoclimate include Box Models, Energy Balance Models (EBMs),

Statistical-dynamical Models (SDMs), Radiative Convective Models (RCMs), Earth System Models of Intermediate Complexity (EMICs) and General Circulation Models (GCMs). This study focuses on GCMs.

1.1. Review on the simulation of influence of orbital parameter change in specific geological period on atmosphere

1.1.1. Numerical simulation of paleoclimate

Numerical simulations are important in paleoclimatological research. The earliest numerical simulations of paleoclimate were carried out in the 1970s (Williams *et al.*, 1974). Since then the simulation time scale has been extended from short-time-scale rapid climate change and orbital-time-scale climate change to climate change at the geological time scale (Kutzbach and Street-Perrott, 1985; Kutzbach, 1989; Rahmstorf, 1994). Simulations have been carried out for the Holocene, Pleistocene, Pliocene, early Cenozoic, and Cretaceous geological periods, extending to the time of the Pangaea supercontinent (Kutzbach and Gallimore, 1989; Barron *et al.*, 1993; Barron *et al.*, 1995; Kutzbach, 1996; Bush and Philander, 1997; Otto-Bliesner and Upchurch Jr, 1997; Ramstein *et al.*, 1997; Cane and Molnar, 2001). Different climate driving factors have been investigated, including changes in orbital parameters, CO_2 concentration and global ice quantity, and thermohaline circulation changes resulting from closed and open ocean channels, plateau uplift, and plate tectonics (Kutzbach and Guetter, 1986; Mitchell *et al.*, 1988; Kutzbach and Gallimore, 1989; Barron *et al.*, 1993; Kutzbach *et al.*, 1993; Barron *et al.*, 1995; Rahmstorf, 1995; Ramstein *et al.*, 1997; Weaver *et al.*, 1998; Cane and Molnar, 2001; Knutti *et al.*, 2004).

1.1.2. Simulation of glacial cycle

The international research effort on glacial cycle simulation has taken the form of a series of large projects, including CLIMAP (Climate Long-range Investigation, Mapping, and Prediction), COHMAP (Cooperative Holocene Mapping Project), and PMIP (Paleoclimate Modeling Intercomparison Project).

CLIMAP is a reconstruction of the earth's environment during the ice age. Carried out in the 1970s, it was led by the marine geologists J. Imbrie and J. Hays, and the geochemist N. Shackleton. Using data related to the reconstructed environment of the ice age, the CLIMAP project simulated the summer climate at the Last Glacial Maximum (LGM). The results show that the global summer climate was drier and colder during the LGM than at present. At the same time, the westerlies over the Northern Hemisphere near the ice cap move southward obviously (Gates, 1976).

In the 1980s, COHMAP, led by J. E. Kutzbach, T. Webb III, and H. E. Wright, was designed to determine and simulate paleoclimate using the land record. It revealed the key role of orbital factors in tropical monsoon climate change, and the bifurcation of the westerly jet stream over North America as a result of the ice cap in North America during the LGM period. Simulations showed that the change of Earth's orbit in the early to middle Holocene led to an increase in seasonality in the Northern Hemisphere and an enhanced monsoon (Kutzbach and Street-Perrott, 1985; Kutzbach and Guetter, 1986; Mitchell *et al.*, 1988; Kutzbach, 1989; Kutzbach and Gallimore, 1989; Barron *et al.*, 1993; Kutzbach *et al.*, 1993; Rahmstorf, 1994; Barron *et al.*, 1995; Rahmstorf, 1995; Kutzbach *et al.*, 1996; Bush and Philander, 1997; Otto-Bliesner and Upchurch Jr, 1997; Ramstein *et al.*, 1997; Weaver *et al.*, 1998; Cane and Molnar, 2001; Knutti *et al.*, 2004). This coincided with generally high lake levels during 9–6 Ka BP (Kutzbach, 1985).

The PMIP, part of Past Global Changes Project (PAGES), in the 1990s was led by S. Joussaume, with contributions from a number of leading paleoclimate scientists including J. E. Kutzbach, A. J. Broccoli, and J. Guiot *et al.*. This research project evaluated the sensitivity and accuracy of models by comparing results from different climate models and the difference between the simulation results and observational record. In general, the models all simulated the main features of climate change since the LGM, including the monsoon enhancement driven by orbital factors in the early to middle Holocene (6 Ka BP) and the enhanced meridional temperature gradient during the LGM period (Joussaume and Taylor, 1995). The simulations significantly underestimated the range of climate change evident in the record. The tropical Atlantic Ocean warming further enhanced the African monsoon in the middle Holocene (Kutzbach *et al.*, 1996; Kutzbach and Liu, 1997).

1.1.3. Simulation before the Pleistocene

Because it is difficult to reconstruct the boundary conditions of the Pleistocene, simulations of the climate before the Pleistocene differ from simulations of glacial cycles in the Quaternary, in that their focus is not on reconstruction, but on sensitivity testing. Prell and Kutzbach (1997) investigated the response of the monsoon to orbital change and its association with the Qinghai–Tibet Plateau height and uplift mode. The results showed a remarkable orbit-driven effect of the uplift mode on monsoon response. Under the mode of stable plateau uplift in the past 15 Ma, monsoon intensity and variable rate index have shown little change, whereas under the mode of large uplift at 11–8 Ma, monsoon intensity and the variable rate index showed a marked increase. These results show that the sensitivity of the orbit-driven monsoon response has strengthened greatly. Under the Quaternary uplift mode, monsoon intensity and variability increased abruptly at 2–3 Ma. This shows that not just the rate of plateau uplift, but also the process and mode of uplift may have an important influence on climate change.

1.2. Review on the simulation of influence of orbital parameter change in non-specific geological period on the atmosphere

Hunt (1979) investigated the effects of a five-fold increase or decrease in rotation rate. The lower rotation rate resulted in phenomena including enhanced mid-latitude westerly jet stream, a decrease in the tropospheric temperature gradient with warming in polar regions, and an expansion of the subtropical arid area. The rapid rotation scenario yielded a significant tropospheric zonal temperature gradient and dry, cold high latitudes. A hemispheric model was used without terrain changes or diurnal processes, considering yearly average cloud, ozone, and ground reflectivity. Because the terrain and the land–sea distribution are different between the Northern and Southern Hemisphere, the anomalies of the atmospheric circulation and climate caused by change in Coriolis force are not hemispherically symmetrical. Therefore, the changes of circulation and climate in the two hemispheres and the differences between the two hemispheres are not depicted well in the model. Furthermore, the monthly and seasonal differences of all sorts of effects cannot be considered in this model.

Kutzbach and Otto-Bliesner (1982) analyzed the effect of orbital parameter change in the Holocene on the African and Asian monsoon. The difference of radiation is largest at the two solstitial points, and two solstitial points are closely related to the monsoon; therefore, the effects of different orbital parameters on the monsoon in summer and winter were analyzed. The results showed that the monsoon in the Holocene was stronger than now and that precipitation in Africa and India was heavier. The seasonal circulation corresponding to different orbital parameters was not discussed in detail.

Even though Kutzbach and Guetter (1986) discussed the effect of different orbital parameters and surface boundary conditions on the climate, the radiation and sea temperature conditions were fixed to January and July conditions. Again, there was no detailed discussion of the effect of different orbital parameters on climate under seasonal cycle. Their conclusion was that the response of the monsoon and tropical precipitation to the change of radiation caused by change in orbital parameter is stronger than the response of the forcing. Similar conclusions were made to those in Kutzbach and Otto-Bliesner (1982).

Using the NCAR Community Climate Model version 0 (CCM0), Jenkins (1993) investigated the effect of a high rotation rate on circulation. Terrain was not considered. The radiative forcing was reduced by 10% from present values, as an approximation of the radiation incident on the early Earth. The carbon dioxide concentration in the atmosphere was higher than present values, and other forcings were annual averages. It was concluded that by reducing the cloud amount, a high rotation rate influences the climate. Using the same model, Jenkins et al. (1993) investigated the effect of the presence or absence of land and of rapid rotation rate on the climate at 2.5–4 billion years ago, yielding the following results:

1. When the day length is 14 hours, the world average cloud coverage drops to 79% of the present value and the global average temperature rises by 2 K.
2. In the absence of land, the global average temperature rises by 4 K compared with the present value.
3. Increasing the rotation rate and reducing the land area offsets the impact of lower solar radiation during early times on the Earth. Therefore the early Earth might not have needed more CO_2 to maintain its temperature above freezing point.
4. The surface wind direction reversed from west to east in mid-latitudes with the high rotation rate of the early Precambrian, and the meridional mean surface wind was

enhanced. Compared with the current single jet stream, it is possible that there were two jet streams in the early Earth.

Jenkins (1996) then used the NCAR CCM1 model to investigate the effect of a high rotation rate on the climate, using the present-day land–sea distribution, CO_2 concentration, and ozone concentration. The climatological January mean sea surface temperature field and radiative forcing were used in the model, yielding the following conclusions:

1. Polar temperatures fall under the high rotation rate and the winter hemisphere is noticeably colder. As rotation rate increases, the permanent ridge and trough features weaken. When the length of day is less than 18 hours, the ridge and trough features disappear completely.
2. Under fast rotation there is strong mid-latitude subsidence, which is more pronounced in the Pacific and Atlantic.
3. While present-day storm trajectories are mainly controlled by large-scale fluctuation, the trajectories under fast rotation consist of small-scale fluctuations. This kind of movement forward to the small-scale fluctuation under the high rotation rate is related to enhanced convective precipitation.

Hall *et al.* (2005) investigated the effect of a change in orbital parameter on the climate in winter in the Northern Hemisphere using an atmosphere–ocean coupled model employing orbital forcings for the past 165,000 years. The results suggest that while global summer temperature differences can be explained by local thermodynamics caused by a change in radiative forcing, the change of winter climate in the Northern Hemisphere cannot be explained completely in this way. The authors concluded that the change in radiative forcing gave rise to an atmospheric circulation anomaly similar to the Northern Hemisphere Annular Mode (NAM). This circulation anomaly perturbed other climatological variables.

1.3. Discussion and future prospective

Research on the effect of orbital parameter change on circulation and climate has deepened our understanding of the paleoclimate, but previous simulations focused on a specific geological period or specific seasons (mostly summer and winter). The influence of a single orbital parameter change on the circulation has never been examined in detail. Moreover, analyses of sediment in the Arabian and South China seas show that variations in global ice-cover have an important influence on the east Asian monsoon, while orbital parameters, especially the change of precession, are the main external driving force for the east Asian summer monsoon and Indian summer monsoon (Clemens *et al.*, 1991; Jian *et al.*, 2001). Previous studies considered the effect of orbital parameters on the monsoon within a limited area, but the effect of orbital parameters on the global monsoon has rarely been considered. Liu *et al.* (2010) and Liu (2011) therefore investigated the effect of rotation rate and obliquity on global circulation, describing the responses of atmospheric elements and three-cell circulation to variations in a single orbital parameter. Furthermore, they discussed the response of the global monsoon to different rotation rates. Their conclusions are that the African monsoon and the monsoon in the temperate and frigid zones weaken when the rotation rate slows, and vice versa, while the Asia–Australia monsoon has no obvious reverse change with a change of rotation rate. The monsoon changes are different with rotation rate in different areas. However, these studies are somewhat preliminary in nature, and future research on the effect of different orbital parameters on circulation could include the following aspects:

1. Consideration of changes in each of the three key elements of the Earth's orbit, using a GCM to simulate the circulation. This will identify the relative effects of the three orbital elements.
2. An atmosphere–ocean coupled model could be used to investigate the effects of different orbital parameters on the circulation. The mechanisms by which the ocean and atmosphere influence each other under the change of orbital parameter should be discussed in more detail in the future.

2. The influence of different rotation rates on the general circulation of the Earth's atmosphere

The observations of celestial optics and modern spatial geodesy have proved that the rotation of the Earth varies on multiple time scales ranging from several hours to geological ages, and fossil analysis has shown that the Earth's rotation rate is currently slowing. Changes in the Earth's rotation can have an important effect on the atmosphere.

Numerical simulations of the influence of change in rotation rate on geological time scales on the atmospheric circulation and climate have greatly improved our understanding of the ancient climate (Hunt, 1979; Kutzbach and Otto-Bliesner, 1982; Kutzbach and Guetter, 1986; Jenkins, 1993, 1996; Jenkins et al., 1993; Hall et al., 2005). However, this research has either focused on a specific geological period or on specific seasons (mostly summer and winter). The influence of a single orbital parameter change on the circulation has not been studied. Moreover, Jenkins (1996) reported a threshold at a day length of 18 hours: when the day length is shorter than 18 hours, the atmospheric circulation will change significantly. The same conclusion can be drawn from numerical simulations in which the meridional circulation changes from two circuits to a single circuit, when the day length increases from 16 to 64 times its present-day length (Del Genio, 1996). What happens when the day length changes by one hour? Is the atmospheric change significant? Based on the above considerations, the present study performed simulations to investigate the atmospheric circulation under different rotation rates.

2.1. Model description and experimental design

The atmospheric general circulation model (AGCM) used in this study is the NCAR Community Atmospheric Model version 2 (CAM2) released in July 2003. CAM2 uses spherical harmonic functions truncated at wave number 42. To facilitate physical computing, the model grids are based on 64 latitude Gaussian points and 128 longitude points (2.8125 degree intervals) in the horizontal plane, with 26 hybrid vertical levels. The model includes detailed radiative forcing, cumulus convection, and a land surface process parameterization scheme. The sea surface temperature (SST) adopted is the climatological average SST field. Present-day terrain is used. A complete description of this model version is available online at http://www.ccsm.ucar.edu/models/atmcam/docs/description/index.html. A control run with 24-hour day length and two sensitivity runs, with 23- and 25-hour day length, are discussed in this section. These two sensitivity runs represent a condition from geological history and some 10,000 years in the future, respectively. The time series of whole layer average dimensionless angular momentum and the integrals of unit quality atmospheric kinetic energy in the troposphere (surface–100 hPa) and stratosphere (100 hPa–stratosphere top) in the sensitivity runs show that the atmospheric conditions reach steady state after two years (figure not shown). The whole-layer averaged dimensionless angular momentum is an index of atmospheric rotation, which is defined as the ratio of the specific angular momentum $a\cos\phi(u+a\Omega\cos\phi)$ to the mean specific angular momentum of the atmosphere at rest $2a^2\Omega/3$ (Hourdin et al., 1995). We run a 32-year simulation. The first two years are considered as spin-up, and only the results of the last 30 years are analyzed.

2.2. The results of climatic annual mean atmospheric circulation under different rotation rates

The wind speed difference can be used as a representation of the strength of the three-cell circulation (Oort and Yienger, 1996; Quan et al., 2004). With increasing day length, the strength changes of the three-cell circulation are different for the two hemispheres and at different latitudes, while the extent of the three-cell circulation does not change significantly. Figure 1 shows that the change under the higher rotation rate is opposite to

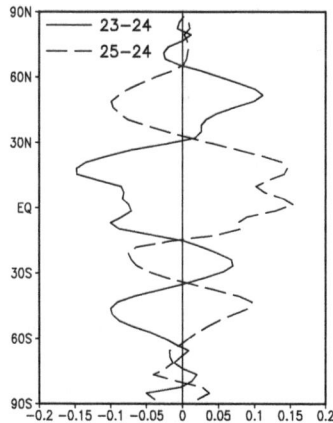

Figure 1. Three-cell circulation index (high-level meridional wind minus low-level meridional wind) obtained by subtracting the control run from the sensitivity run for the climate annual mean. The solid line is the 23-hour day sensitivity run minus the control run. The dashed line is the 25-hour day sensitivity run minus the control run. Units: m s⁻¹.

that under the lower rate. The three-cell circulation is stronger with slower rotation than with faster rotation. It should be pointed out that the changes of circulation at latitudes south of 80°S and between 10°S–10°N do not behave in this way. In these regions the circulation strength decreases under the slow rotation, and increases under faster rotation compared with the control run. In general, the main characteristics are that the circulation strength increases globally when the rotation slows, and vice versa.

Results for the temperature field are as follows (Fig. 2). There are cold anomalies in the Northern Hemisphere and warm anomalies in the Southern Hemisphere for the lower rotation rate, while the opposite applies for the higher rate. The boundary between cold and warm anomalies is at 15°S. The trend of geopotential height change is the same as for the temperature field between different rotation rates (figure not shown).

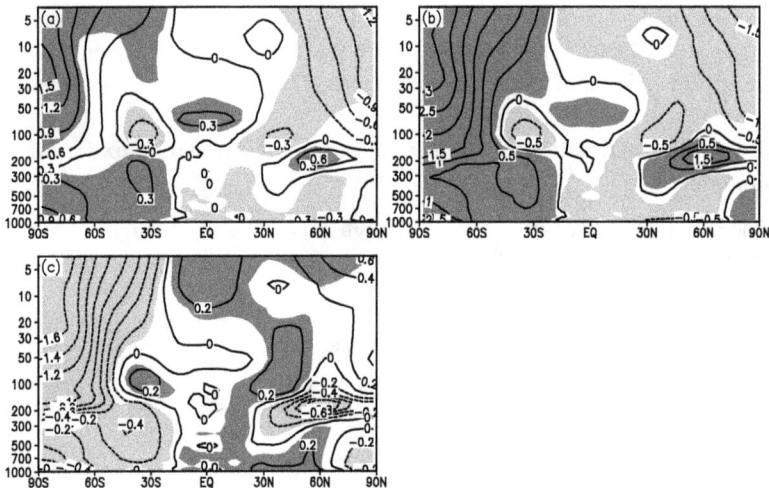

Figure 2. Latitude–pressure cross-section of climatological annual temperature anomaly field, which is the sensitivity run minus the control run. a) 25-hour day minus 24-hour day; b) Same as a) but for 25-hour minus 23-hour; c) Same as a) but for 23-hour minus 24-hour. The shaded areas indicate significant difference at the 95% level according to Student's t-test. Positive anomalies are shown by dark shading, negative ones by light shading. Units: K.

The sign of the annual mean zonal wind field anomalies changes when the rotation rate is changed (Fig. 3). The positive and negative anomalies reverse between higher and lower rotation rates. Westerlies are strengthened in the middle-higher latitudes (40°S–60°S and 40°N–60°N) under the lower rotation rate, and weakened elsewhere. The reverse situation applies under the higher rotation rate, but at slightly different latitudes. Compared with the control run, with slow rotation the easterlies strengthen and westerlies weaken at 15°S–30°S and below 700 hPa over the region 0°–30°N. Westerlies strengthen and easterlies weaken above 500 hPa at 30°S–70°S and 15°S–60°N. The situation reverses under the higher rotation rate. As the rotation rate becomes even slower, the extent of the positive and negative areas also increase. Convergence and divergence of the tropospheric wind field are enhanced when the rotation rate decreases (figure not shown). Vertical velocities are enhanced in the troposphere and north of 30°S in the stratosphere with a lower rotation rate, while the vertical velocity weakens south of 30°S in the stratosphere. The situation with a higher rotation rate reverses, but the change is not completely symmetrical (figure not shown).

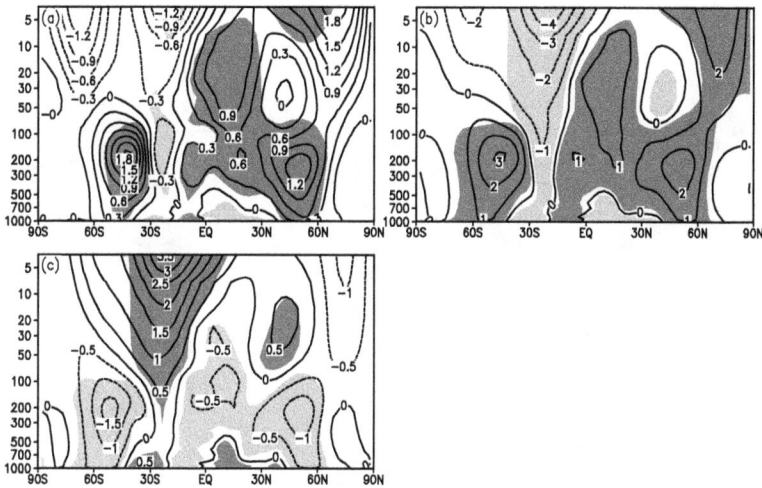

Figure 3. Same as Fig. 2 but for zonal wind. Units: m s^{-1}.

2.3. The results of climatic seasonal mean atmospheric circulation under different rotation rates

Figure 4 shows the changes in the climatological seasonal mean three-cell circulations with different rotation rates. The changes of three-cell circulation in the Northern and Southern Hemispheres are not consistent under different rotation rates. In the Northern Hemisphere the three-cell circulation strengthens under the slower rotation and weakens under the faster rotation. Enhancement of the three-cell circulation is not obvious in the Southern Hemisphere. Autumn changes in the ascending branches of the Hadley cell over the low-latitude (0°–25°S) Southern Hemisphere and in the Ferrel cell over high latitudes (60°S–90°S) are not consistent with changes in other seasons, which show weakening under the lower rotation rate and strengthening under the higher rate. The changes in strength of the three-cell circulation in autumn are the most obvious. For the Hadley cell, the change in strength in winter is the second most obvious. The changes in the climatological annual mean three-cell circulation are affected by those of the climatological seasonal mean three-cell circulation.

The magnitudes of change vary significantly with season, with the change in autumn being the largest. The largest changes in strength and extent of geopotential height under different rotation rates are seen in the autumn. The changes in spring under different rotation rates of geopotential height field, temperature field, meridional wind field in the stratosphere and vertical velocity field are opposite to those in summer and winter (and to the annual mean). Westerlies in mid-latitudes are strengthened in all four seasons with slow rotation. Changes in the zonal wind in the two hemispheres are opposite in spring and autumn.

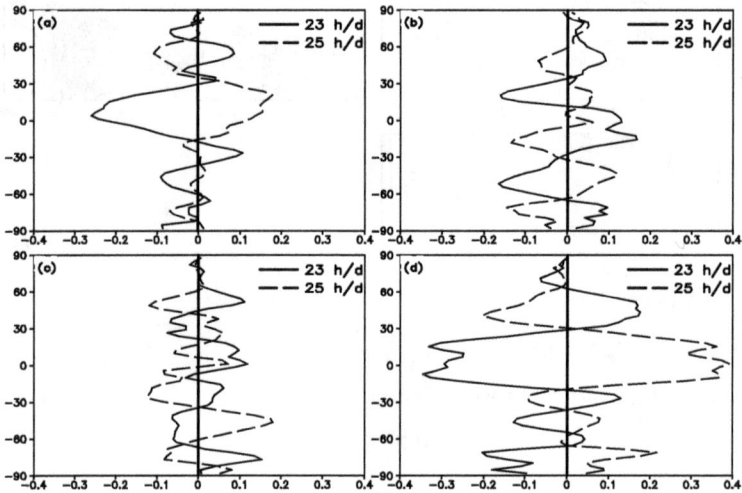

Figure 4. Same as Fig. 1 but for seasonal mean. The solid line and the dashed line are 23 h/d and 25 h/d, respectively. a)–d) are the situations in winter, spring, summer, and autumn, respectively. Units: m s^{-1}.

2.4. Concluding remarks

The general circulation of the Earth's atmosphere has been modeled under different rotation rates using the NCAR CAM2 model. The results indicate that the intensity of the three-cell circulation strengthens when the rotation rate slows. Slower rotation gives cold anomalies in the annual mean temperature field in the Northern Hemisphere and warm anomalies in the Southern Hemisphere, with the boundary between cold and warm anomalies located at 15°S. The annual mean zonal wind field has positive and negative anomalies when the rotation rate is changed. The positive and negative anomalies reverse between higher and lower rotation rates. In spring, the changes of geopotential height field, temperature field, meridional wind field in the stratosphere, and vertical velocity field are opposite to those in summer and winter (and the annual mean) under different rotation rates. Westerlies in mid-latitudes strengthen in all four seasons with slow rotation. Trends in the zonal wind changes in the two hemispheres are opposite in spring and autumn. Quantitative changes have significant seasonal differences, with the largest changes in autumn.

3. The influence of different obliquity on the general circulation of the Earth's atmosphere

The obliquity of the Earth ranges from 21.6° to 24.5° over a 41,000-year cycle. The obliquities of other planets and dwarf planets with atmospheres in our solar system range from 3° to 120°. Each atmosphere has its own characteristics. How does a change in obliquity influence atmospheric circulation?

Determining the influence of obliquity change over geological time scales involves the reconstruction and numerical simulation of paleoclimate. Kutzbach and Otto-Bliesner (1982) analyzed the effect of orbital parameter change in the Holocene on the African and Asian monsoons. The effects of different orbital parameters on the monsoon in summer and winter were analyzed in detail, revealing that the monsoon in the Holocene was stronger than now and that precipitation in Africa and India was heavier than now. The effect on seasonal global circulation of changes in the orbital parameter was not discussed in detail. Although Kutzbach and Guetter (1986) discussed the effect of different orbital parameter and surface forcing conditions on the climate, the radiation and sea temperature conditions were fixed at January and July values. The authors provided no detailed discussion of the effect of different orbital parameters on the seasonal cycle. Hall et al. (2005) investigated the effect of change in orbital parameter on the climate in winter in the Northern Hemisphere using an atmosphere–ocean coupled model with orbital parameters from the past 165,000 years. Their results suggest that while differences in global summer temperature can be explained by local thermodynamics caused by changes in radiative forcing, the change of winter climate in Northern Hemisphere cannot be explained in this way. They proposed that the change of radiative forcing has generated an atmospheric circulation anomaly similar to the NAM. This circulation anomaly perturbed other climatological variables. Tuenter et al. (2003) investigated the effects of procession and obliquity on the African monsoon. Past research on the effect of orbital parameter changes over geological scales on circulation and climate has deepened our understanding of paleoclimate, but these simulations have either focused on a specific geological period or a specific season (mostly summer and winter). The influence of a single orbital parameter change on the circulation has never been examined in detail. Moreover, there has been little work on the effect on the seasonal general circulation. The influence of different rotation rates on the atmospheric circulation was studied in the last section; in this section, we consider the influence of obliquity on the general circulation.

3.1. Model description and experimental design

In this section we use the same model as in Section 2. In addition to the control run, sensitivity runs with obliquities of 0°, 20°, 30°, and 60° are discussed. The same methods are used to judge whether the model is stable. We run a 32-year simulation; again, only the last 30 years are analyzed.

3.2. The results of climatic annual mean atmospheric circulation under different obliquities

Figure 5 shows that the intensity of the three-cell circulation weakens with increasing obliquity. The ascending branch of the Hadley cell in the Southern Hemisphere is enhanced with an obliquity of 60°. With obliquity increased, the Hadley cell in the Southern Hemisphere expands, and the Hadley cell in the Northern Hemisphere and the Ferrel cell in the Southern Hemisphere contract. It should be noted that the three-cell circulation in the Southern and Northern Hemispheres is not symmetric at an obliquity of 0°, possibly because of the asymmetry between the terrain in the Southern and Northern Hemispheres, or because the temperature forcing used in the model is the annual climatological mean. This latter factor can also cause the asymmetry of the three-cell circulation between the Northern and Southern Hemispheres at an obliquity of 0°. Compared with the control run, the distributions of anomalies are very similar for the sensitivity runs of 0° and 60°, and those of 20° and 30°, while the anomaly magnitudes of 0° and 60° are larger than those of 20° and 30°. There is a linear weakening in the three-cell circulation with increasing obliquity.

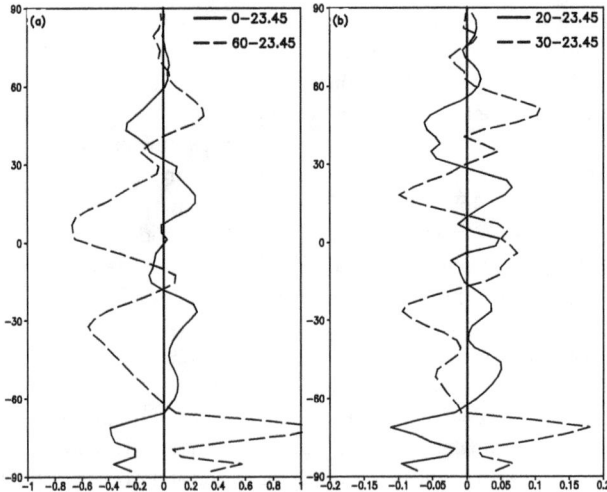

Figure 5. Three-cell circulation index (high-level meridional wind minus low-level meridional wind) for different obliquities, obtained by subtracting the climate annual mean control run from the sensitivity run. a) Solid line represents 0° minus 23.45°. The dashed line represents 60° minus 23.45°; b) Same as a) but for 20° minus 23.45° (solid line) and 30° minus 23.45° (dashed line). Units: m s⁻¹.

These variations in the three-cell circulation can be explained as follows. As obliquity changes, the highest latitude that sunlight can reach at winter solstice and summer solstice will change. The larger the obliquity, the higher the latitude that sunlight can reach. The radiation received throughout the year at high latitudes increases, while that received at low latitude decreases. This pattern necessarily weakens ascent over the equator and descent over the polar regions, so that the Hadley and anti-Hadley cells are weakened. Furthermore, the Ferrel cell is weakened.

Latitude–pressure cross-sections of the climate-mean annual geopotential height anomaly (figure not shown), which is the sensitivity run minus the control run, show that the positive anomaly mainly lies at 30°S–30°N (reaching a maximum of 205 gpm and 60 gpm, respectively) with smaller obliquity. The negative anomaly lies south of 30°S and north of 30°N, and the anomaly in the Northern Hemisphere is larger than that in the Southern Hemisphere. The situation reverses when obliquity is increased. There are increases in geopotential height at 30°S–30°N, and decreases south of 30°S and north of 30°N, with the obliquity decreased. The change of geopotential height with the obliquity change is larger in the Northern Hemisphere than in the Southern Hemisphere.

Figure 6 shows latitude–pressure cross-sections of the climate-mean annual temperature anomaly field. Consistent with the changes in geopotential height field, a reduction in obliquity to 0° and 20° gives positive anomalies at 30°S–30°N (reaching a maximum of 4 K and 1.2 K, respectively), with negative anomalies south of 30°S and north of 30°N. The situation reverses under larger obliquity. At an obliquity of 0°, there is a zone of positive temperature anomaly at 10 hPa over the South Pole region, extending northward and downward, reaching 45°S near the surface. At an obliquity of 60°, there is a region of positive anomaly between 100 to 200 hPa over the equator.

Latitude–pressure cross-sections of climate-mean annual zonal wind anomaly field (not shown) show that the extent of easterlies in the stratosphere over the equator with an obliquity of 0° is smaller than that in the control run, while with an obliquity of 20°, it is close to that in the control run. The extent of easterlies in the stratosphere over the equator with an obliquity of 30° is slightly larger than in the control run and is largest in these simulations for an obliquity of 60°. The velocity of the easterlies in the stratosphere over the equator increases with increasing obliquity. Easterly velocities can reach 80 m s^{-1} when the obliquity is 60°. The extent of the westerlies decreases with increasing obliquity. At the same time the extent of the jet stream is reduced. The jet stream at middle latitudes in the Northern Hemisphere decreases from 30 to 10 m s^{-1} when the obliquity increases from 0° to 60°. The strongest winds at high latitudes in the Northern and Southern Hemispheres decrease from 20 to 10 m s^{-1} and from 50 to 40 m s^{-1}, respectively.

Figure 7 shows the situation with obliquity reduced from 23.45° (Fig. 7a and b). The easterlies strengthen and westerlies weaken below 500 hPa over 30°S–45°S and 10°S–20°S, and below 300 hPa over 10°N–20°N and 60°N–90°N. At the same time, easterlies weaken and westerlies strengthen in the troposphere over 45°S–80°S, 20°S–30°S, 10°S–10°N, and 35°N–60°N. Easterlies weaken and westerlies strengthen south of 45°S in the stratosphere

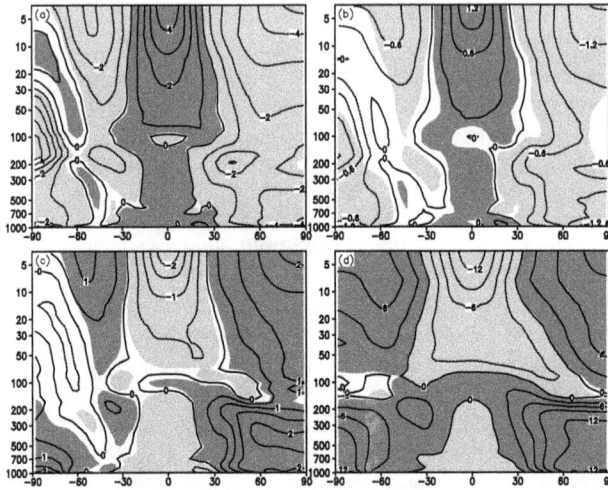

Figure 6. Latitude–pressure cross-section of climate-mean annual temperature anomaly field, which is the sensitivity run minus the control run. a) 0° minus 23.45°; b) 20° minus 23.45°; c) 30° minus 23.45°; d) 60° minus 23.45°. Units: K. The shaded areas are significant at the 95% level according to Student's *t*-test. Dark shading indicates positive anomaly, light shading indicates negative anomaly.

and over 40°S–60°N. With obliquity greater than 23.45° (Fig. 7c and d), easterlies weaken and westerlies strengthen over 80°S–90°S below 500 hPa, 30°S–45°S, 10°S–20°S below 300 hPa, 10°N–20°N below 300 hPa, and over 60°N–90°N. The easterlies strengthen and westerlies weaken over 45°S–80°S in the troposphere, and over 20°S–30°S, 10°S–10°N, and 35°N–60°N. The easterlies strengthen and westerlies weaken south of 45°S in the stratosphere and over 40°S–60°N.

To summarize, the effect of increased obliquity is as follows. The extent of easterlies in the stratosphere increases and wind velocity strengthens. The extent of westerlies in the Southern and Northern Hemispheres is reduced. The jet stream in the Northern Hemisphere weakens, while that in the Southern Hemisphere strengthens. Except for the strengthening of easterlies over 10°S–10°N, the wind fields over other regions near the surface weaken. The pattern with reduced obliquity is reversed.

For the annual-mean meridional wind field (not shown), as obliquity is increased the strengths of the three-cell circulation, convergence, and divergence all weaken. With increasing obliquity, southerlies weaken and northerlies strengthen south of 50°S in the stratosphere and north of 40°N. Northerlies weaken over 50°S–0° below 5 hPa in the stratosphere, while southerlies weaken over 0°–25°N. Northerlies above 5 hPa over 50°S–0° and southerlies above 5 hPa over 0°–25°N strengthen.

For the vertical wind field (figure not shown), the results in the troposphere are consistent with the change of three-cell circulation and meridional wind field. With increasing obliquity the vertical wind velocity decreases. In the stratosphere, the ascent strengthens over 45°S–65°S, 25°S–25°N, and 60°N–90°N, and weakens over 65°S–90°S. Descent strengthens over 25°S–45°S and 25°N–45°N.

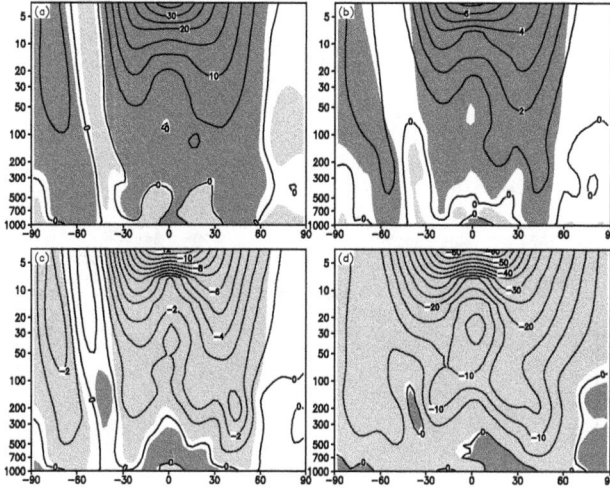

Figure 7. Same as Fig. 6 but for zonal wind field. Units are m s⁻¹.

3.3. The results of climatic seasonal mean atmospheric circulation under different obliquities

For the three-cell circulation, the results of increased obliquity are as follows (Fig. 8). The three-cell circulation in the Northern Hemisphere strengthens in winter, and that in the Southern Hemisphere weakens. The three-cell circulation in the Northern Hemisphere weakens in spring, and the Hadley cell in the Southern Hemisphere strengthens. The other two cells weaken in the Southern Hemisphere in spring. In summer, the three-cell circulation weakens in the Northern Hemisphere, and the Hadley cell in the Southern Hemisphere strengthens. At the same time, the changes of the other two cells are not obvious. In autumn, the global three-cell circulation weakens. The trend of changes in the three-cell circulation in the Northern Hemisphere in winter, and the Hadley cell in the Southern Hemisphere in spring and summer are opposite to the corresponding trends of the annual mean. The changes in strength of the three-cell circulation in winter and summer are larger than in the other two seasons. With the exception of the Northern Hemisphere three-cell circulation in winter and the Southern Hemisphere Hadley cell in spring and summer, which strengthen with increased obliquity, the general trend is for cells to weaken.

The results for the geopotential height field with increased obliquity are as follows (figure not shown). In spring the geopotential height decreases over 60°S–20°N, and increases south of 60°S and north of 20°N. In summer the geopotential height increases north of 30°N and over 40°S–30°N at 100–20 hPa, while it decreases below 100 hPa over 30°S–30°N, above 5 hPa at 30°S–30°N, and south of 30°S. In autumn the geopotential height decreases over 30°S–

30°N and south of 50°S at 200–10 hPa, while it increases north of 30°N, over 30°S–50°S, and below 500 hPa south of 30°S. In winter the geopotential height decreases north of 20°S and increases south of 20°S.

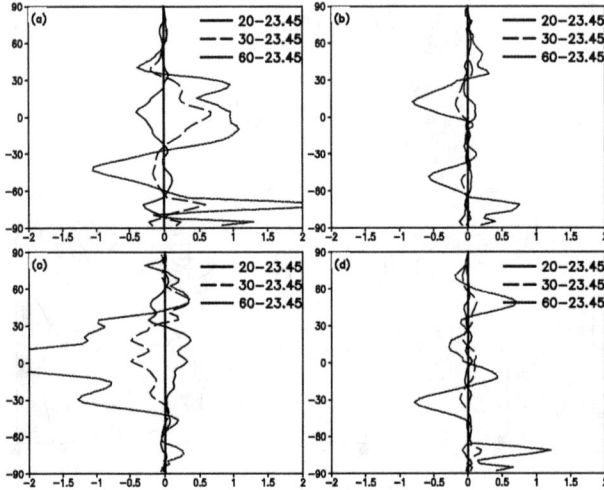

Figure 8. Same as Fig. 5 but for the seasonal mean. The solid line, dashed line, and short dotted line are for obliquities of 20°, 30°, and 60° minus 23.45°, respectively. a)–d) are the situations in winter, spring, summer, and autumn, respectively. Units: m s^{-1}.

The results for the temperature field with increased obliquity are as follows (figure not shown). In spring, there are cold anomalies (magnitude approximately 1 K) over 45°S–20°N, and warm anomalies south of 45°S and north of 20°N (the maximum warm anomaly can reach 2–3 K). In summer there are warm anomalies north of 10°N, and at 200–50 hPa over 30°S–10°N, while cold anomalies exist south of 10°N. In autumn the temperature decreases over 30°S–30°N and south of 30°S at 300–20 hPa, while it increases north of 30°N, above 30 hPa at 30°S–60°S, and south of 30°S below 300 hPa. In winter the temperature decreases north of 25°S and increases south of 25°S.

The zonal wind field responds to increased obliquity as follows (figure not shown). The extent of easterlies increases over the equator and easterly velocity strengthens, while the extent of westerlies decreases at middle latitudes. The jet stream strengthens at middle latitudes in the Southern Hemisphere while that in the Northern Hemisphere weakens. In spring, the wind strengthens in the stratosphere over the equator, while the easterlies weaken in the troposphere over the equator. The westerlies strengthen over 30°S–40°S below 200 hPa. In summer the extent and strength of easterlies increase in the stratosphere over the equator, but the easterlies weaken in the troposphere over the equator. The westerlies weaken in the Northern Hemisphere. The westerlies weaken at low latitudes (0°–30°S) in the Southern Hemisphere, strengthen at middle latitudes (30°S–60°S), and weaken at high latitudes (south of 60°S). In autumn the situation is similar to summer, with increased extent and strength of easterlies in the stratosphere over the equator. Easterlies weaken in the troposphere over the equator, and in the Northern Hemisphere the westerlies weaken at middle–low latitudes (south of 60°N) and strengthen at high latitudes (north of 60°N). In winter the extent and velocity of easterlies increase in the stratosphere over the equator, while over the equator and globally, westerlies weaken.

Changes in the seasonal-mean vertical wind field are consistent with that of the three-cell circulation, and only the seasonal anomalies in the stratosphere resulting from increased obliquity are described here. In spring, ascending motion weakens over 0°–30°S, 80°S–90°S, and 70°N–80°N, while it strengthens over 0°–30°N. In summer, ascending motion weakens over 0°–30°S, south of 60°S, and north of 70°N, while it strengthens over 0°–50°N, and descending motion strengthens over 45°S–60°S. In autumn, ascending motion strengthens over 0°–30°S and weakens over 0°–30°N. In winter, descending motion strengthens over 0°–30°N and ascending motion strengthens over 0°–40°S. The descending motion over 40°S–60°S and ascending motion south of 70°S weaken.

3.4. Concluding remarks

The general circulation of the Earth's atmosphere has been simulated with different obliquities using the NCAR CAM2 model. The results indicate that the three-cell circulation weakens for large obliquity except in the Northern Hemisphere in winter, and the Hadley circulation in the Southern Hemisphere in spring and summer, which strengthen. The intensity of the annual mean three-cell circulation weakens for large obliquity. The extent of the Hadley circulation in the Southern Hemisphere increases, while that of the Hadley circulation in the Northern Hemisphere and the Ferrel circulation in the Southern Hemisphere decreases for large obliquity. The ascending branch of the Hadley circulation in the Southern Hemisphere strengthens significantly under the 60°-obliquity condition compared with the normal obliquity of the Earth. Both the annual mean extent and velocity of easterly winds in the stratosphere over the equator increase for large obliquity, whereas the extent of westerly winds decreases. The jet stream in the Northern Hemisphere weakens for large obliquity, while that in the Southern Hemisphere strengthens. In all four seasons, the easterlies in the troposphere and westerlies in the Northern Hemisphere weaken for large obliquity, as does the Northern Hemisphere mid-latitude jet stream. The differences in seasonal response for large obliquity are as follows. The mid-latitude jet stream in the Southern Hemisphere strengthens in spring, the westerly winds at middle and high latitudes change in opposite senses in summer and autumn in the Southern Hemisphere, and the global westerly winds are weaker in winter.

4. The influence of different orbit parameter on the monsoon

Generally the change of orbital parameter can modify the monsoon system by changing solar radiation. For example, Kutzbach and Otto-Bliesner (1982) analyzed the effect of orbital parameter change in Holocene on the African and Asian monsoon. The difference of radiation at two solstitial points are biggest the two solstitial points are closed related with the monsoon, therefore the effects of different orbital parameters on the monsoon in summer and winter are analyzed. The results show that the monsoon in the Holocene was stronger than today and that precipitation in Africa and India were greater than today. In the 1980s, COHMAP, led by J. E. Kutzbach, T. Webb III, and H. E. Wright, was designed to determine and simulate paleoclimate using the land record. It revealed the key role of orbital factors in tropical monsoon climate change, and the bifurcation of the westerly jet stream over North America as a result of the ice cap in North America during the LGM period. Simulations show at the same time that the change of Earth's orbit in the early to middle Holocene led to a seasonal increase in the Northern Hemisphere and an enhanced monsoon (Kutzbach and Street-Perrott, 1985; Kutzbach and Guetter, 1986; Mitchell *et al.*, 1988; Kutzbach, 1989; Kutzbach and Gallimore, 1989; Barron *et al.*, 1993; Kutzbach *et al.*, 1993; Rahmstorf, 1994; Barron *et al.*, 1995; Rahmstorf, 1995; Kutzbach *et al.*, 1996; Bush and Philander, 1997; Otto-Bliesner and Upchurch Jr, 1997; Ramstein *et al.*, 1997; Weaver *et al.*, 1998; Cane and Molnar, 2001; Knutti *et al.*, 2004).

Can a change in rotation rate change the monsoon system? What is the effect of altering the obliquity on the monsoon system? In this section we seek to answer these questions using a unified dynamical index of the monsoon—the normalized seasonality and dynamical normalized seasonality (DNS), proposed by Li and Zeng (2000, 2002, 2003). The DNS is computed using the results in the previous two sections, to investigate the influence of the rotation rate and obliquity on the monsoon system. The normalized seasonality is given by

$$\delta = \frac{\| \overline{V_1} - \overline{V_7} \|}{\| \overline{V} \|} - 2,$$

where $\overline{V_1}$ and $\overline{V_7}$ are the mean January and July climatological wind vectors, respectively. $\overline{V}$ is the mean climatological wind vectors. The region of $\delta > 0$ represents the monsoon region. The DNS is given by

$$\delta^*_{m,n} = \frac{\| \overline{V_1} - \overline{V}_{m,n} \|}{\| \overline{V} \|} - 2,$$

where $\overline{V}_{m,n}$ is the monthly wind vector for year n and month m.

4.1. The influence of different rotation rates on the global monsoon system

Figure 9 shows model output for the global normalized variability at 850 hPa. Compared with the control run (Fig. 9b), there is no obvious change in the extent of the tropical monsoon region and subtropical monsoon region, while the monsoon regions in the temperate and frigid zones change for both higher (Fig. 9a) and lower rotation rate (Fig. 9b).

The vertical distribution of the monsoon (Figs 10–14) shows no obvious change except for a slight movement of the edge of the monsoon region, but the strength of the monsoon clearly changes. Because of the importance of the monsoon in the low level of the troposphere, Student's *t*-test was used to assess the statistical significance of the strength change of the monsoon at 850 hPa (Fig. 15). With faster rotation, the south and north African monsoons strengthen, while the tropical African monsoon weakens (Fig. 15c). In general, the African monsoon strengthens. Monsoons strengthen from the Arabian Sea to India, north of the Bay of Bengal, and from the southeast coast of China to the middle and lower reaches of the Yangtze River. At the same time, the monsoons in the Far East and North Pole region strengthen, while those in northeast Asia and from the south of the Bay of Bengal to the South China Sea and the East China Sea weaken. With slower rotation (Fig. 15a and b) the African monsoon, European monsoon, monsoon in the Far East, and monsoon in the North Pole region all weaken. Monsoons in the South China Sea and the East China Sea are still weakened. Monsoons in the Arabian Sea, Bay of Bengal, and north of Africa all strengthen. The African monsoon and the monsoons in the temperate zone and frigid zone weaken when the rotation rate slows, a reversal of the response to faster rotation. The Asian–Australian monsoon does not exhibit this reverse relationship. Globally, the monsoons do not change consistently with rotation rate.

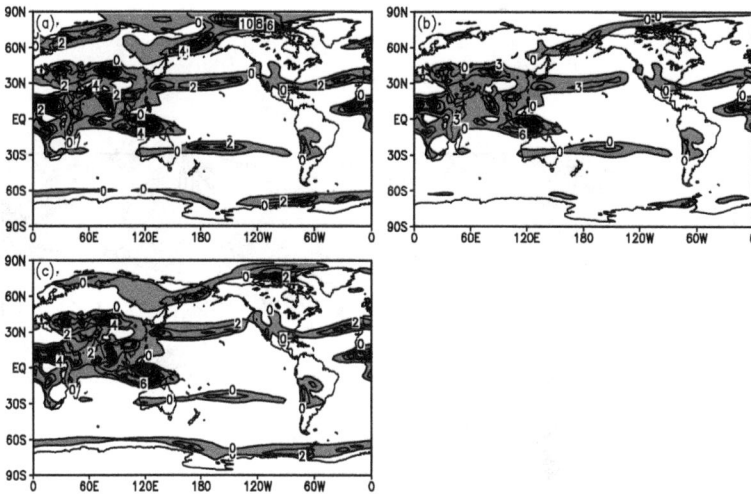

Figure 9. Global normalized variability at 850 hPa obtained from simulations. a) 23-hour day length; b) 24-hour day length; c) 25-hour day length. Shading indicates regions in which the normalized variability is greater than zero.

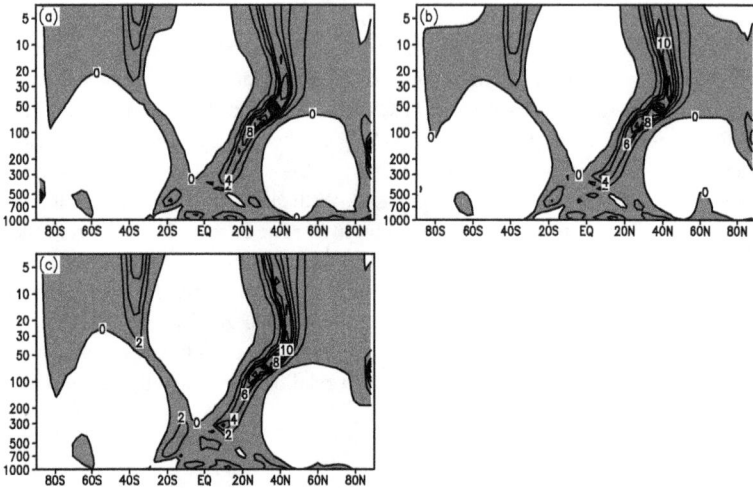

Figure 10. Latitude–pressure cross-section of zonal mean normalized variability. a) 23-hour day length; b) 24-hour day length; c) 25-hour day length. Shading indicates regions in which the normalized variability is greater than zero.

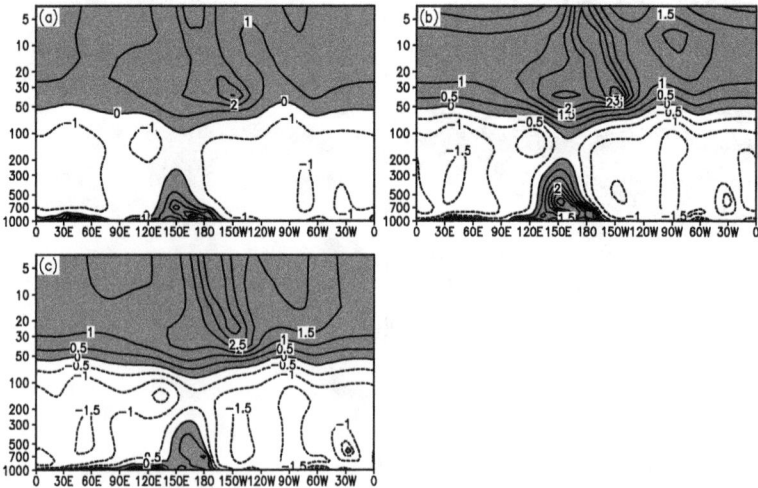

Figure 11. As Fig. 10, but for 50°N–70°N.

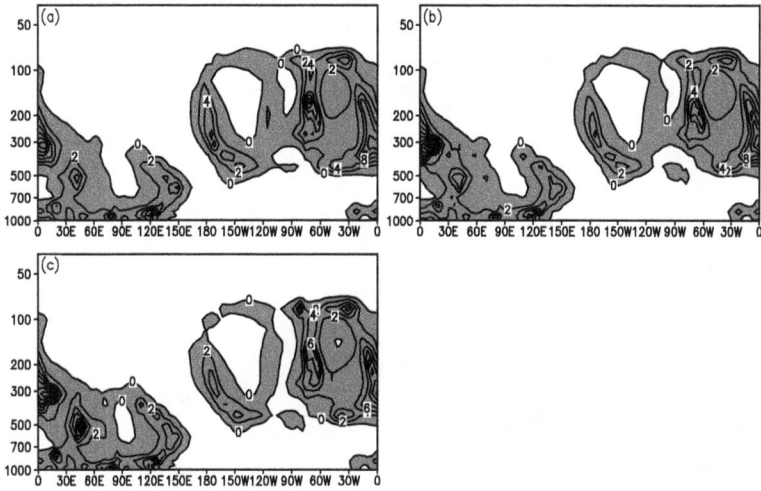

Figure 12. As Fig. 10, but for the tropics (10°S–10°N).

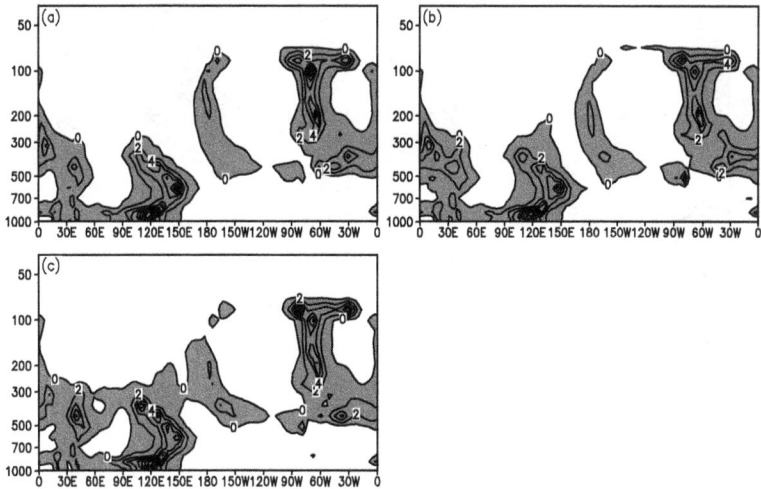

Figure 13. As for Fig.10, but for the tropical Southern Hemisphere (10°S–0°).

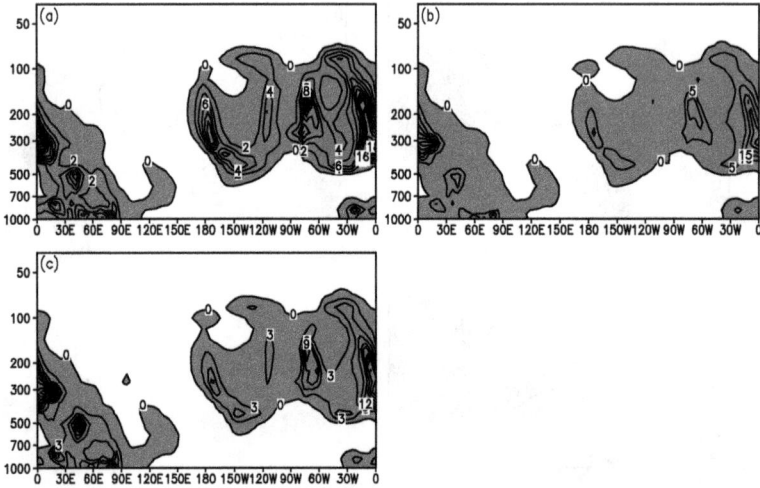

Figure 14. As for Fig. 10, but for the tropical Northern Hemisphere (0°–10°N).

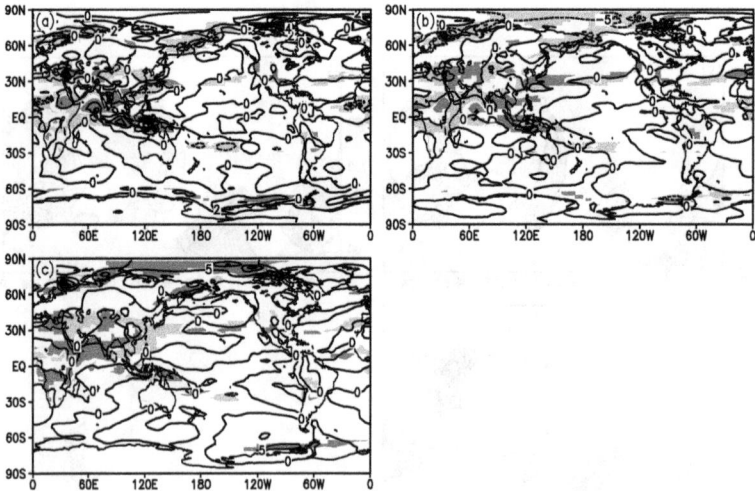

Figure 15. Normalized variability anomaly field, which is the sensitivity run minus the control run. (a) 25-hour day length minus 24-hour day length; (b) 25-hour day length minus 23-hour day length; (c) 23-hour day length minus 24-hour day length. Shading indicates statistically significant difference at the 95% level according to Student's *t*-test. Dark and light areas indicate positive and negative anomalies, respectively.

4.2. The influence of different obliquity on the global monsoon system

A change in obliquity will inevitably change the distribution of radiation; indeed, there is no seasonal variation with an obliquity of zero. Does the meridional wind reverse between January and July seasonally in the other sensitivity runs? Figure 16 shows the monthly-mean meridional winds over the domain 0°–10°N, 0°–120°E The meridional wind in January is opposite to that in June and July. Therefore, it is reasonable to use the DNS proposed by Li and Zeng (2000, 2002, 2003) and used in Section 4.1 to investigate the influence of obliquity on the monsoon system.

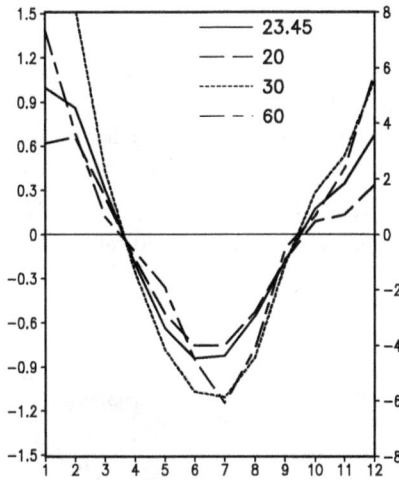

Figure 16. Monthly-averaged meridional winds over the domain 0°–10°N, 0°E–120°E. The horizontal axis shows the month of the year. The scale for meridional wind velocity for obliquities of 23.45°, 20° and 30° is on the left, and for obliquity of 60° on the right. Units are m s⁻¹.

Figure 17 shows the global normalized variability at 850 hPa obtained from the numerical simulations. Compared with the control run (Fig. 17d), the extent of the global monsoon region increases significantly with increased obliquity. The vertical distribution of the monsoon for different obliquities (Fig. 18) shows that the edge of the monsoon above 20 hPa in the stratosphere retreats to high latitudes with increased obliquity. At the same time, the monsoon in the stratosphere expands downward. It links up with the monsoon in the troposphere, which expands upward south of 60°S with increased obliquity. The monsoon in the stratosphere in the Northern Hemisphere also expands downward to the whole stratosphere. The monsoon in the Northern Hemisphere troposphere also expands.

In the temperate and frigid zones (50°N–70°N), the monsoon in the stratosphere extends down to the troposphere with increased obliquity, while that in the low troposphere extends slightly upward (Fig. 19). The tropical monsoon (10°S–10°N) (Fig. 20) expands at high levels with increased obliquity east of 180°E so that it covers the whole region from 500 hPa to 100 hPa east of 180°E. The part of the monsoon west of 180°E but east of 90°E expands westward and shrinks downward, while the part west of 90°E shrinks westward. The monsoon near the surface extends eastward to east of 150°E. For the tropical monsoon in the Southern Hemisphere (Fig. 21), the monsoon east of 90°W expands slightly horizontally. The monsoon over 180°E breaks up, and the western part shrinks horizontally and vertically. In particular, the tropical monsoon in the Northern Hemisphere (Fig. 22) expands with increased obliquity.

Because of the importance of the low-level monsoon, the significance of the t-test for the changes in monsoon strength at 850 hPa is also assessed here (Fig. 23). Figure 23 shows that the African, South American, North Pacific, and East Asian monsoons all strengthen with increased obliquity, and that the extent of the global monsoon increases. Furthermore, with decreased obliquity, the African, South American, North Pacific, and East Asian monsoons all weaken, and the extent of the global monsoon decreases. There is no obvious linear relationship with obliquity for monsoon strength in other regions. Therefore, the influence of obliquity on the monsoon in other region is perhaps indirect and non-linear.

In general, increased obliquity results in an increase in the extent of the global monsoon. This kind of change has a complex horizontal and vertical structure.

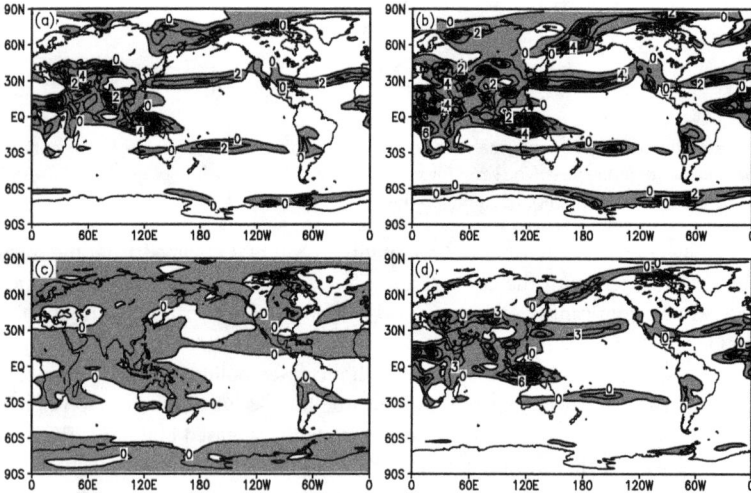

Figure 17. As for Fig. 9 but for different obliquity conditions. (a) 20° obliquity; (b) 30° obliquity; (c) 60° obliquity; (d) 23.45° obliquity.

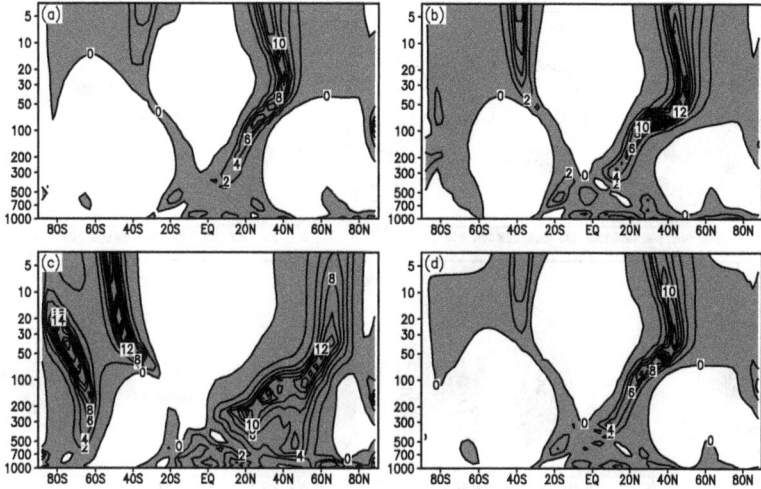

Figure 18. Zonal mean latitude–pressure cross-section. (a) 20° obliquity; (b) 30° obliquity; (c) 60° obliquity; (d) 23.45° obliquity.

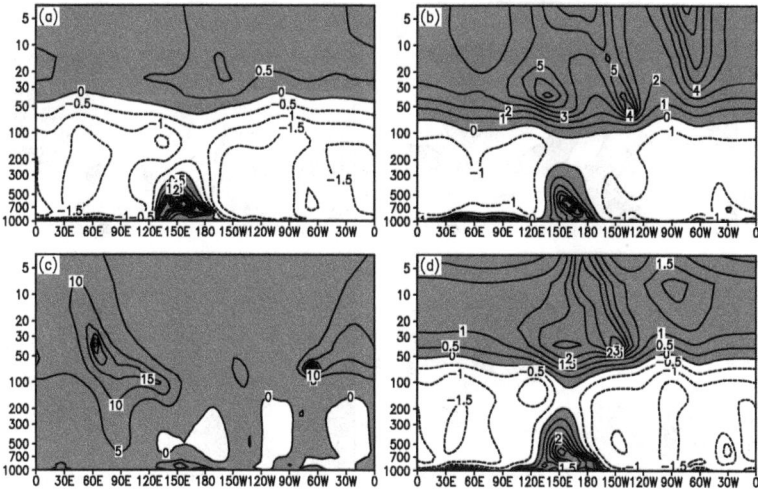

Figure 19. As for Fig. 18 but zonal cross-section for 50°N–70°N.

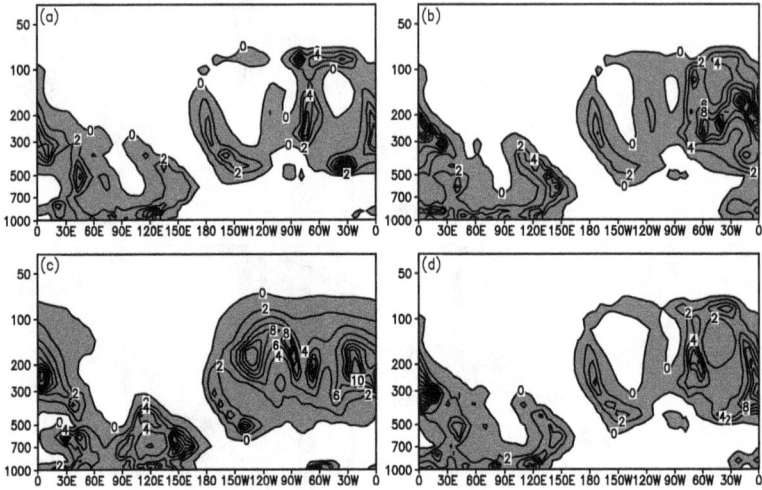

Figure 20. As for Fig. 18 but a tropical (10°S–10°N) latitude–pressure cross-section.

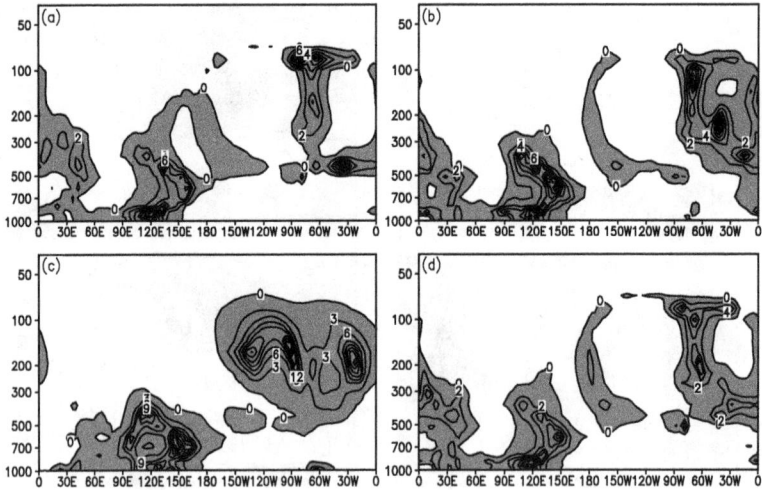

Figure 21. As for Fig. 18 but a Southern Hemisphere tropical latitude–pressure cross-section (10°S–0°).

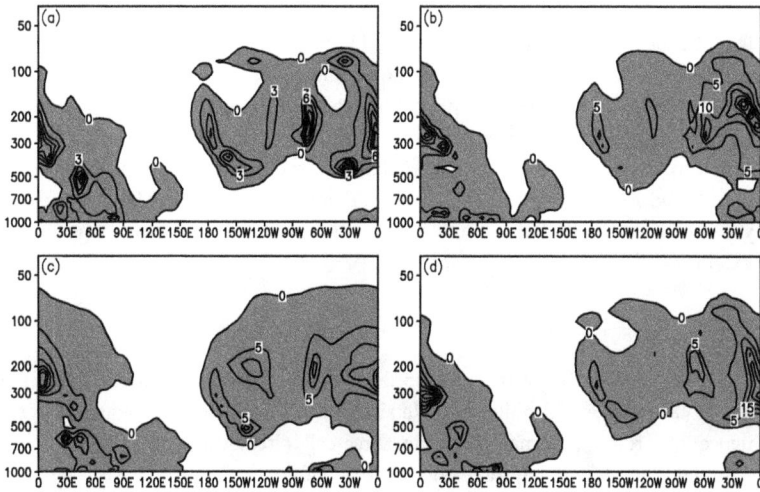

Figure 22. As for Fig. 18 but a Northern Hemisphere tropical latitude–pressure cross-section (0°–10°N).

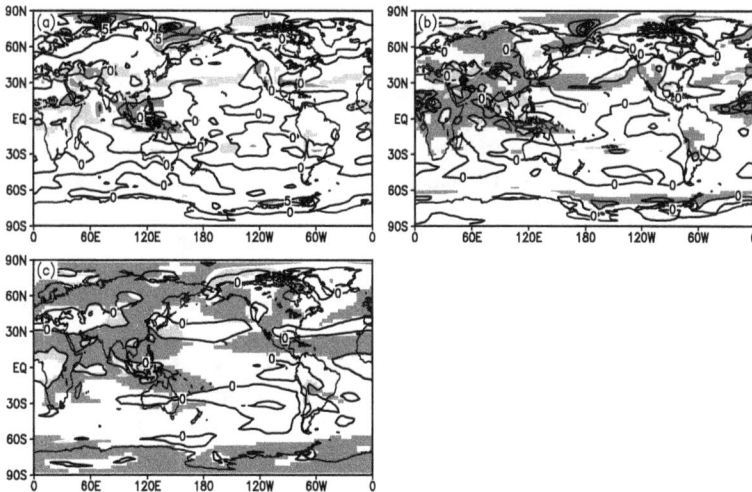

Figure 23. As for Fig. 15 but for different obliquity conditions. (a) 20° obliquity minus that of 23.45°; (b) 30° obliquity minus that of 23.45°; (c) 60° obliquity minus that of 23.45°. The shaded areas are significant at the 95% level according to Student's t-test. The dark and light areas indicate positive and negative anomalies, respectively.

4.3. Concluding remarks

There are three regions of linear response to the effect of rotation rate on the monsoon: Africa, the Middle East, and the temperate and frigid zones. For example, the African monsoon and the monsoon in the temperate and frigid zones weaken when the rotation rate slows. In general, with increased obliquity, the extent of the global monsoon is increased. This change also has a complex horizontal and vertical structure. At 850 hPa, the African, South American, North Pacific, and East Asian monsoons all strengthen with increased obliquity, and the extent of the global monsoon increases. At 850 hPa, with decreased obliquity, the African, South American, North Pacific, and East Asian monsoons all weaken, and the extent of the global monsoon decreases.

5. Results and discussion

The general circulation of the Earth's atmosphere has been simulated under different orbital parameters (rotation rate and obliquity). Retaining the same sea/land positions, we study the influence of orbital parameter on the three-cell circulation, potential height field, temperature field, wind field and monsoon on different timescales.

The results for the mean annual atmospheric circulation under different rotation rates are as follows:

1. The strength of the three-cell circulation is increased with the slower rotation compared with the higher rate, except for latitudes south of 80°S and 10°S–10°N, where the circulation weakens with slower rotation and strengthens with faster rotation, compared with the control run.
2. There are negative anomalies in the Northern Hemisphere annual mean temperature and positive anomalies in the Southern Hemisphere with the lower rotation rate, while the opposite is true for the faster case. The boundary between negative and positive anomalies is 15°S. Geopotential height changes in the same way as temperature for the different rotation rates.
3. Westerlies strengthen in the regions corresponding to 40°S–60°S and 40°N–60°N in the case of slow rotation, and weaken in other regions. The situation is reversed for faster rotation, but at slightly different latitudes. Westerlies strengthen in middle–high latitudes in both hemispheres for slow rotation. Compared with the control run, the easterlies strengthen and westerlies weaken over 15°S–30°S and below 700 hPa over 0°–30°N. Westerlies strengthen and easterlies weaken above 500 hPa over 30°S–70°S and 15°S–60°N. The situation reverses under faster rotation. Vertical velocities are enhanced in the troposphere and north of 30°S in the stratosphere for the lower rotation rate, while the vertical velocity weakens south of 30°S in the stratosphere. The situation under the higher rotation rate reverses, but the change is not completely symmetrical.

The results for the mean seasonal atmospheric circulation under different rotation rates are as follows:

1. The three-cell circulation strengthens for slow rotation and weakens for faster rotation, but the strengthening is not obvious in the Southern Hemisphere. Changes in autumn over the Southern Hemisphere, in the ascending branches of the Hadley cell over low latitudes (0°–25°S) and the Ferrel cell over high latitudes (60°S–90°S), which weaken for low rotation rate and strengthen for rapid rotation, are not consistent with those of other seasons. The largest changes of strength of the three-cell circulation occur in autumn, and for the Hadley cell, the second largest occur in winter. The changes in the mean annual three-cell circulation are driven by those in the seasonal mean three-cell circulation.

2. The change in autumn is the largest. The strength and extent of changes of geopotential height in autumn under different rotation rates are the largest. In spring, the changes with rotation rate in the geopotential height field, temperature field, and meridional wind field in the stratosphere and in the vertical velocity field are the opposite of the changes in summer and autumn (and annual mean). Westerlies in mid-latitudes strengthen in all four seasons with slower rotation. Changes in the zonal wind in both hemispheres are opposite in spring and autumn.

Results for the mean annual atmospheric circulation under different obliquity are as follows:

1. The intensity of the three-cell circulation weakens with increased obliquity. The ascending branch of the Hadley cell in the Southern Hemisphere strengthens with an obliquity of 60°. With increased obliquity, the Hadley cell in the Southern Hemisphere expands, and the Hadley cell in the Northern Hemisphere and the Ferrel cell in the Southern Hemisphere contract. Compared with the control run, the anomaly distributions are very similar between the sensitivity runs of 0° and 60° and those of 20° and 30°, while the anomaly magnitudes of 0° and 60° are larger than those of 20° and 30°. With increased obliquity, the three-cell circulation weakens linearly.

2. With decreased obliquity the geopotential height increases over 30°S–30°N, and decreases south of 30°S and north of 30°N. The change of geopotential height with obliquity change is asymmetrical in the Southern and Northern Hemispheres. Changes in the Northern Hemisphere are larger than in the Southern Hemisphere.

3. With decreased obliquity, the warm temperature anomalies mainly lie between 30°S and 30°N (reaching maximum values of 4 K and 1.2 K for 0° and 20° obliquity, respectively), while the cold anomalies lie south of 30°S and north of 30°N. The situation reverses for increased obliquity. With an obliquity of 0°, a zone of warm temperature anomaly at 10 hPa over the South Pole region extends northward and downward, reaching near the surface at 45°S. With an obliquity of 60°, the region of warm anomaly lies from 100 to 200 hPa over the equator.

4. With large obliquity, the extent of easterlies in the stratosphere increases and wind velocity strengthens. The extent of westerlies in both Southern and Northern Hemispheres decreases. The jet stream weakens in the Northern Hemisphere and strengthens in the Southern Hemisphere. Winds over other regions near the surface weaken, except over 10°S–10°N where the easterlies strengthen. The situation reverses with decreasing obliquity. In the mean annual meridional wind field, with increased

obliquity, southerlies weaken and northerlies strengthen south of 50°S in the stratosphere and north of 40°N. The northerlies weaken over 50°S–0° below 5 hPa in the stratosphere. The southerlies weaken over 0°–25°N. Northerlies strengthen above 5 hPa at 50°S–0° and southerlies strengthens above 5 hPa at 0°–25°N. Changes in the vertical wind in the troposphere are consistent with changes in the three-cell circulation and meridional wind field. With increased obliquity the vertical winds weaken. In the stratosphere, ascent strengthens over 45°S–65°S, 25°S–25°N, and 60°N–90°N, and weakens over 65°S–90°S. Descent strengthens over 25°S–45°S and 25°N–45°N.

The mean seasonal atmospheric circulation responds to changes in obliquity as follows:

1. In winter, with increased obliquity, the three-cell circulation in the Northern Hemisphere strengthens, and that in the Southern Hemisphere weakens. The three-cell circulation in the Northern Hemisphere weakens in spring, and the Hadley cell in the Southern Hemisphere strengthens. The other two cells weaken in the Southern Hemisphere in spring. In summer, the three-cell circulation weakens in the Northern Hemisphere, and the Hadley cell in the Southern Hemisphere strengthens. In this season, the changes of the other two cells are unclear. In autumn, the global three-cell circulation weakens. Changes in the three-cell circulation in the Northern Hemisphere in winter and the Hadley cell in the Southern Hemisphere in spring and summer are opposite to the corresponding changes in the annual mean. The changes in strength of the three-cell circulation in winter and summer are larger than in the other two seasons.

2. With increased obliquity, in spring the geopotential height decreases over 60°S–20°N, and increases south of 60°S and north of 20°N. In summer the geopotential height increases north of 30°N and over 40°S–30°N at 100–20 hPa, while it decreases below 100 hPa over 30°S–30°N, above 5 hPa both at 30°S–30°N and south of 30°S. In autumn, the geopotential height decreases over 30°S–30°N and south of 50°S at 200–10 hPa, while it increases north of 30°N, over 30°S–50°S, and below 500 hPa south of 30°S. In winter, geopotential height decreases north of 20°S and increases south of 20°S.

3. With increased obliquity, in spring there are cold temperature anomalies (magnitude approximately 1 K) over 45°S–20°N, and warm anomalies south of 45°S and north of 20°N (with maximum values of 2–3 K). In summer there are warm anomalies north of 10°N, and at 200–50 hPa over 30°S–10°N, while cold anomalies exist south of 10°N. In autumn the temperature decreases over 30°S–30°N and south of 30°S at 300–20 hPa. Temperatures increase north of 30°N, over 30 hPa at 30°S–60°S, and south of 30°S below 300 hPa. In winter the temperature decreases north of 25°S and the temperature increases south of 25°S.

4. With increased obliquity, in spring the zonal wind strengthens in the stratosphere over the equator. The easterlies weaken in the troposphere over the equator, while westerlies strengthen over 30°S–40°S below 200 hPa. In summer the extent of easterlies expands in the stratosphere over the equator, and the velocity of easterlies strengthens. The easterlies weaken in the troposphere over the equator, and the westerlies weaken in the Northern Hemisphere. Westerlies weaken at low latitudes (0°–30°S) in the Southern Hemisphere, strengthen at middle latitudes (30°S–60°S), and weaken at high latitude

(south of 60°S). In autumn the situation is similar to the summer, so that with increased obliquity, the extent and velocity of easterlies increase in the stratosphere over the equator. At the same time, easterlies weaken in the troposphere over the equator, and westerlies weaken at middle low latitudes (south of 60°N) and strengthen at high latitudes (north of 60°N) in the Northern Hemisphere. In winter, the extent and velocity of easterlies increase in the stratosphere over the equator, while the velocity over the equator and the global westerlies weaken.

5. Changes in the mean seasonal vertical wind field are consistent with those of the three-cell circulation. Responses in the stratosphere to increased obliquity are as follows. In spring, ascending motion weakens over 0°–30°S, 80°S–90°S, and 70°N–80°N, and strengthens over 0°–30°N. In summer, ascending motion weakens over 0°–30°S, south of 60°S, and north of 70°N, and strengthens over 0°–50°N while the descending motion strengthens over 45°S–60°S. In autumn, ascending motion strengthens over 0°–30°S and weakens over 0°–30°N. In winter, descending motion strengthens over 0°–30°N and ascending motion strengthens over 0°–40°S. The descending motion over 40°S–60°S and ascending motion south of 70°S are weakened.

Results for the monsoon system are as follows:

1. There is no obvious change in the extent of the tropical and subtropical monsoon regions under different rotation rates, while the monsoon region in the temperate and frigid zones clearly changes. The vertical distribution of monsoon does not appear to change except for a slight movement of the edge of the monsoon region. The monsoon strength clearly changes with different rotation rates. Both the southern and northern parts of the African monsoon strengthen with rapid rotation, while the tropical African monsoon weakens. Monsoons strengthen from the Arabian Sea to India, north of the Bay of Bengal, and from the southeast coast of China to the middle and lower reaches of the Yangtze River. At the same time, the monsoons in the Far East and North Polar Region strengthen, while the monsoons in northeast Asia and from south of the Bay of Bengal to the South China Sea and the East China Sea weaken. With slow rotation the African and European monsoons, as well as the monsoon in the Far East and in the North Polar Region all weaken. Monsoons in the South and East China Sea still weaken. Monsoons in the Arabian Sea, Bay of Bengal, and north of Africa are all strengthened. The African monsoon and the monsoon in the temperate and frigid zones weaken when the rotation rate reduces. The situation is reversed with faster rotation. The Asian–Australian monsoon does not exhibit this reverse relationship. Globally, the monsoons do not change consistently with rotation rate.

2. With increased obliquity, the extent of the global monsoon increases. This change has complex horizontal and vertical structure. With increased obliquity, at 850 hPa, the African, South American, North Pacific, and East Asian monsoons all strengthen, and the global monsoon increases in extent. With decreased obliquity, at 850 hPa, the African, South American, North Pacific, and East Asian monsoons all weaken, and the extent of the global monsoon decreases.

The present results indicate that the influence of obliquity or rotation rate change on general circulation can be evaluated under the assumptions that the land–sea distribution does not change and that the terrain height remains the same. The results can help us to discover which factors, obliquity or rotation rate, are responsible for the circulation anomalies in simulations of paleoclimate. At the same time, analyzing the influence of individual orbital parameters on geophysical fields for each of the four seasons helps us to better understand the laws governing components of climate change in geological periods. Investigation of the influence of variation in individual orbital parameters on the monsoon can deepen our knowledge of the monsoon.

6. Conclusion

Our results indicate that the three-cell circulation strengthens when the Earth's rotation rate slows. Furthermore, there are cool anomalies in the annual mean temperature field in the Northern Hemisphere and warm anomalies in the Southern Hemisphere for the lower rotation rate. The boundary between regions of cool and warm anomalies is located at 15°S. The sign of the annual mean zonal wind field anomalies changes when the rotation rate is changed. The positive and negative anomalies reverse between lower and higher rotation rates. The changes in the geopotential height field, temperature field, and meridional wind field in the stratosphere and in the vertical velocity field in spring are opposite to those in summer and autumn (and in the annual mean) under different rotation rates. Mid-latitude westerlies strengthen in all four seasons with a low rotation rate. Changes in the zonal wind in the two hemispheres reverse in spring and autumn. Quantitative changes show significant seasonal variation, with the largest changes in autumn. There are three regions of linear monsoon response to changing rotation rate: Africa, the Middle East, and the temperate and frigid zones.

Our results suggest that the three-cell circulation weakens for large obliquity, except for the three-cell circulation in the Northern Hemisphere in winter, and the Hadley circulation in the Southern Hemisphere in spring and summer, which strengthen with increased obliquity. The annual mean three-cell circulation weakens for large obliquity. The annual mean Hadley circulation in the Southern Hemisphere expands for large obliquity while the Hadley circulation in Northern Hemisphere and the Ferrel circulation in Southern Hemisphere shrink for large obliquity. The ascending branch of the Hadley circulation in the Southern Hemisphere strengthens significantly when the obliquity is increased from its normal value to 60°. The annual mean extent and velocity of easterly winds in the stratosphere over the equator increase for large obliquity, but the extent of the westerly winds decreases. The jet stream weakens in the Northern Hemisphere and strengthens in the Southern Hemisphere. In all four seasons, the easterlies in the troposphere and westerlies in the Northern Hemisphere weaken, and the mid-latitude jet stream in the Northern Hemisphere weakens for large obliquity. The strength of the mid-latitude jet stream in the Southern Hemisphere strengthens in spring, the westerly wind at middle and high latitudes changes in opposite senses in summer and autumn in the Southern Hemisphere, while the westerly wind weakens globally in winter.

The global monsoon expands with increased obliquity, and the African, South American, North Pacific and East Asian monsoons strengthen.

From the above, we can conclude that rotation rate and obliquity can have a significant effect on the general circulation of the Earth's atmosphere and on the monsoon system. This may partly explain changes in the atmosphere that have occurred as the Earth has evolved. By this means, we can determine where the responses to changing orbital parameters are linear or nonlinear. Of course these experiments are not enough by themselves. Additional experiments are required in which we change the land–sea distribution, the topography, and make simultaneous changes to more than one orbital parameter, to further determine the linearity of the response. These experiments will require extensive numerical simulations. In addition, the thermal contrast between land and sea can have a significant effect on the monsoon and atmospheric circulation. Further simulations with air–sea coupled models will deepen our knowledge of the monsoon and atmospheric circulation. All of these studies can help us to design further simulations to investigate and understand the Earth's changing climate.

Author details

Xinhua Liu
National Meteorological Center (NMC), Beijing, China
Key Laboratory for Semi-Arid Climate Change of the Ministry of Education, Lanzhou University, Lanzhou, China

Acknowledgement

The author of this chapter is grateful to the editor and the reviewer for their helpful comments and suggestions that greatly improved the manuscript. This study is jointly supported by the National Science Foundation of China (41105026) and the Foundation of Key Laboratory for Semi-Arid Climate Change of the Ministry of Education in Lanzhou University (the Fundamental Research Funds for the Central Universities Programs for Science and Technology Development of China, 223-860011). I also thank Professor Jianping Li and Athena Coustenis for their instructions and guidance.

7. References

Barron, E. J., P. J., Fawcett, D., Pollard, et al., 1993: Model simulations of Cretaceous climates: the role of geography and carbon dioxide. Philos. Trans. R. Soc. London B, 341, 307-316.

Barron, E. J., P. J., Fawcett, W. H., Peterson, et al., 1995: A 'simulation' of mid-Cretaceous climate. Paleoceanography, 10, 953-962.

Berger A., J. Imbrie, J. Hays, et al., 1984: Milankovitch and Climate. D Reidel Publishing Company, Dordrecht, Boston, Lancaster.

Bush, A. B. G., S. G. H., Philander, 1997: The late Cretaceous: Simulation with a coupled atmosphere-ocean general circulation model. Paleoceanography, 12, 495-516.

Cane, M. A., P., Molnar, 2001: Closing of the Indonesian seaway as a precursor to east African aridification around 3-4 million years ago. Nature, 411, 157-162.

Clemens, S. C., W., Prell, D., Murray, et al., 1991: Forcing mechanisms of the Indian Ocean Monsoon. Nature, 353, 720-725.

Collins, W.D., J.J., Hack, B.A., Boville, P.J., Rasch, D.L., Williamson, J.T., Kiehl, B., Briegled, J.R., Mccaa, C., Bitz, S-J., Lin, R.B., Rood, M.H., Zhang, Y.J., Dai, 2003: Description of the NCAR Community Atmosphere Model (CAM2). Boulder, Colorado, http://www.ccsm.ucar.edu/models/atm-cam/docs/cam2.0/description/index.html.

Del Genio, A. D., W., Zhou, 1996: Simulations of superrotation on slowly rotating planets: sensitivity to rotation and initial condition. Icarus, Vol 120, pp.332-343.

Gates, W. L., 1976: Modeling the ice age climate. Science, 191, 1138-1144.

Hall, A., A., Clement, D. W. J., Thompson, A., Broccoli, C., Jackson, 2005: The Importance of Atmospheric Dynamics in the Northern Hemisphere Wintertime Climate Response to Changes in the Earth's orbit. J. Clim., 18, 1315-1325.

Hourdin, F., O., Talagrand, R., Sadourny, R., Courtin, D., Gautier, C.P., McKay, 1995: Numerical simulation of the general circulation of the Titan. Icarus, Vol 117, pp.358-374.

Hunt, B.G., 1979: The influence of the earth's rotation rate on the general circulation of the atmosphere. J. Atmos. Sci., 36, 1392-1407.

Jenkins, G. S., 1993: A general circulation model study of the effects of faster rotation, enhanced CO2 concentrations and reduced solar forcing: Implications for the Faint-Young Sun Paradox. J. Geophys. Res., 98, 20803-20811.

Jenkins, G. S., 1996: A sensitivity study of changes in Earth's rotation with an atmospheric general circulation model. Global Planet. Change, 11, 141-154.

Jenkins, G. S., H. G., Marshall, W. R., Kuhn, 1993: Precambrian climate: The effects of land area and Earth's rotation rate. J. Geophys. Res., 98, 8785-8791.

Jian, Z., B., Huang, W., Kuhnt, et al., 2001: Late quatemary upwelling intensity and East Asian monsoon forcing in the South China Sea. Quatemary Research, 55, 363-370.

Joussaume, S., K. E., Taylor, 1995: Status of the paleoclimate modeling intercomparison project (PMIP). Proceedings of the first international AMIP scientific conference (WCRP Report, 1995), 425-430.

Kutzbach, J. E., B. L., Otto-Bliesner, 1982: The sensitivity of the African-Asian monsoonal climate to orbital parameter changes for 9000 years B.P. in a low-resolution general circulation model. J. Atmos. Sci., 39, 1177-1188.

Knutti, R., J., Fluckiger, T. F., Stocker, et al., 2004: Strong hemispheric coupling of glacial climate through fresh water discharge and ocean circulation. Nature, 430, 851-856.

Kutzbach, J. E., F. A., Street-perrott, 1985: Milankovitch forcing of fluctuations in the level of tropical lakes from 18~0 kyr BP. Nature, 317, 130-134.

Kutzbach, J. E., G., Bonan, J., Foley, et al., 1996: Vegetation and soil feedbacks on the response of the African monsoon to orbital forcing in the Early to Middle Holocene. Nature, 384, 623-626.

Kutzbach, J. E., P. J., Guetter, 1986: The influence of Changing orbital parameters and surface boundary conditions on climate simulations for the past 18000 years. J. Atmos. Sci., 43, 1726-1759.

Kutzbach, J. E., P. J., Guetter, W. F., Ruddiman, et al., 1989: The sensitivity of climate to late Cenozoic uplift in south-east Asia and the American southwest: Numerical experiments. J. Geophys. Res., 94, 18393-18407.

Kutzbach, J. E., R. G., Gallimore, 1989: Pangean climates: Megamonsoons of the megacontinent. J. Geophys. Res., 94 (D3), 3341-3357.

Kutzbach, J. E., W. L., Prell, W. F., 1993: Ruddiman, Sensitivity of Eurasian climate to surface uplift of the Tibetan plateau. The Journal of Geology, 101, 177-190.

Kutzbach, J. E., Z., Liu, 1997: Response of the African monsoon to orbital forcing and ocean feedbacks in the middle Holocene. Science, 278, 440-443.

Li, J., and Q. Zeng, 2003: A new monsoon index and the geographical distribution of the global monsoons. Adv. Atmos. Sci., 20, 299-302.

Li, J., and Q. Zeng, 2002: A unified monsoon index. Geophys. Res. Lett., 29(8), 1274, doi:10.1029/2001GL013874.

Li, J., and Q. Zeng, 2000: Significance of the normalized seasonality of wind field and its rationality for characterizing the monsoon. Sci. China (Ser. D), 43(6), 646-653.

Liu, Xinhua, Jianping Li, Qingliang Zhou and Yi Yang, 2010: Numerical simulation of the influence of changing rotation rate on the general circulation of the Earth's atmosphere. Information Science and Engineering (ICISE), 2010 2nd International Conference on, 5203-5206, 10.1109/ICISE.2010.5691888.

Liu. Xinhua, 2011: Numerical simulation of atmospheric general circulation under different obliquity of Earth. 2011 International Conference on Remote Sensing, Environment and Transportation Engineering, RSETE 2011 - Proceedings, p 5115-5118.

Mitchell, J. F. B., N. S., Grahame, K. H., Needham, 1988: Climate simulation for 9 000 years before present: Seasonal variations and the effects of Laurentide ice sheet. J. Geophys. Res., 93, 8283-8303.

Oort, A. H., J. J., Yienger, 1996: Observed interannual variability in the Hadley circulation and its connection to ENSO. J. Climate., Vol 9, pp.2751-2767.

Otto-bliesner, B. L., G. R., Upchurch Jr, 1997: Vegetation induced warming of high-latitude regions during the Late Cretaceous period. Nature, 385, 804-807.

Prell, W. L., J. E., Kutzbach, 1997: The impact of Tibet-Himalayan elevation on the sensitivity of the monsoon climate system to changes in solar radiation. Ruddiman, W. F., Tectonic uplift and climate change. New York: Plenum Press, 171-201.

Qian, W. H., 1995: The observational study and numerical experiment on the effect of the variation of the earth's rotation on the global sea surface temperature anomaly. Chinese J. Atmos. Sci., Vol. 19, pp.654-662.

Quan, X. W., H. F., Diaz, M. P., Hoerling, 2004: Change in the tropical Hadley cell since 1950, in the Hadley Circulation: Past, Present, and Future, edited by H. F. Diaz and R. S. Bradley, Cambridge Univ. Press, New York.

Rahmstorf, S., 1994: Rapid climate transitions in a coupled ocean-atmosphere model. Nature, 372, 82-85.

Rahmstorf, S., 1995: Bifurcations of the Atlantic thermohaline circulation in response to changes in the hydrological cycle. Nature, 145-149.

Ramstein, G., F., Fluteau, J., Besse, et al., 1997: Effect of orogeny, plate motion and land-sea distribution on Eurasian climate change over the past 30 million years. Nature, 386, 788-795.

Tuenter, E., S. L., Weber, F. J., Hilgen, L. J., Lourens, 2003: The response of the African summer monsoon to remote and local forcing due to precession and obliquity. Global and Planetary Change, 36, 219-235.

Weaver, A. J., M., Eby, A. F., Fanning, et al., 1998: Simulated influence of carbon dioxide, orbital forcing, and ice sheets on the climate of the last glacial maximum. Nature, 394, 847-853.

Williams, J. R. G., R. G., Barry, W. M., Washington, 1974: Simulation of the atmospheric circulation using the NCAR global circulation model with ice age boundary conditions. J. Appl. Meteorol., 13, 305-317.

Williams, G. E., Megacycles, 1981: Long-Term Episodicity in Earth and Planetary History. Hutchinson Ross Publishing Company, Stroudsburg, Pennsylvania.

Zheng, D. W., G., Chen, 1994: Relation between equatorial oceanic activities and LOD changes. Science in China (A), 37, 341-347.

Chapter 3

Study of Climate Change in Niger River Basin, West Africa: Reality Not a Myth

Juddy N. Okpara, Aondover A.Tarhule and Muthiah Perumal

Additional information is available at the end of the chapter

1. Introduction

Before the beginning of 21st, when the subject of global warming and climate change became a frontline issue among the climate scientists, it was thought by a lot of people, especially the decision- and policy-makers that climate change was a myth and not real. This underscores the reason why many countries even till today are not convinced yet why they should ratify the Kyoto protocol. But today, hardly is there any part of the globe that has not experienced the impacts of climate change either positively or negatively. As a result, addressing climate change issue has become one of the humanity's most pressing and difficult environmental challenges of our time, requiring urgent and concerted efforts. It is a complex, long-term problem, two centuries in the making. Climate change is ubiquitous-there are only, but a few human activities that do not contribute to it. Its effects are already being felt all over through weather and hydrological extremes (floods and droughts) in the Niger River basin and will only worsen, seriously affecting in particular sustainable development, with adverse impacts on the economic development of developing countries, as well as social welfare, the environment, natural resources and physical infrastructure. It would certainly most likely affect in general, the way of life in all countries, especially the developing countries with low resilience(depending on river basin's resources), as well as fragile ecosystems and even threaten global security through migratory pressures and resources conflicts. Again, the existing mismatch between the primary culprits of climate change and the primary victims of its impacts often bedeviled international efforts to address the problems at global level; and there is no other place this climatic impact is more obvious and critical than in countries of Africa, particularly the West African region; even though they have contributed just a little or nothing to the factors responsible for the global warming. Even though, there is no consensus amongst the Global Climate Models (GCMs) over the future climate of the West Africa [1], the region's experience so far, starkly demonstrates the development setbacks and high level of vulnerability of the area to impacts of climate change; ranging from recurrent droughts, Niger River zero flow of 1984 and 1985 at Malaville and Niamey (Benin and Niger) respectively, to shrinkage and disappearance of Lake Chad, devastation of Abuja National stadium velodrome, collapse of connecting Nukkai and Sokoto bridges, in 2003, 2005 and 2010 respectively (all in Nigeria). This is because; the current climate of the region is strongly characterized by climatic variability and extreme events (floods and droughts) that already have serious implications on economic development of the region.

It is now very obvious too that an inherent characteristic of climate is change, and a period of change is already underway which has the potential to threaten the fabric of human society and its development across the globe both at present and in the future. This is because climate, development and the world's water-resource systems of a river basin have a unique relationship insofar as water resources depend on the hydrological cycle which is itself part of the Earth climate system driven by the sun's energy. What came to be known as global warming (i.e. the rising of average air temperature of the Earth's atmosphere and ocean) that began to increase in the late 19th century (Fig. 1) and is projected to continue rising is human's making; through human's deliberate actions to conquer nature, as against living in harmony with natural systems. Global warming is the main culprit of climate change. While humans have affected weather and climate, weather and climate have in turn affected humans through extreme events (Fig. 2).As a result, there has been growing needs to study, understand and quantify the potential impacts of climate change on climate sensitive sectors of national economies and the hydrologic regimes. Also, decision-makers in many climate–sensitive sectors: water, energy, agriculture, fisheries, health, forestry, transport, tourism, disaster risk management - are now more than ever before, increasingly concerned by the growing adverse impacts of climate change associated risks, because they are ill-equipped to adequately tackle these challenges.

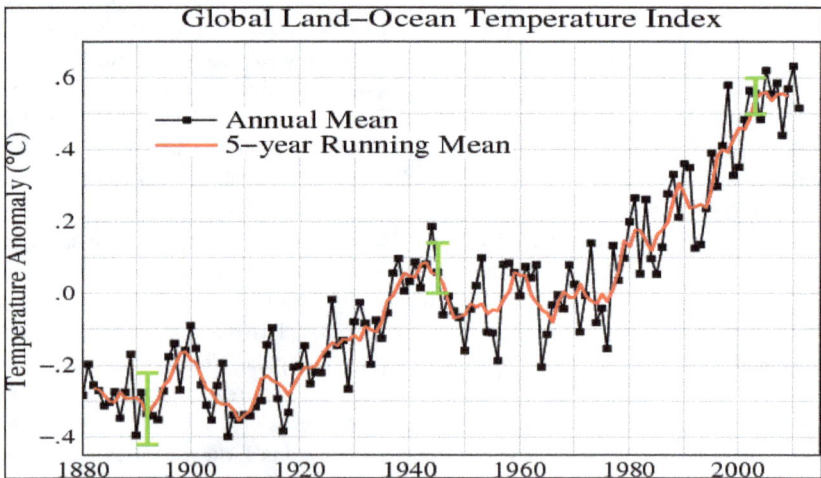

Figure 1. Global Temperature Anomaly 1880-2010 (Sources NASA, 2011)

Figure 2. Reaping Extreme Hydrological disasters due to climatic Variability and changing climate (Nigeria experience in the lower Niger Sub-basin)

Water is an indispensable element of life; the water resources of the region's river basins are highly dependable and sensitive to climate variability and change; due to inter-connection between the climate system, hydrological cycle and water resources system. Thus, if the trends in climate contexts that took place over the last three decades continue to prevail unabatedly, West Africa will no doubt experience decreased freshwater availability. Also, compared to previous decades, it is observed that since the early 1970s, the mean annual rainfall has decreased by 10% in the wet tropical zone to more than 30% in the Sahelian zone while the average discharge of the region's major river systems dropped by 40 to 60%. This sharp decrease in water availability will be complicated by greater uncertainty in the spatial and temporal distribution of rainfall and surface water resources [2, 3]. Again, it is important to note that Niger River basin is not just simply water; it is an origin of identity for the region, a route for migration and commerce; but also seriously threatened by man-made climate change. The region's recent experience is a demonstration of the fact that climate change is real and not a myth. It is against this backdrop that this paper attempts to highlight the various experiences of the level of vulnerability of Niger River basin and its inhabitants to the impacts of climate change including climatic variability and extremes; as well as provide scientific evidence to substantiate the characterization of the current climatic variability and the future impacts of climate change on the region. So the paper will try to distinguish between climate variability and climate change. We adopt the Inter-governmental Panel on Climate Change (IPCC) definitions for both terms. Accordingly, climate change is referred to as statistically significant variation in the mean state of the climate or the long-term changes in climate conditions observable over several decades or longer [4]. Climate variability on the other hand is the deviations of climate statistics over a given period of time from the long-term climate statistics relating to the corresponding calendar period or short-term variations in climate over periods of days, months, years and decades [4].

2. Intellectual merit

For a very long period of time there has been harmonious and balanced use of water of Niger River, such that it is even limited in its natural functions and services. But over time, this trend changed, as the basin suffered for many decades from human pressure and new uses such as construction of dams that have been disturbing the characteristics, structure and functioning of the river basin's ecosystems. This intensive use of these natural resources added to the growing population and climate changes impacts, among which severe droughts and their impacts have had severe consequences on the status of the Niger River and its tributaries, biodiversity, landscapes, key habitats and floodplains. Aside this, climate change and freshwater resource systems are interconnected in a complex ways, with rainfall patterns, evaporation and water demand or use influencing the availability of both surface and groundwater resources in the region.

Of course, the climatic future of the Niger River Basin may remain uncertain, due to inconsistency of the global climate models over the area; but climate change is expected to have a major influence not only on water resources, but on food and human society at large through its impacts on climatic variability and extremes. Again, it is impossible to rule out the occurrence of other possible indirect influences of climate change such as higher temperatures leading to high evaporative-and greater-demand for water. Undeniably, such disagreement between the climate models may be interpreted that nothing could be said with certainty about the future evolution of rainfall in the basin; but very high degree of climatic variability is projected to continue, which could even become more pronounced on seasonal, annual and decadal timescale. West African rivers are mainly strongly seasonal and humid with fairly modest inter-annual rainfall variability. As shown in figure 3, the rivers display very strong relationships with rainfall that accounts for about 60% - 70% of river flow variability [5]. In all the cases, river flows show much greater coefficient of variability than rainfall mainly because of heterogeneity and nonlinear response of runoff to changes in rainfall; especially to the variations in rainfall intensity. The impacts of climate change on freshwater resources systems can sometimes be direct, stemming from the relationship between temperature and/or precipitation, and the abundance and quality of the available water resources; while the indirect impacts occur by causing shifts in temperature, lifestyle, population, economy or technology, which in turn may trigger shifts in demand for water. Thus, it is expected that climate change could further amplify and entrench water resources anomalies such as local drought or flooding at the lake / river level.

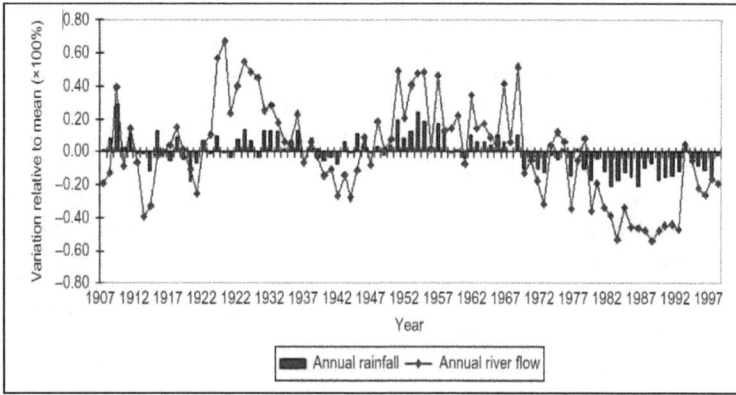

Figure 3. Rainfall-Runoff interactions in Sudano-Guinean zone of Niger Basin in Mali and Guinea (source: Mahel *et al.*, 2009)

According to Intergovernmental panel on climate change (IPCC), climate change is expected to result in severe water stress over much of Africa; particularly, in the agro-pastoralist region of West Africa [6], where Niger River basin is located. Consequently, not only will freshwater become scarcer in already dry regions, but changing freshwater temperatures could affect natural ecosystems and water quality. In fact, the West African region of the continent is characterized by extreme climatic variability with extreme weather events; the last 40 years since 1969 have witnessed dramatic reductions in mean annual rainfall (Fig. 4) throughout the region [7, 8, 9 and 10]. A rainfall decrease of 29 – 49 percent has been observed in the 1968 – 1997 period compared to the 1931 – 1960 baseline period within the Sahel region [6].

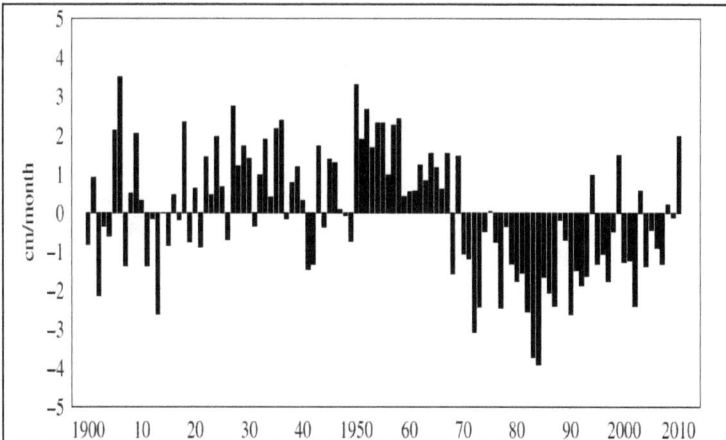

(Source: NOAA NCDC Global Historical Climatology Network Data)

Figure 4. Typical Rainfall Anomalies (1900 -2011) over West African Sudano-Sahel zone (11-18N and West of 10E)

The projected warming climate caused by increasing concentration of greenhouse gases is very likely to exacerbate the present climatic variability and extremes in the region; which already have implications for water resources availability and food production not only in the Niger River basin, but over the entire West African region. For example, the main livelihood of the people in Niger River basin is traditional, low input, rain-fed farming and nomadic pastoralism; any increase in volatility of summertime temperatures will therefore, have serious effects in grain-growing regions of the basin. Again, climate change is expected to lead to intensification of the hydrological cycle; and by implication will ensue in increasing drought or flooding episodes.

Aside this, the Niger River is not just simply water; it is an origin of identity for the region, a route for migration and commerce, as well as a catalyst for potential conflict and cooperation too. Hence, the rising concern on the adequate management and assessment of the water resources in the face of the changing climate is quite important. The river and its tributaries are the lifeline for the teeming human population, with annual growth rate of 2.8 per cent; as well as the major sources of hydropower to most of the riparian countries within the basin. Much of the population of the basin suffers from extreme, chronic poverty and vulnerable to droughts and increasing malnutrition rate, due to increasing food and water insecurity resulting from climatic, demographic and land use changes.

Climate change is indeed real, not a myth. The region's recent past experience of unusual vagaries of weather and climate is a clear demonstration of reality of climatic variability and change. In fact, climate change represents a shock to the rural farmers' with low resilience to climate change impacts. It in fact poses a pincer threat that reveals how vulnerable the basin and its inhabitants are to hydrological extremes (droughts floods landslides etc.). Today, droughts and floods have become the most common natural disasters in the Niger River basin frequently accompanied by loss of lives, properties and croplands. With climate variability and change, changes in the onset and cessation dates of rainy season are very likely; as presently being experienced in Nigeria (Fig. 5). Almost all the droughts that occurred in the region are associated with late starts of rainy season and early cessation of the rains, resulting in drastic reductions in the length of rainy season and invariably the length of growing season as well.

Under such a context, only a rational mobilization and adequate management of water resources in the River Niger basin seem to be the most adequate answer and the catalyst for the development of growth and gradual alleviation of poverty. Therefore, sustainable water resources and agricultural development has become an absolute necessity for food security and health in 21st century; if poverty eradication is to become a reality in Africa and Nigeria in particular. Additionally, with climate variability affecting the agricultural sector badly, coupled with uncertainty about the future climate that confounds planning among the smallholder and commercial farmers, climate change could aggravate this already tensed situation, triggering unfathomable impacts such as crop failures, floods, droughts and malaria epidemics. This will no doubt compromise the region's ability of achieving the Millennium Development Goals (MDGs) related to poverty, hunger and human health. Climate change will further pose serious challenge to the attainment of other MDGs related to reduce child mortality, improved maternal health, ensuring environmental sustainability and combating HIV/AIDS, malaria and other diseases as a result of malnutrition ensuing from food insecurity. Thus, the need for anticipatory strategies for adaptation to climate variability and change has become even more urgent as resources demands increase through population growth and development.

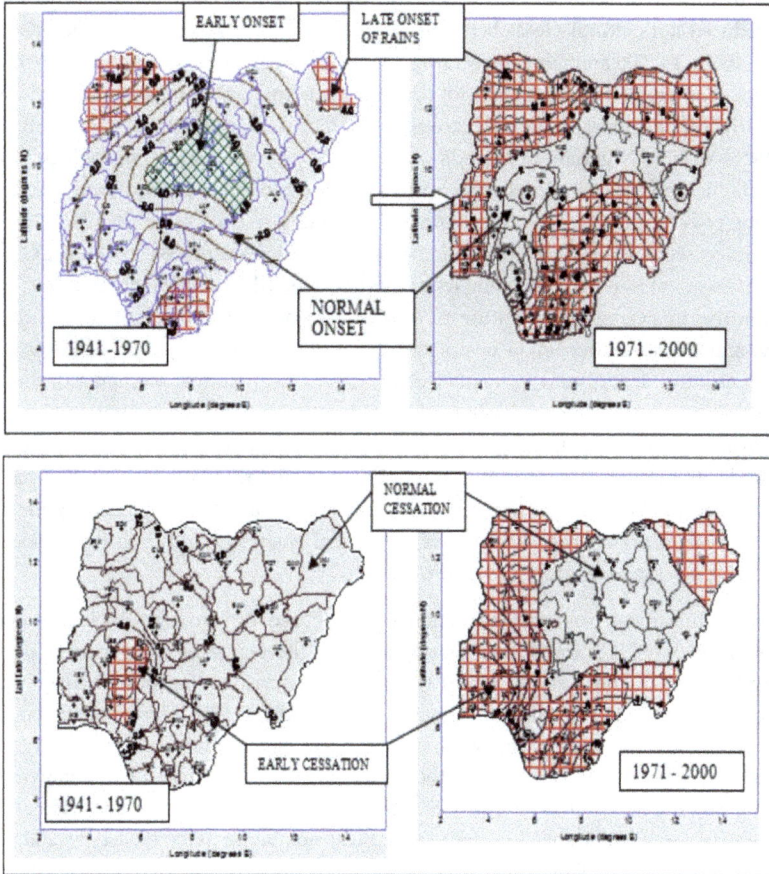

(Source: Nigeria Meteorological Agency (NIMET, 2007))

Figure 5. Fig. 5: Changes in Onset (top) and Cessation (bottom) Dates of Rainy Season (Nigeria)

Number of studies in the Niger basin have shown significant trends towards a false onset (a situation where the rainy season starts normally and then ceases abruptly, creating a dry period between the false onset and the true onset), late or delayed onset (a situation where the expected start of the rainy season is delayed) and early cessation (a situation where the rainy season stops far ahead of the expected time of the summer rains) [8]. For example, prolonged drought occurred from 1738-1756, centred in the area of the Niger River bend that induced famines which killed half of the population of the city of Timbuktu. More recently too, there has been persistent drought in the Sahelian regions of the basin since the late1960s that has resulted in a decrease in the availability of freshwater resources [6]. Up to 250,000 people and millions of herds of cattle are reported to have perished. In the lower Niger sub-basin(Nigeria) alone, the impacts of the drought episodes of 1968 and 1973/74 which reduced agricultural yields between 12% and 14% of annual average and ensued in death of about 300,000 animals representing 13% of livestock population was still being felt several years after[11]. So, the impacts of drought are extremely serious and often dramatic particularly for the most vulnerable groups - women and children. Drought forces the inhabitants of the Nigerian dry lands to resort to survival strategies, which further exacerbate the desertification problems, with associated reduction in land productivity and worsening poverty problems. Some other major famines also occurred from 1983 - 1985 and 2007 in large sections of the Niger basin. This was well documented in the IPCC's 2007 Fourth Assessment Report, confirming its earlier findings, which shows that trends in Africa include a rise in average temperature of 0.7°C for most of the continent during the 20th century, and decreases in rainfall of up to 30 percent over large portions of the Sahel. In Niger River basin, it further documents that the rivers mean annual discharge declined by 40-60 per cent; and in future, the basin could see changes in rainfall, evaporation and runoff of approximately 10 per cent. Also, major changes in rainfall in terms of annual and seasonal trends and extreme events of flood and drought have been documented by [12, 13, 14 and 6]. Other parts of the West Africa is not left out; for example, annual hydrological regime of the Nakambe River, Burkina Faso has shown substantial changes too during the period1955-1998 with a shift occurring around1970[15].

Moreover, while most research and action have focused on drought challenges and its impacts in the Sahel it is important to also state that the region also experiences periodic flooding. In 1953 or there about, heavy rainfall leading to flooding destroyed crops and resulted in famine that lingered on for the first nine months of 1954. This affected about five million people in both western and south-central Niger and northern Nigeria and northern Cameroun [16]. Since the recovery of the Sahel rainfall in the mid-1990s following the prolong drought periods of 1970s and 1980s[17],floods associated with intense rainfall have again become more pronounced in the region, most notably in 1995, 1998-1999, 2002-2003, 2006-2008 and 2010[18]. This resurgent of flood phenomenon has been associated with a number of factors; including anomalous heating in the tropical Atlantic Ocean and La-Nina event in the tropical Pacific Ocean[19]. This argument on whether such intense rainfall can lead to flooding was further substantiated by linking the occurrence of the floods to accumulative rainfall in the days prior to heavy rain event [20]. Also two distinct flood events occur every year in the Niger basin, especially in the lower sub-basin in Nigeria. The first is the 'black flood' that originates from the high rainfall area at the headwaters; arriving at Kainji (Nigeria) every November and lasting till March at Jebba (Nigeria) after attaining a peak of about 2000 m³/s every February. The second type of flood is called 'white flood', which becomes prominent only downstream of Sabongari (Nigeria), soon after Niger River enters Nigeria; usually laden with silt and other suspected particles. The flood derives its flow from the local tributaries and reaches Kainji every August and attaining peak rate of 4000 – 6000 m3/s between September and October in Jebba every year [21]. How all these flood and droughts events will evolve in future in the face of the changing climate still remains a subject of research till today.

Till today, there has been increasing menace of frequency of flood in the lower Niger basin, due to intense falls of short duration particularly for the years - 2005, 1999, 1994 and 1988 and the resulting casualties are all still very fresh in mind. For instance, in August, 2005, the old Nukkai Bridge in Jalingo State, Nigeria, collapsed and sank into the overflowing Jalingo River, killing more than 100 people (ThisDay newspaper, 2005). Most recently too is the collapse of Sokoto bridge, Nigeria in 2010 in the same manner the Minneapolis Bridge collapsed and sank into Mississippi River in USA; on 2nd August, 2007 (http://www.washingtonpost.com/wp-srv/photo/gallery/070801/GAL-07Aug01-833303/ind ex.html), all due to massive flooding from intense rainstorms ensuing from a changing climate, because these bridges were initially designed for passage of specific discharge of flood water that may be far less than the discharge that it is presently able to convey. These bridges were designed on the assumption of stationarity of hydrological series and return period. With climate change, the assumption of stationarity of series in hydraulic design of water resources system is dead [22]. With hydrological cycle projected to intensify in the face of the changing climate, these hydrological extremes (droughts and floods) are expected to be on the increase [6].

With increasing climatic variability, climate change will impose additional pressures on the water availability, water accessibility and water demand in the region; although the scant available data in the region make it presently difficult to predict these changes with recognizable certainty [6]. Also, observed is the consequent collapse of the region's ecological zones from 6 (Table 1) to 5 (Fig. 6), as a result of decline in rainfall; there has been 200km southward shift in isohyets (Fig. 7). Following the decline in average annual rainfall, before and after 1970, with ranges from 15% to over 30% depending on the location within the Niger basin[23], the savanna zone (interface of desert and forest) is resultantly pushed further south with the desert advancing at a fast rate of 700m per annual on the average. Hence, we now have the Sudano/Sahel extending to about lat: 10.5N from lat: 12.5N, covering about 35% of the landmass of the country.

Furthermore, evidence of changing climate and its effects on local hydrology can already be seen in the historic stream flow records of the Niger River. Records have shown substantial decrease in observed flows across the basin over a time period from 1907 to 2000, due primarily to increasing temperature. A minimum zero flow condition was observed in 1985 over Niamey (Niger) Gauging station (Table 2), at the upstream of Nigeria [24]. Even in the humid lower Niger sub-basin, the average river flow of the recent time slice (1982-2000) in table 3 is far lower than the previous reference time period (1960-1981). The logical consequence of a decline in precipitation and streamflow is a change in the timing and magnitude of the precipitation and streamflow pattern.

Ecological Zones	Altitude in(m)	Mean Monthly Temperature(°C)	Mean Annual Rainfall(mm)	Type of Rainfall Distribution	Length of Rainy Season (days)
Mangrove Forest and/ Freshwater Swamp Forest	< 100	28 - 25	> 2000	Extended Modal	300 - 360
Rain Forest	100	28 - 24	1200 - 2000	Bimodal	250 - 300
Derived Savanna and /Southern Guinea Savanna	< 500	30 - 26	1100 - 1400	Bimodal	200 - 250
Northern Guinea Savanna	400-500	30 - 23	1000 - 1300	Unimodal	150 - 200
Sudan Savanna	<300- >600	31 - 21	600 - 1000	Unimodal	90 - 150
Sahel Savanna	300-400	32 - 25	400 - 600	Unimodal	90

Table 1. Characteristics of the ecological zones over West Africa

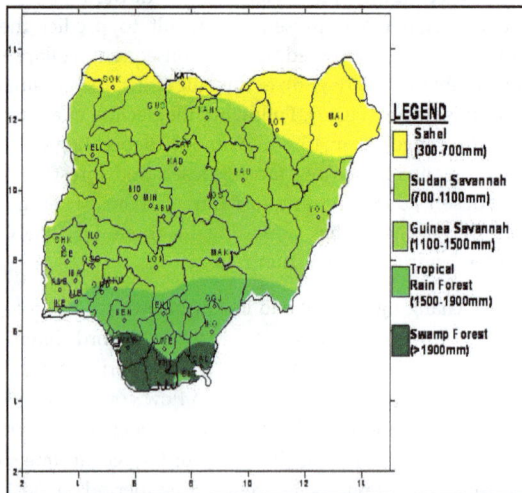

Figure 6. Collapsed Ecological Zone of Nigeria from 6 in table 1 to 5 zones due to changing climate

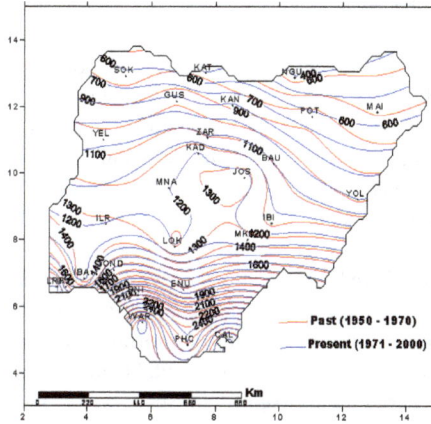

Figure 7. Isohyets shift due to southward advancing of aridity

Station	River	Period	Qmean (m³/s)	Qmax (m³/s)	Year	Qmin (m³/s)	Year
Koulikoro (Mali)	Niger	1907-2000	1385	9670	1925	13	1973 and 1982
Niamey (Niger)	Niger	1928-2000	870	2360	1968	0	1985
Lokoja (Nigeria)	Niger	1915-2000	5590	26,300	1956	599	1974

Table 2. Discharge Characteristics of River Niger (Source: Archives of HYDRONIGER)

Station	River	Period	Qmean (m³/s)	Qmax (m³/s)	Year	Qmin(m³/s)	Year
Lokoja	Niger	1960-1981 1982-2003	68936.5 58646.9	94790 (1960-2003)	1969	25760	2003
Onitsha	Niger	1960-1981 1982-2003	71136.64 59721.14	87810 (1960-2003)	1999	25760	2003
Makurdi	Benue	1960-1981 1982-2000	41369.21 34660.84	61869 (1960-2000)	1969	20378	1983
Jebba	Niger	1960-1976 1977-1997	18122.53 11275.74	23377 (1960-1997	1963	6253	1991

Table 3. Lower Niger River flow Characteristics

Again, the need for development and investment in the region is evident too, and the Niger River holds tremendous development potentials. Development opportunities range from those directly related to the river, such as power, irrigation, and navigation, to those "beyond the river," such as increases in trade, communication investments, and enhanced labor flows [25]. However, despite the rich potentials of the basin, the basin has not been meeting the rising water demands of the region occasioned by high population growth rate, which is projected to double by 2050 as shown in Figure 8[26]. This clearly reveals the high vulnerability of the basin to climatic variability and potential climate change. This is because water availability in Niger basin is highly variable, aside the growing concern on land and water degradation occasioned by climate change and human activities. Also threatened is groundwater, which is found to be safer than surface water, especially concerning pollution vulnerability. Consequently, there has been threat of tension between the Member States of the basin, underscoring the need for equitable sharing of water resources [27]. Hence, studies of this kind is conceptualized as a means of informing and improving knowledge of the availability of water for food production and equitable management and sharing of the natural resources, since water is the source of food security. To worsen the already fragile situation, deforestation too is progressing at an alarming rate due to urbanization and population pressure. By 2000, tropical forest and woodland covered less than 15% of the land mass [28].

Another issue of great concern is the gas flaring, which contributes to global warming, the main culprit of climate change, apart from causing other environmental degradations. The flaring of gas has been practiced in the Niger Delta region in the lower Niger sub-basin for over four decades (Fig. 9). Today there are about 123 flaring sites in the region (Energetic Solution Conference, 2004), making Nigeria one of the highest emitter of greenhouse gases in Africa. For example, some 45.8 billion kilowatts of heat are discharged into the atmosphere of the Niger Delta from flaring 1.8 billion cubic feet of gas every day [29].

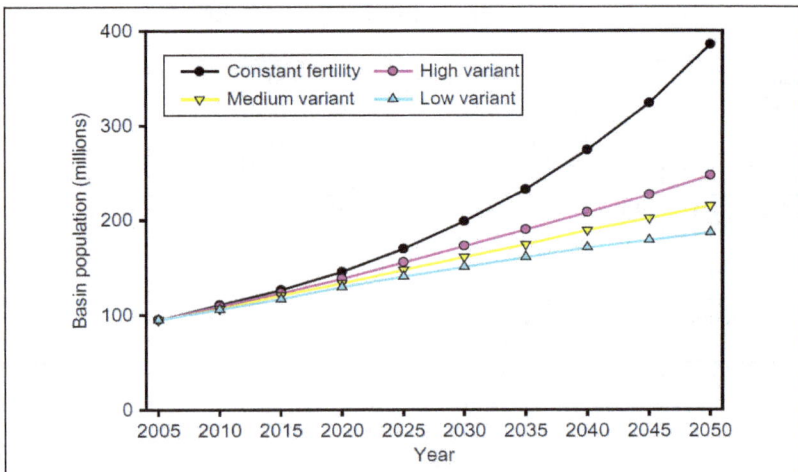

Figure 8. The evolution of Niger River Basin population 2005-2050(Source: based on UN population Division (2006)

Gas flaring has raised temperatures and rendered large areas uninhabitable. Between 1970 and 1986, a total of about 125.5 million cubic meters of gas was produced in the Niger Delta region, about 102.3 (81.7%) million cubic meters were flared while only 2.6 million cubic meters were used as fuel by oil producing companies and about 14.6 million cubic meters were sold to other consumers [30]. Gas flaring and other oil exploration and exploitation activities have contributed significantly to the degradation of the environment in the region. Gas flaring leads to acid rains; and the concentration of acid in rain water appears to be higher in the Niger Delta region and decreases further away from the region. Though there is need to do more research on this. It has altered the vegetation of the area, replacing local vegetation with "stubborn" elephant grasses, as it is called locally, a grass plant that can grow in very harsh environment. Unfortunately, in spite of the negative implications of gas flaring to the environment, the multi-national oil firms operating in Nigeria have continued in these bad environmental practices unabatedly.

Although, recent studies suggest that increase in atmospheric concentration of GHG on West Africa from current elevated levels up to about 550ppm, may make West Africa rainfall regime more robust and drought less frequent and persistent (Brook, 2004). So there is need to assess how much impacts the future change would have on the local and regional available water resources.

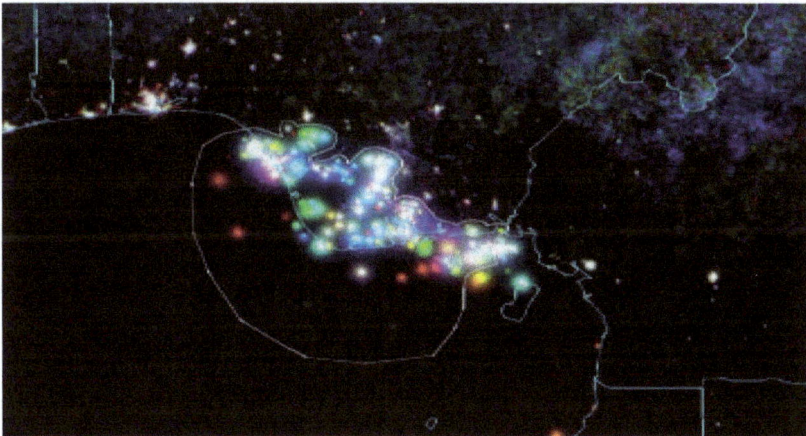

Figure 9. Satellite image shows Nigeria's coastline burning bright with gas flares at night. The red dots represent gas flared in 2006, the green dots represent 2000 and the blue dots represent 1992. The white line encircles the flares associated with Nigeria.

Drawing on a review of published literature on Niger Basin, it is observed that though the assessment of the impacts of climate change on hydrology and water resource is given accelerated attention in most parts of world today, not much of research works have been done to assess quantitatively the impact of the projected climate change on water resources, socio-economic activities and hydrological extremes in the region. Also, since the majority of the basin lies in the Sahel zone (between lat. 10°N and 20° N and long. 20°W and 10°E), researches on characterization of current climatic variability and future climate change have been lopsided focusing more on the Sahel. Less published information is available for the more humid portions of the basin [17], underscoring the reason for focusing on the lower Niger sub-basin in this study. However, amongst the few studies available include the work of reported in [31], 32, 33, 34 and 35]. Also, an average temperature increase of 0.4°C has been observed within Nigeria over the 20 years [36]; while climatic variations were observed over Benin – Owena River Basin, southwestern, Nigeria that appears as fluctuations of wet and dry periods every 2-3 years in terms of rainfall and stream flow [37], with a positive temperature trend rising at the rate of 0.37°C/ decade. Also, observed is a decreasing trend of rainfall over Lake Chad basin[38] and it was opined that climatic change apart from some other human activities (irrigation) was the main key factor responsible for the shrinkage of the Lake size from 25,000km^2 in the 1960s to presently barely 2,000km^2.

3. Application area

The case study area being investigated is the Niger River basin with much focus on the lower Niger sub-basin area, a much humid portion of the Niger River basin where Nigeria is domiciled and less researched by climate change scientists. River Niger is located between 5°N and 23°N of latitude, and 12°W and 17°E of longitude. It rises in Guinea high grounds and flows for a total length of about 4,100 km through Mali, Niger and Nigeria before reaching the Atlantic Ocean. Niger River basin is the largest trans-boundary basin in West Africa, and the second largest river in Africa by discharge volume (5,700 m^3/s; 1948-2006) after Congo River (42,000 m^3/s) and the third longest (4, 100 km). The total drainage area of Niger River (2.2 million km^2) [39], with hydrologically active area (1.5 million km^2) [40] covers fully 7.2 per cent of the continent Africa with a total population of over 100 million people distributed among the nine riparian countries [40] that share the basin's resources of which 71 per cent live in Nigeria. The nine countries presently sharing the active catchment area are; Benin, Burkina Faso, Cameroon, Chad, Cote D'Ivoire, Guinea, Mali, Niger and Nigeria (Fig.10), Seventy-six per cent of the basin area is located within Mali (Upper Niger), Niger (Middle Niger) and Nigeria (Lower Niger) sub-basins.

About 44.2% of the basin area is located within Nigeria, which constitute about 61.5 % of human population of the basin [26]. According to the lowest climate change scenario, demographers estimated that the population of the basin will double by 2050, but if the present fertility rates remain constant, the population could even increase fourfold by 2050 (Fig. 8). The choice of lower Niger as the focus stems from the fact that Nigeria is strategically located at the downstream of the basin and more vulnerable to the adverse impacts of environmental changes at the upstream, apart from gas flaring and other forms of environmental degradation taking place in Nigeria The region is also less researched in terms of climate change vulnerability and impacts studies. Moreover, Nigeria' ecological strata truly represents the climatic profile of the basin. As shown in figure 11, the climatic zones of Niger Basin varies from hyper-arid in the north to sub-equatorial and annual rainfall fluctuates from about 4000mm in the southern/Cameroun to less than 400mm (with

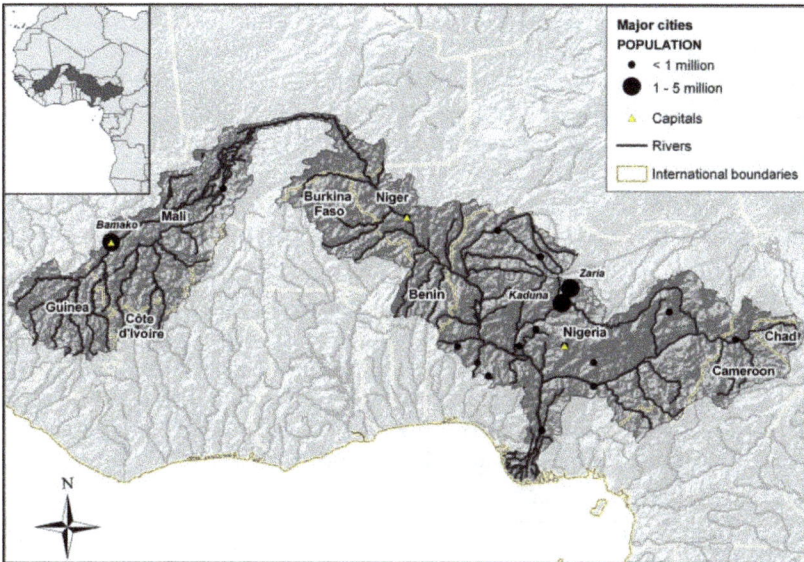

Figure 10. Niger River Basin Catchment Area and Member countries

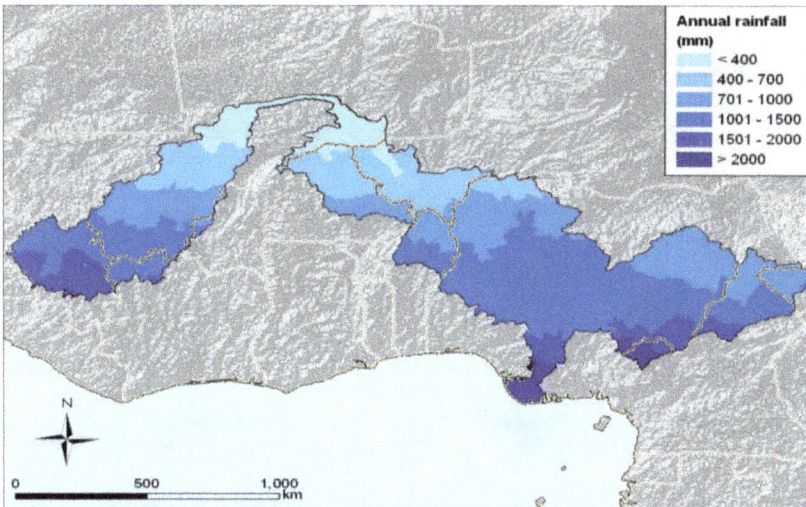

Figure 11. The Niger Basin annual rainfall (source: Mahe et al., 2009a)

no rain in some years) on the fringes of the Sahara desert in northern Mali and Niger [26]. Spatio-temporal variability of rainfall is high in the basin, causing water stress and droughts, which are very problematic for agricultural planning than low annual rainfall [41] also cited in [26]. Rain-fed agriculture remains a common practice in the region. Total rainfall provides a measure of water supplied to rain-fed agriculture on a given area; while evapotranspirable water corresponds to the fraction of rainfall actually available to the plants and excludes rain that falls when the plants cannot exploit it or in excess of the demand. That said, northwards from the very humid, eastern coastal locations, to the boundary with the desert, the vegetation profile includes Moist Evergreen Rain Forests, Dry Semi-Evergreen Rain Forests, Derived Savannah, Southern Guinea Savannah, Northern Guinea Savannah, Sudan Savannah, and Sahel Savannah [42, 43]. Rainfall in Niger River basin depends on the Atlantic West African monsoon (WAM) between May and November each year and gives dry and wet seasons respectively.

Another important index of climate change posing serious challenge in the basin is the land use/ land cover changes. Land cover just like in any other part of the global environments results mostly from combination of natural and anthropogenic influences. The main natural force of the change remains rainfall changes induced by climatic variability. This has been found to often reduce the natural regeneration rate of land resources in the area. Table4 shows the percentage of changes that took place between 1976 and 1995 in Nigeria.

S/n	Ecology	Area km2 (1976)	Area km2 (1995)	Change 1976-1995	Percent change
1	Agricultural Tree Crop Plantation	824.15	1,656.88	832.73	101.0
2	Alluvial	523.61	282.38	-241.23	-46.1
3	Discontinuous grassland dominated by grasses and bare surfaces	7,614.72	12,517.23	4,902.51	64.4
4	Disturbed Forest	14,677.70	19,491.29	4,813.59	32.8
5	Dominantly grasses with discontinuous shrubs and scattered trees	13,053.77	12,487.62	-566.15	-4.3
6	Dominantly shrubs and dense grasses with a minor tree component	118,529.55	85,020.98	-33,508.57	-28.3
7	Dominantly trees/woodlands/shrubs with a subdominant grass component	154,933.40	83,281.15	-71,652.25	-46.2
8	Extensive (grazing, minor row crops) Small Holder Rainfed Agriculture	170,837.55	192,892.33	22,054.77	12.9
9	Extensive Small Holder Rainfed Agriculture with Denuded Areas	4,417.88	10,118.47	5,700.58	129.0
10	Floodplain Agriculture	9,671.81	21,576.03	11,904.21	123.1
11	Forest Plantation	1,000.85	1,581.24	580.39	58.0
12	Forested Freshwater Swamp	18,564.71	16,696.51	-1,868.20	-10.1
13	Graminoid/Sedge Freshwater Marsh	5,882.74	1,136.51	-4,746.22	-80.7
14	Grassland	1,196.74	8,146.74	6,950.00	580.7
15	Gullies	125.35	19,070.48	18,945.13	15,113.2
16	Intensive (row crops	329,227.97	373,481.34	44,253.37	13.4
17	Irrigation Project	148.85	1,008.86	860.01	577.8
18	Livestock Project	51.02	139.65	88.63	173.7
19	Major Urban	1,102.58	1,362.37	259.79	23.6
20	Mangrove Forest	10,157.12	10,067.31	-89.81	-0.9
21	Minor Urban	958.69	4,022.98	3,064.29	319.6
22	Montane Forest	7,900.02	8,053.76	153.74	1.9
23	Montane grassland	2,502.27	3,898.15	1,395.88	55.8
24	Natural Waterbodies: Ocean	6,766.53	15,588.36	8,821.83	130.4
25	Rainfed Arable Crop Plantation	15.92	521.38	505.46	3,174.9
26	Reservoir	1,331.41	2,901.16	1,569.75	117.9
27	Riparian Forest	7,506.46	5,330.46	-2,176.01	-29.0
28	Rock Outcrop	1,445.15	2,647.96	1,202.81	83.2
29	Saltmarsh/Tidal Flat	18.84	596.92	578.08	3,067.5
30	Sand Dunes/Aeolian	1,032.77	5,428.30	4,395.53	425.6
31	Shrub/Sedge/Graminoid Freshwater Marsh/Swamp	17,749.63	10,251.68	-7,497.95	-42.2
32	Teak/Gmelina Plantation	624.44	1,156.43	531.99	85.2
33	Undisturbed Forest	28,022.42	13,477.90	-14,544.52	-51.9
34	Canal		30.76	30.76	
35	Mining Areas		61.15	61.15	

Table 4. Typical Land use/land cover changes in Niger River Basin from 1976-1995 (lower Niger Basin) (Source: Fasona and Omojola, 2005)

4. Study design and methodology

4.1. The characterization of current climatic variability in Niger River Basin

There are many different ways by which changes in hydro-climatological series can take place, either abruptly (step change) or gradually (trend) or may even take more complex dimension. The characterization of current climate of the basin is based on the data available in the archives of the Nigerian Meteorological Agency (NIMET), generated from the synoptic weather stations. The length of data used is 62 years (1941-2002), which is within the WMO recommendation of >50 years for change detection [44] Using these data, the characterization of the spatio-temporal variability of the basin was examined based on parametric (Regression) and non-parametric (Kendall Rank correlation, Thie and Sens) approaches. Before this is done, the regionalization of the point climatic data into areally integrated climate data using ArcGIS Thiessen Polygon method was done. The regional index employed in the study is calculated as the average of the standardized climatic variables of the stations included in the region. The index is calculated at the monthly and yearly time-scales. The uses of standardized values are important to allow comparison of time-series whenever climatic variables present significant spatial gradient throughout the area of study as in the case in the Niger basin. The standardized variable is expressed as:

$$Z = \left(\overline{X} - \mu \right) / \sigma \qquad (1)$$

where $\overline{X}$ is sample annual rainfall mean for the rainfall station, μ is the long term mean and σ is the standard deviation of long term annual rainfall.

The baseline data used for all the computation is WMO recommended period 1961 – 1990. Apart from ensuring compliance with World Meteorological Organization [45] standard, this analysis is necessary to eliminate part of the local variability factors associated with a specific station and not reflected by a regional change. Due to the relevance of such study in water resources management, the year considered in the analysis is standard hydrological year applicable in Nigeria (i.e. 1st June – 31st May). Following the characterization of the current climatic variability, the potential impacts of future climate change on the hydrology and water resources of the region was evaluated using the Thornthwaite water balance and Artificial Neural Networks (ANNs). Given the huge size of the basin within Nigeria (562,372 km²), its heterogeneous nature in terms of agro-ecological zoning and diverse hydroclimatic variability, the whole study area was be divided into five(5) sub-basins, namely: Upper Niger Sub-basin (131,600 Km²), Lower Niger Sub-basin (158,100 Km²), Niger south Sub-basin (53,900Km²), Upper Benue Sub-basin (158,900Km²), and Lower Benue Sub-basin (73,000Km²).

4.2. Hydrological modelling of potential impacts of Future climate change on Lower Niger River Basin

Water balance model description and data source

The water balance model used was the one developed by Thornthwaite in 1948 and later revised in 1955[46]. The method is basically a book-keeping procedure, which estimates the balance between the inflow and outflow of water. The main inputs into the model are precipitation and potential evaporation, while the main outputs are actual evaporation and water surplus or runoff. The model estimates the potential evaporation using the Priestley-Taylor method. Estimation of evaporation is based upon knowledge of the potential evapotranspiration, available water-holding capacity of the soil, and a moisture extraction function. The general structure of this model is often represented as below to include the monthly time scale:

$$S(t + 1) = S(t) + P(t) - E(t) - Q(t) \tag{2}$$

where $S(t)$ represents the amount of soil moisture stored at the beginning of time interval t; $S(t + 1)$ represents the storage at the end of that interval; flow across the control surface during the interval consists of precipitation $P(t)$, actual evapotranspiration $E(t)$ and soil water surplus $Q(t)$.

The water balance model, developed to work with GIS-Arcview Avenue programs (ArcView's object-oriented programme language) interface, is being employed to estimate the water surplus (runoff) that indicates the available water resources in the basin. The study required and obtained from ftp.crwr.utexas.edu in the directory /pub/crwr/gishydro/Africa; gridded climatic dataset interpolated to a 0.5 degree grids. The climate dataset, are mean monthly values for the period of record. Generated water surpluses are excess rainfalls that are available for streamflow generation. In place of using the usual water surplus routing processes like Muskingum-Cunge method or response function approach, or the two-step flow routing approach for the transfer of the water surplus into the subwatersheds of the Niger basin, an artificial neural networks (ANNs) is being employed.

5. ANNs model structure and weight distribution

The neural network structure and weight distribution used in the study for the training of the networks during the calibration simulation are shown in figure 12. By definition, the regression of a dependent variable y on an independent variable x, estimates the most probable value for y, given x and a training set. The regression method will produce the estimated value of y which minimizes the mean-squared error. The simulation is terminated when a reasonable coefficient of correlation (R), say 0.96 and above and a reasonably low value of RMSE are achieved

ANNs being an empirical and black-box model has an in-built capability that takes into consideration the watershed characteristics during simulations. The ANNs model parameters were estimated through model calibration known as the training of the networks. The network connection type is multi-layer normal feed forward, while the total number of layers is 15, the transfer function is sigmoid and root mean square errors (RMSE) set as the objective function. The model parameters were optimized by minimizing the values of the objective function. There are number of ways of modelling or linking the relationship between the causative factor, rainfall, to runoff observed at a particular site of a basin by using equations which describe the major physical processes involved in the transformation or by representing these component processes in a conceptual manner, or alternatively using the neural networks approach. So considering the impossibility of representing the component processes of transformation process using a physics-based approach, one may either opt for the conceptual representation or the neural network modelling approach. While calibrating the conceptual model or the neural network model using the past recorded input with the corresponding outflows observed at a specific location of a river, it is implicit that when the calibrated model is applied for future predictions, the input that would be used in the model is in the same range of input used for the calibration. It is on this premise that a simpler approach of model-to-model calibration technique, i.e. using one model to calibrate another model, has been adopted in the study. Ninety six monthly data has been used for calibrating the ANNs and twenty four months data were used for verification. Areally averaged temperature and precipitation changes from formulated climate change synthetic scenarios were imposed on each sub-basin for assessing climate change impacts on the generated water surplus (runoff).

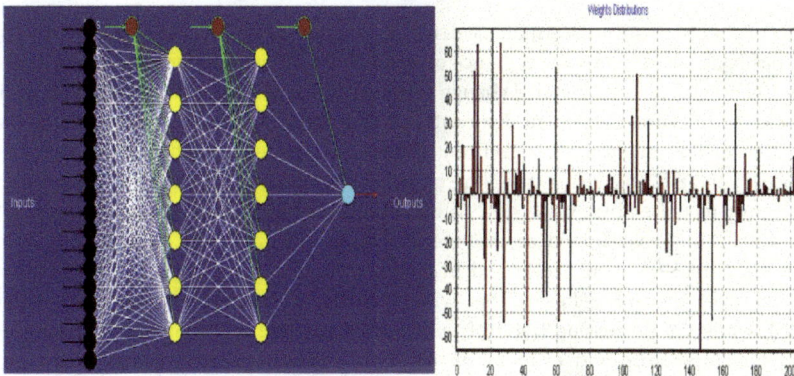

Figure 12. (a) typical architecture of the neural networks. (b) Typical weight distribution of neural networks structure

6. Application of climate change scenarios

Climate change scenarios are plausible indications of how future climate of a place will evolve. They are not predictions, as we know it in weather forecasting. They are generated using various global climate models (GCMs). There are three generic types of climate change scenarios: synthetic scenarios, analogue scenarios and scenarios based on output s from GCMs. All the three types have been used in climate change impacts research. In this study, the synthetic or hypothetical type of scenario is being applied in order to avoid the complexities of downscaling of the scenario from the outputs of GCM, which is most often used. The choice is on the premise that all the scenarios are not predictions of the future in the way that weather forecast are, but plausible indicators. Synthetic scenarios describe techniques where particular climatic (or related) elements are changed by a realistic but arbitrary amount, often according to a qualitative interpretation of climate model simulations for a region. Most studies have adopted synthetic scenarios of constant changes throughout the year [47] but some have introduced seasonal and spatial variations in the changes [48] and others have examined arbitrary changes in inter-annual, within-month and diurnal variability as well as changes in the mean [49, 50]. The selected scenario will be applied in the same manner it was used by Jiang *et al.*, 2007. The synthetic climate changes scenarios adopted in the study are shown in Table 5.

Scenario no.	1	2	3
ΔTemp (°C)	2	2	2
Δ Precip (%)	-20	0	20

Table 5. Synthetic climate change scenario

7. Results and discussion

Results of analyses show discernible evidence of climatic variability and change in the Niger River basin, as adjudged by the presence of trends in the series. The practical significance of a trend is judged by a percentage change of the sample mean over an observation period. The field significance assessment demonstrates that annual temperature, precipitation, and river flow in the region show significant change as adjudged by the obtained results in Tables 6, 7, and 8. Considering the entire basin holistically and sub-basin-wise, temperature significantly increased by a value < 3.79% over the entire basin; while rainfall decreased by a value < 10.0% and river flow decreased is in the range of 14.24 % - 40.8 % (Table 6). Further results show that the increasing trend in temperature is at the rate of 0.001°C/month and 0.02°C/yr over the entire basin. This will invariably create high evaporative demands (Fig.13), while the decreasing trend of rainfall is at the rate of 2.45mm/yr. The increasing trends of temperature and evaporation observed over the basin are due to the global warming known to be the culprits of climate change.

This indeed is in agreement with the report of IPCC over the region [6], projected to have an increasing rate of temperature of 0.2 to 0.5°C per decade. As further evidence from the interactions of Niger River flows and the sub-basins rainfall in Fig 14, the whole sub-basins are highly sensitive to the climate variability and changes in the region, a decrease in runoff observed everywhere in the basin coincides with decrease in rainfall and with hardly a time lag of > 2 years. The actual starting periods of the trends were substantiated using WMO [45] recommended 5-year average smoothening and this revealed that the present trends began since the post-civil war, economic development and population growth of 1970. These observed trends also strongly agree with that of Yue and Hashino [51], because a trend > 10% particularly in rainfall and discharge is quite significant in water resources management and planning.

The statistical significance of these trends was further explored by Thie and Sen's technique. Results obtained from this Thie and Sen approach are strongly supported by results obtained by regression test for linear trend and Kendall's Rank Correlation test (Tables 7 and 8). This said, it is important to also assert that in some cases a well-defined rainfall trends over Upper Niger, Lower Niger and Upper Benue Sub-basins could only be established when the length of data series was increased from 60 years to a number ≥70 years data. These observed hydroclimatic characteristics and trends were significant at 95% confidence level and the concomitant effect resulted in the downward trends of the river discharge over the basin, which portends danger for water resources of the region. Above results indicate the presence of change as adjudged by the trends in the series, also strongly supported by exploratory analyses shown in figures 15-17. Figures 17 and (18a and 18b) are typical of the river flow behavior at the upstream (Niger) and downstream (Nigeria). Niger River at the headwater in Niamey is obviously being threatened by the changing climate as indicated by these results and this has serious implications on the downstream flow over Nigeria. Further results characterizing the temporal climatic variability of the region reveal an increasing variability in areal temperature with coefficient of variability (CV) of 1.43% on long-term periods. The magnitude of the variability of recent time slice (1972 - 2002) was higher relative to the 1940-1971 reference periods; with CV of 1.21% (Table 9). On the contrary, the magnitude of the temporal variability of rainfall was higher during the reference period of 1940-1971, with coefficient of variability of 8.1%, even though the recent time slice (1972 – 2002) showed the greatest tendency towards aridity or drying condition, with lesser mean rainfall of 1427.8mm (Table 10).

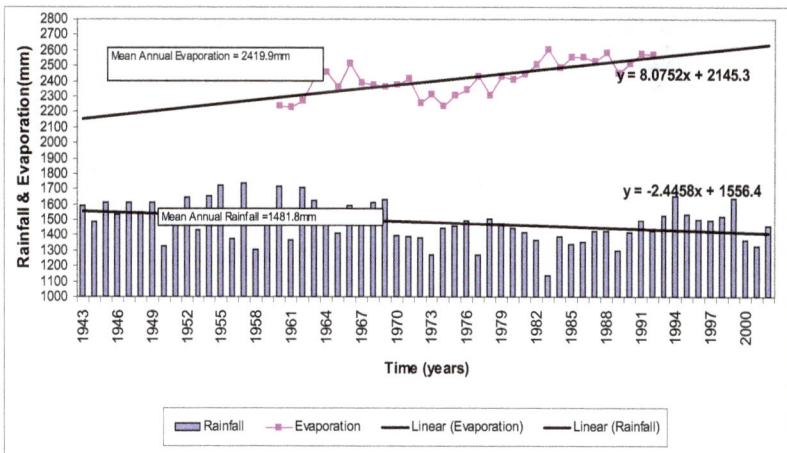

Figure 13. Annual Areal Average Rainfall and Evaporation Relationship

Testing the Significance of Temperature Trends Slope				
Sub-basins	Area Coverage (Km²)	Median slope (°C / yr)	Median Slope %	Remarks
Upper Niger	131,600	0.0076	1.3212	Significance
Lower Niger	158,100	0.0163	2.9826	Significance
Upper Benue	158,900	0.0093	1.6667	Significance
Lower Benue	73,000	0.0181	3.2871	Significance
Niger South	53,900	0.0171	3.2708	Significance
Entire Basin	594,000	0.0169	3.7905	Significance
Testing the Significance of Rainfall Trends Slope				
Sub-basins	Area Coverage (Km²)	Median Slope	M.Slope %	Remarks
Upper Niger	131,600	-3.57	-25.19	Significant
Lower Niger	158,100	-6.01	-4.47	Not Significant
Upper Benue	158,900	-2.22	-5.48	Not Significant
Lower Benue	73,000	-5.65	-16.66	Significant
Niger South	53,900	-13.89	-12.32	Significant
Entire Basin	594,000	-2.45	-10.00	significant
Testing the Significance of Niger River Flow Trends Slope				
Stations	Area Coverage(Km²)	Median Slope	M.Slope %	Remarks
Kainji(Upper Niger)	131,600	-3.62	-24.72	Significant
Jebba(Lower Niger)	158,100	-438.39	-40.80	Significant
Lokoja(Lower Niger)	158,100	-419.45	-28.93	Significant
Onitsha Niger South)	53,900	-217.50	-14.24	Significant
Numan(Upper benue)	158,900	-319.68	-31.10	Significant
Makurdi(Lower Benue)	73,000	-258.75	-27.85	Significant

Table 6. THIE AND SEN'S Median Slope Computation

Station	Slope	Variance	T-Statistics	t-Test	Error	Nature of Trend at 5% Sig. Level
Upper Niger	0.0087	0.00000635	3.4514	2.001	0.002519	Positive Trend
Lower Niger	0.0170	0.00000419	8.3190	2.001	0.002047	Positive Trend
Niger South	0.0179	0.00000388	9.0740	2.001	0.001972	Positive Trend
Upper Benue	0.0097	0.00000822	3.3708	2.001	0.002867	Positive Trend
Lower Benue	0.0197	0.00000739	7.2478	2.001	0.002719	Positive Trend
Entire Basin	0.0172	0.00000364	9.0271	2.001	0.00191	Positive Trend

a

Station	Slope	Variance	T-Statistics	t-Test	Error	Nature of Trend at 5% Sig. Level
Upper Niger	-1.904	0.695	-2.284	2.003	0.834	Negative Trend
Lower Niger	-10.01	13.088	-2.775	2.003	3.617	Negative Trend
Niger South	-11.837	21.453	-2.56	2.003	4.632	Negative Trend
Upper Benue	-6.640	4.745	-3.048	2.003	2.178	Negative Trend
Lower Benue	-8.564	10.136	-2.690	2.003	3.184	Negative Trend
Entire Basin	-2.446	0.8377	-2.672	2.003	0.915	Negative Trend

b

Table 7. a: Results of Regression Test for Linear Trend of Temperature over Niger River Basin
b: Results of Regression Test for Linear Trend of Rainfall over Niger River Basin

Station	Test Statistics	P-Values	Trend	Year Trend Begins	Nature of Trend	Warmest Year of Period & value
Upper Niger	6.569	1400	Positive	1972	Sig. between 1 and 5% Sig. Level	1987(2.45°C)
Lower Niger	5.868	1345	Positive	1979	Sig. between 1 and 5% Sig. Level	1987(3.0°C)
Niger South	7.233	1246	Positive	1973	Sig. between 1 and 5% Sig. Level	1998 (2.54°C)
Upper Benue	4.605	1452	Positive	1979	Sig. between 1 and 5% Sig. Level	2002(2.81°C)
Lower Benue	6.163	1167	Positive	1979	Sig. between 1 and 5% Sig. Level	1998(2.53°C)
Entire Basin	6.722	1412	Positive	1979	Sig. between 1 and 5% Sig. Level	1998(2.6°C)

a

Station	Test-Statistic	P-Values	Trend	Year Trend Begins	Nature of Trend
Upper Niger	-2.755	963	Negative	1970	Sig. between 1 and 5% Sig. Level
Lower Niger	-3.261	914	Negative	1982	Sig. between 1 and 5% Sig. Level
Niger South	-2.80	715	Negative	1972	Sig. between 1 and 5% Sig. Level
Upper Benue	-1.962	1173	Negative	1967	Sig. at 5% Sig. Level
Lower Benue	-2.375	750	Negative	1973	Sig. at 5% Sig. Level
					b

Table 8. a: Results of Kendall's Rank Correlation Test of Temperature over Niger Basin
b: Results of Kendall's Rank Correlation Test of Rainfall over Niger River Basin

Period	Mean	Std. Dev.	CV (%)	Max	Year	Min	Year	Max as % of Ave.	Min as % of Ave.
Long-term Period									
1943-2002	319.60	4.58	1.43	330.70	1987/88	312.52	1961/62	103.47	97.78
Reference Periods									
1943-1971	316.39	2.75	0.87	327.22	1972/73	312.52	1961/62	103.42	98.78
1972-2002	322.60	3.87	1.20	330.70	1987/88	315.81	1974/75	102.51	97.9

Table 9. Temporal Areal Temperature Variability over Niger River Basin in Nigeria

Period	Mean	Std. Dev.	CV (%)	Max	Year	Min	Year	Max as % of Ave.	Min as % of Ave.
Long-term Period									
1943-2002	1481.8	128.01	8.64	1738.6	1957/58	1139.7	1983/84	117.33	76.91
Reference Periods									
1943-1971	1539.6	125.26	8.14	1738.6	1957/58	1303.2	1958/59	112.92	84.65
1972-2002	1427.8	106.48	7.46	1656.0	1994/95	1139.7	1983/84	115.99	79.82

Table 10. Temporal Areal Rainfall Variability over Niger River Basin in Nigeria

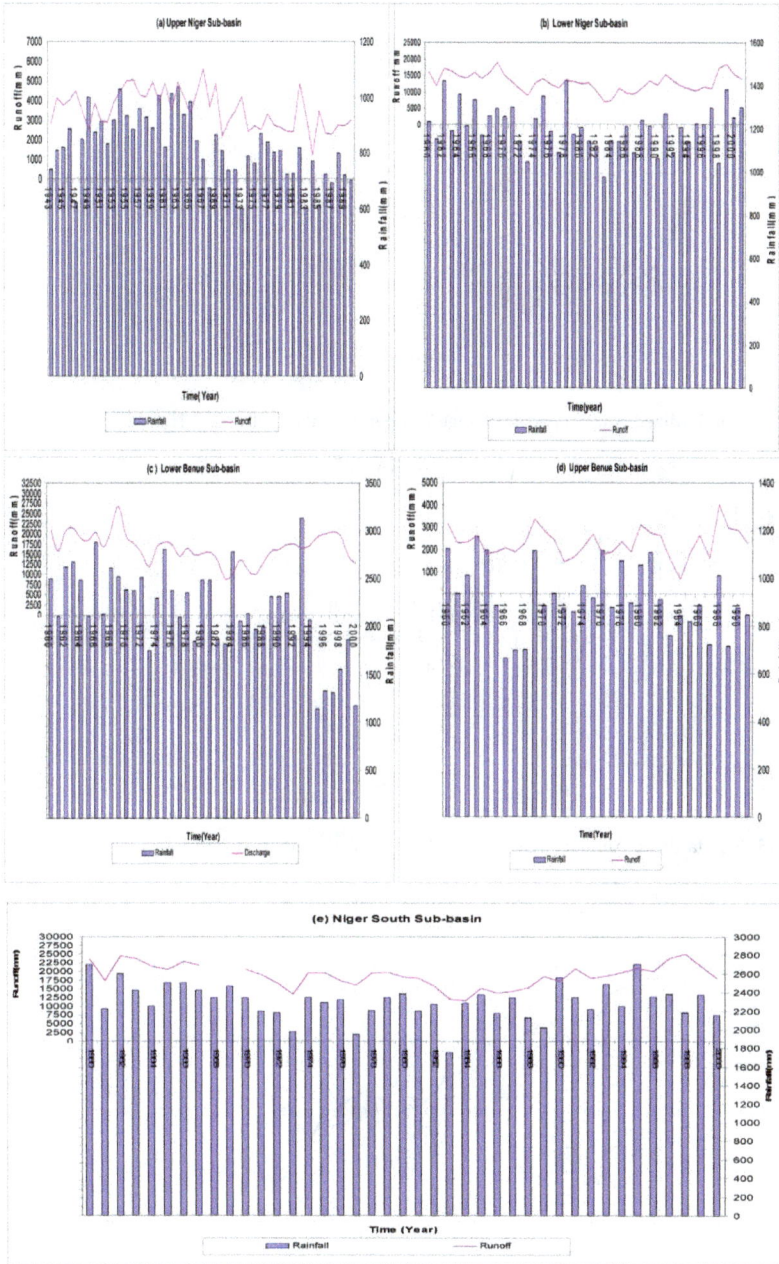

Figure 14. River Flows Interactions with Sub-basins Rainfall

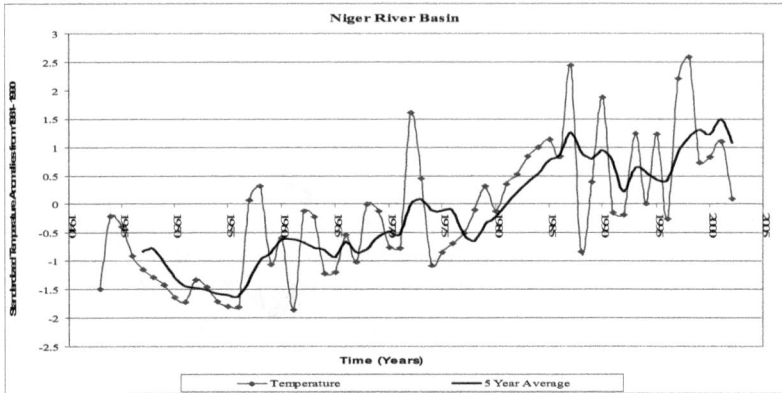

Figure 15. Standardized Annual Areal Average Temperature Variability and Trend

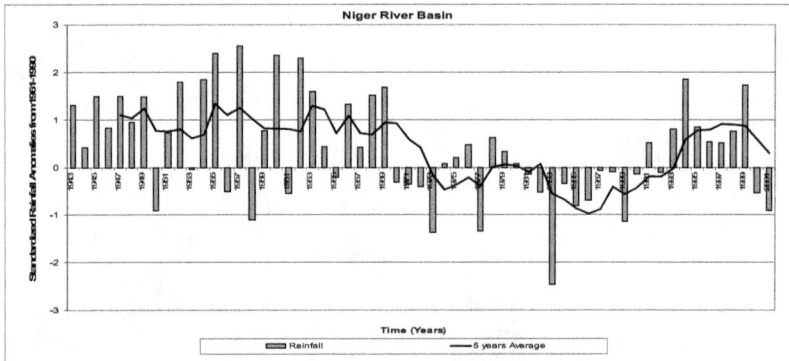

Figure 16. Standardized Annual Areal Average Rainfall Variability and Trend

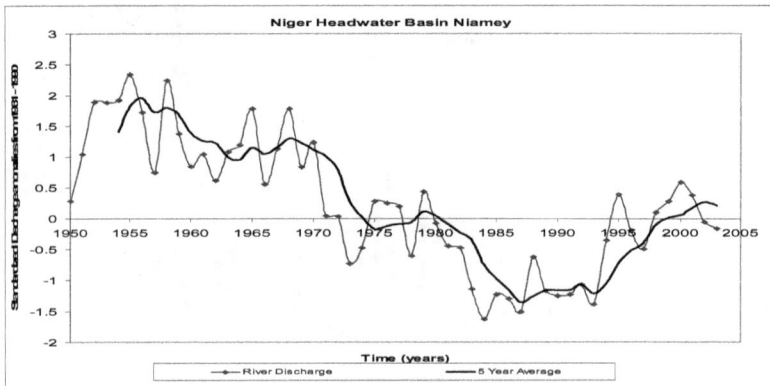

Figure 17. Typical Temporal Variability and Trend of Niger River over Niamey (Headwater Basin)

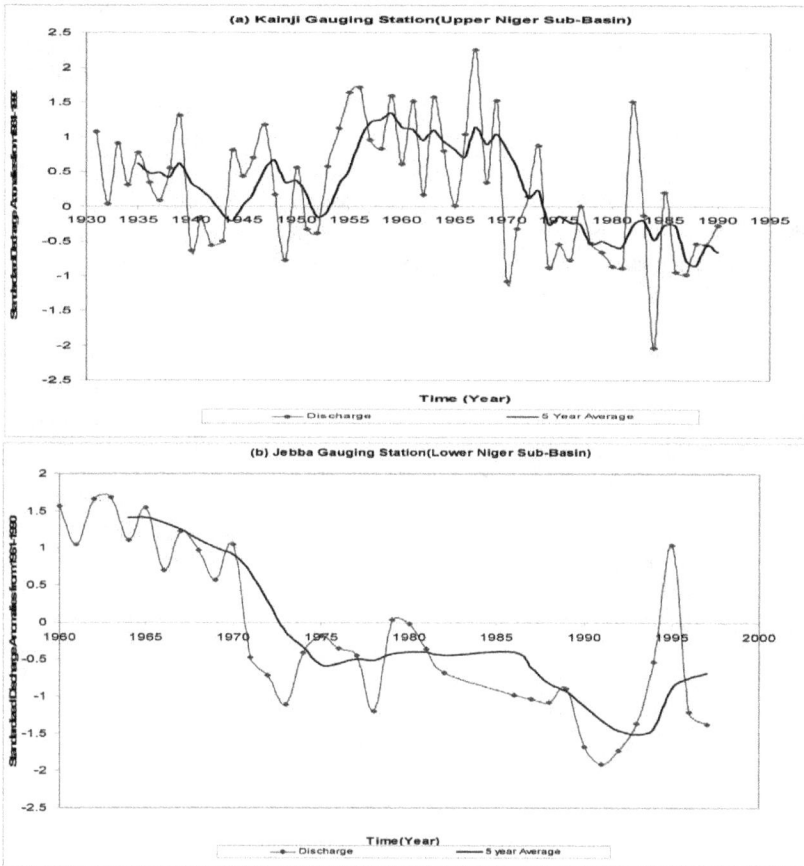

Figure 18. Typical Temporal Variability and Trends of annual River Flow over the Niger Sub-basins

7.1. Future climate change in the Niger River Basin

Although, there are no consensus among the Global Climate Models(GCMs) on whether the West African region will become drier or wetter over the course of the 21st century, because half of the 21 GCMs used by IPCC predict increased rainfall, while the remaining half predict decreased rainfall [1]. But most of the models do agree that climatic variability will increase over the region. Subsequently, climate change is expected to have a key influence not only on water resources, but on food and human security through its impacts on climate variability and extremes. Again, it is important to state that major portion of the Niger River basin located outside the Sahelian region lies within the humid tropical zone of southern Nigeria; a region already experiencing high temperature, evapotranspiration and rainfall. While climate models differ however, modeling done in this study reveals the high level of varying sensitivity of the Niger River basin, particularly the lower Niger sub-basin to the changing climate, which justifies the various experiences of impacts of climate change in the region. Results show direct link between rainfall and runoff (water surplus) generated over the Niger River basin. The peak of the rainfall coincides with the same month as the peak of water surplus and *vice versa* (Fig. 19). Since water surplus indicates the available water resources of the region, this reveals how sensitive the available water resources are to the climate of the region. Over the Upper Niger sub-basin, a peak rainfall of 265 mm in the month of August, produced a runoff or water surplus of 160.5 mm; while over Lower Niger and Niger South sub-basins, peak rainfall of 294 mm and 347 mm yielded water surplus of 185 mm and 238 mm, respectively. The annual water surplus or runoff was highest over the Niger south (1241.2mm) closer to Atlantic ocean, followed by Lower Benue (973.6mm), Lower Niger (729.4mm), Upper Benue (495.3mm) sub-basins and the least value is observed over Upper Niger Sub-basin (360.7mm)in the Sudano-Sahel zone. It is further observed that the water surplus is much more sensitive to the accuracy of potential evaporation estimate (that depends on temperature) in the humid climate than the arid climate.

The calibration and optimization results show that the models performance has been quite good, as adjudged by the values of RMSE and Nash Sutcliff Efficiency (E). Generic example of such outcomes are Upper Niger (12.6909 /and 0.9362), Lower Niger (8. 4665 and 0.9578) and Niger South (4.1315 and 0.9987). The highest RMSE values and lowest E values were observed over Niger South and Lower Niger sub-basins respectively.

Figure 20 is the visual output of the calibration simulation of the ANNs. It suffices to say from the result that the models have shown good capabilities to reproduce historical monthly runoff series with an acceptable accuracy proved by verification results (Figs. 20). However, the necessary assumptions made and the absence of intermediate component processes of rainfall- runoff transformation should be kept in mind; hence estimated observed values obtained through ANNs model has been used in place of actual observed runoff for the water balance calibration. Therefore, there may be need to further verify these results with actual field data.

Additionally, in order to evaluate the seasonal and inter-annual changes, differences in mean monthly runoff simulated by the ANNs using the climate change scenarios were compared with a baseline runoff values. Results show various changes in runoff expected over the Niger basin in the face of the projected climate change (fig. 21). It is further observed that when temperature increases by 2°C, the mean monthly runoff on the average changed by -10 to -50%, -5 to -40% and 15 to 60% respectively for precipitation changes of -20%, 0% and 20% (Table 11). Hence, if the current prevailing climatic conditions of increasing temperature trend and decreasing rainfall trends continue unabatedly, projected climate change may exacerbate its impacts on the water resources of the region, resulting in water stress condition. By implications then, freshwater, hydro-power, health and food security of the country may be seriously threatened, unless rainfalls turnout to be on the increase and adequate management strategy put in place.

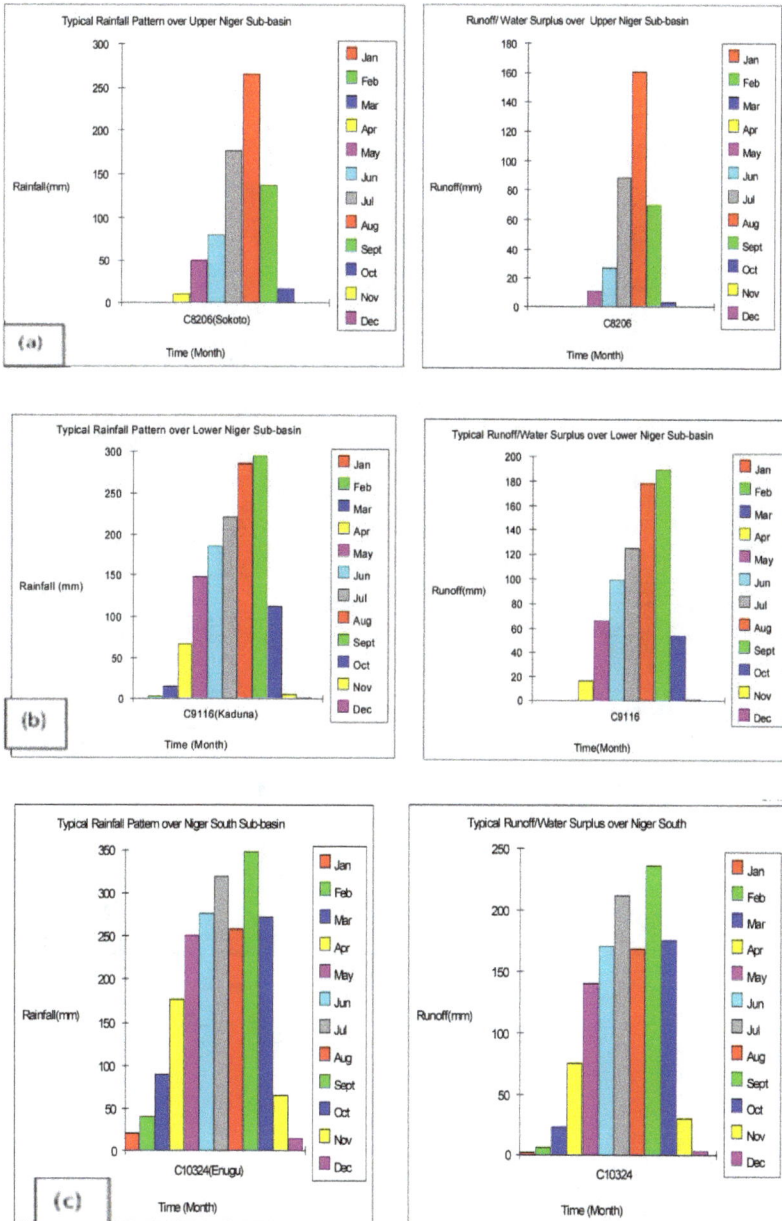

Figure 19. Typical Simulated Runoff and Climatic Conditions over (a) Upper Niger Basin (b) lower Niger and(c) Niger south sub-basins

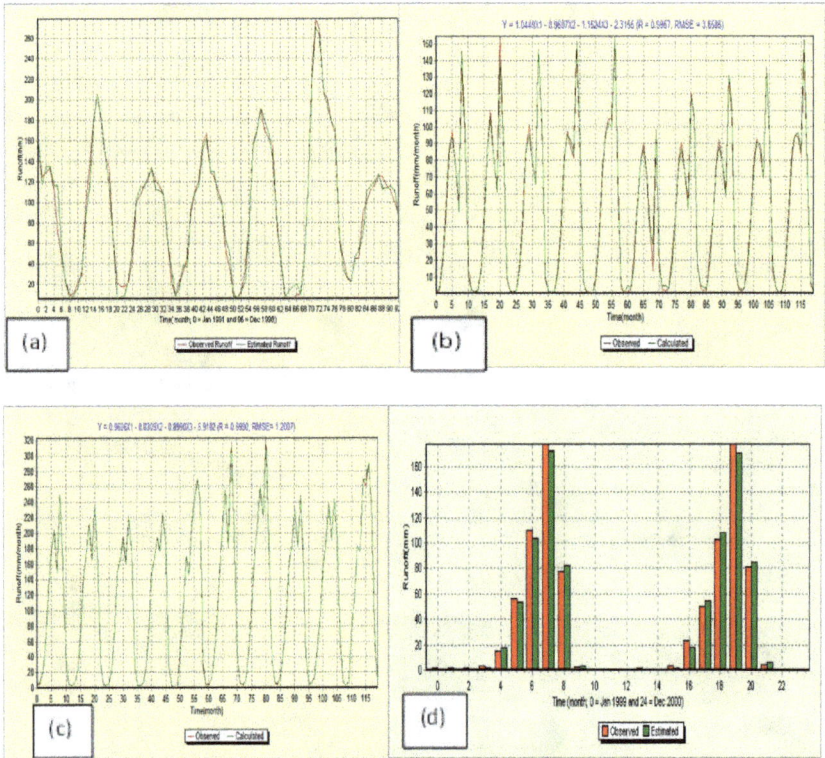

Figure 20. Typical ANNs Model Calibration Simulation Output over (a) Upper Niger (b) Lower Niger (c) Niger South Sub- basins and (d) model verification output (Upper Niger)

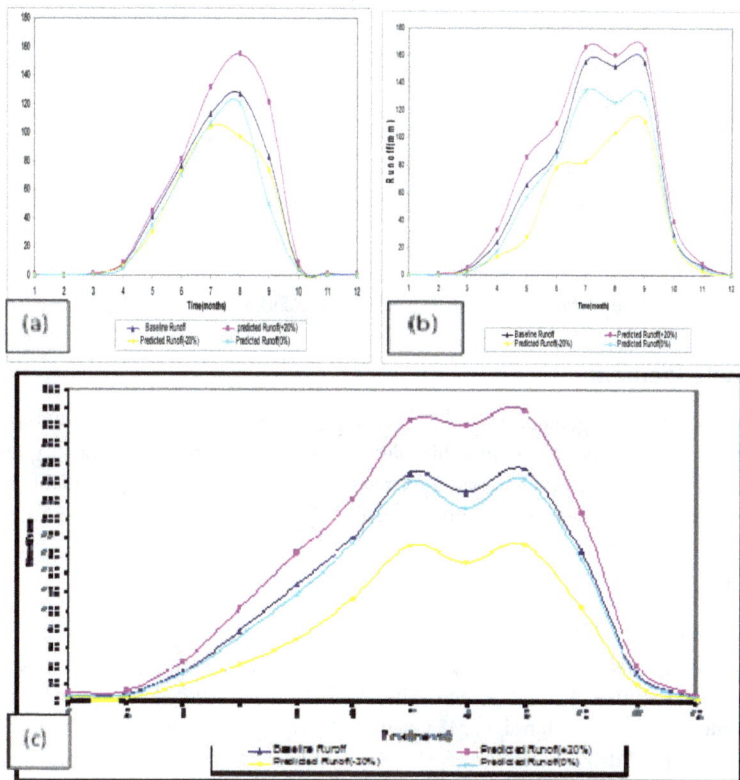

Figure 21. Typical Changes in Runoff under Climate Change Scenarios of Temperature = +2°C and Precipitation = + 20% and 0% over (a) Upper Niger,(b) Lower Niger and (c)Niger South Sub-basins

8. Conclusion

This study explored and assessed the potential impacts of projected climate change on water resources of Nigerian sector of Niger River basin using both parametric and non-parametric approaches and simulation models i.e. Thornthwaite water balance accounting scheme and Artificial neural networks (ANNs). There is discernible evidence of climate change in Nigeria, adjudged by the observed changes in the in the onset and cessation dates of seasonal rains and the presence of trends in the hydo-climatic series. It is further observed that apart from the Sahel region, Sudano-Guinean region or the humid portion in the lower Niger sub-basin are also vulnerable to the changing climate and its impacts. The observed changes are not unlikely to be connected to the long time variability in the climatic variables of the region, and land use changes due to increasing anthropogenic activities and gas flaring and population pressure. Also, a future drier climate is expected to impact negatively on the runoff and invariably on the available water resources of the region. Hence, proactive and aggressive management strategy is seriously needed to match any unfathomable impact.

Author details

Juddy. N. Okpara and Aondover. A.Tarhule
Department of Geography and Environmental Sustainability, University of Oklahoma, Norman, USA

Muthiah Perumal
Department of Hydrology, Indian Institute of Technology, Roorkee, India

Acknowledgement

The greatest acknowledgement goes to God Almighty, the author and finisher of our life for the inspiration and wisdom to put this piece of work together. We are also grateful to Professor Maidment and his team and the University of Texas for making available the GIS-ArcView avenue script used for the modeling, as well as the gridded data.

9. References

[1] Christensen, J.H., Hewitson, B., Busuioc, A., Chen, A., Gao, X., Held, I., Jones, R., Kiolli, R.K., Kwon, W.-T., Laprise, R., Magaña Rueda, V., Mearns, L., Menéndez, C.G., Räisänen, J., Rinke, A., Sarr, A. and Whetton, P. (2007). 'Regional climate projections' in Solomon, S., Qin, D., Manning, M., Chen, Z., Marquis, M., Averyt, K.B., Tignor, M. and Miller, H.L. (Eds.) *Climate Change 2007: The Physical Science Basis*. Contribution of Working Group I to the Fourth Assessment Report of the Intergovernmental Panel on Climate Change. Cambridge, UK and New York, US: Cambridge University Press.

[2] Oyebande L. ,Amani A., Mahe G., Diop I. N.(2002) Climate change, water and wetlands in West Africa: Building linkages for their integrated Management.Special Comminssioned paper. IUCN-BRAO. Ouagadougou, April, 2002

[3] Niasse, M., Afouda, A. and Amani, A. (2004) Reducing West Africa's Vulnerability to Climate impacts on water resources, wetlands and desertification: Elements for a Regional Strategy for preparedness and Adaptatio. International Union for Conservation of Nature and Resources (IUCN), Cambridge, 84 pp

[4] Watson, R.T. (Ed.) (2001). *Climate Change 2001: Synthesis Report*, Contribution of Working Groups I, II and III to the Third Assessment Report of the Intergovernmental Panel on Climate Change. Published for the Intergovernmental Panel on Climate Change. Cambridge: Cambridge University Press. Available at
http:// www.ipcc.ch/pdf/climate-changes-2001/synthesis-syr/english/front.pdf (79Kb).

[5] Conway, D., Persechino, A., Ardoin-Bardin, S., Hamandawana, H., Dieulin, C. and Mahe, G. (2009). 'Rainfall and river flow variability in Sub-Saharan Africa during the twentieth century', Journal of Hydrometeorology, Vol. 10, pp. 41–59.

[6] IPCC (2007): Summary for policy-makers. In: Climate Change 2007: The physical Science Basis. Contributions of Working Group I to the Forth Assessment Report of the Intergovernmental Panel on Climate Change.

[7] Hulme, M. (2001). 'Climatic perspectives on Sahelian desiccation: 1973–1998', *Global Environmental Change*, Vol. 11, pp. 19–29.

[8] Camberlin, P. and Diop, M. (2003): Application of Rainfall principal Component Analysis to the assessment of rainy season characteristics in Senegal

[9] Giannini, A., Saravanan, R. and Chang, P. (2003). 'Oceanic forcing of Sahel rainfall on interannual to interdecadal timescales', *Science*, Vol. 302, pp. 1027–1030.

[10] Dai A., Lamb, P.J., Trenberth, K.E., Hulme, M., Jones, P.D. and Xie, P.(2004) The recent Sahel drought is real. Int. J. Climatol., 24, 1323-1331

[11] Okorie, F.C. 2003. Studies on Drought in the sub- Saharan Region of Nigeria using remote sensingand precipitation data, JNCASR- Costed Fellowship Programme, University of Hyderabad, India,January- April.

[12] Olaniran, O. J. (1991). Evidence of climatic change in Nigeria based on annual rainfall series 1919-1985. *Climate Change*, 19, 507-520. doi:10.1007/BF00140169, http://dx.doi.org/10.1007/BF00140169

[13] Hulme, M. (1992): Rainfall Changes in Africa: 1931-1960 to 1961-1990, Int. J. Climatology 12:685-699

[14] Anyadike, R. (1993) Seasonal and annual rainfall variations over Sudan. International Journal of Climatology, 13, 567-580.

[15] Mah´e G, Paturel JE, Servat E, Conway D, Dezetter A. 2005. Impact of land use change on soil water holding capacity and river modelling of the Nakambe River in Burkina-Faso. *Journal of Hydrology* 300: 33–43.

[16] Grolle, J. (1997) Heavy rainfall, famine, and cultural response in the West African Sahel: the "Muda" of 1953–54', *GeoJournal*, Vol. 43, No. 3, pp. 205–214.

[17] Goulden, M. and Few, R.(2011) Climate Change, Water and Conflict in the Niger River Basin, USAID, Reliefweb report-*http://reliefweb,int/node/474471*

[18] Cook, S., Fisher, M., Tiemann, T. and Vidal, A. (2011). 'Water, food and poverty: Global-and basin-scale analysis', *Water International*, Vol. 36, pp. 1–16.

[19] Paeth, H., Fink, A. and Samimi, C. (2009). 'The 2007 flood in sub-Saharan Africa: Spatio-temporalcharacteristics, potential causes, and future perspective', *EMS Annual Meeting Abstracts*, Vol. 6, EMS2009-103.

[20] Tarhule, A. (2005). 'Damaging rainfall and flooding: The other Sahel hazards', *Climatic Change*, Vol. 72, pp. 355–377.

[21] Oyebande, L.(1995):Global Climate Change and Sustainable Water Management for Energy production in the Niger Basin of Nigeria: Proceedings of Global Climate Change: Impact on energy development conference, 1995(ed. Engr J.C. Umolu P.E. and published by DAMTECH Nigeria Limited Anglo-Jos, Plateau State , Nigeria).

[22] Milly, P.C.D.,Betancourt, J. Falkenmark, Hirsch, R. M. Kundzewich Z.W. Lettenmaier, D.P. and Scouffer,R.J. (2008) Stationarity is dead: Whither water management? Science Vol.319, 1st February, 2008, Pp 573-574

[23] Oyebande, L. and Odunuga, S. (2010) Climate Change on water resources at the Transboundary level in West Africa: The cases of the Senegal, Niger and Volta basins, The open Hydrology journal, 2010, 4,163-172

[24] Olomoda, I. A. (2004): Impact of climatic change on river Niger hydrology. Proceedings of International Workshop on Managing Shared Aquifers Resources in Africa, held in Tripoli Libya, 2nd to 4th June 2002, pgs 190-197

[25] Dessouassi, R. (2000). Revue des perspectives dans le Bassin du fleuve Niger, *ABN*, *Niger*. Instititut National de la Statistique: INS, Niamey, Niger, 2010.

[26] Ogilvie, A., Mahe G., Ward, J., Serpantie, G., Lemoalle, J.,Morand, P. Barbier, B., Diop, A.T., Caron, A., Namarra, R., Kaczan D., Lukasiewicz, A., Paturel, J.E., Lienou, G. and Clanet, J. C.(2010) Water, agriculture and poverty in the Niger River basin, Water International vol. 35, No. 5, September, 2010, 549-622

[27] Descroix, L.,G. Mahé, T. Lebel,G. Favreau, S. Galle, E. Gautier, J.C. Olivry, J. Albergel,O. Amogu,B. Cappelaere,R. Dessouassi,A. Diedhiou, E. Le Breton, I. Mamadou, D. Sighomnou (2009). Spatio-Temporal variability of hydrological regimes around the boundaries between Sahelian and Sudanian

[28] Salami, A.T. and Akinyede, J. (2006).: Space technology for monitoring and managing forest in Nigeria, a paper presented at the International Symposium on Space and Forests, United Nations Committee on Peaceful Uses of Outer Space (UNOOSA), Vienna, Austria.

[29] Agbola, T. and Olurin, T. A. (2003): *Landuse and Landcover Change in the Niger Delta*. Excerpts from a Research Report presented to the Centre for Democracy and Development, July, 2003

[30] Awosika, L. F. (1995). Impacts of global climate change and sea level rise on coastal resources and energy development in Nigeria. In: Umolu, J. C. (ed). Global Climate Change: Impact on Energy Development. DAMTECH Nigeria Limited, Nigeria

[31] Casenave L. (2004) Hydro-climatic variability: comparison of different global circulation model in western Africa. Master thesis. University of Chalmers, Sweden. 52 p.

[32] Nicholson, S.E. (2000) Land surface processes and Sahel climate', *Reviews of Geophysics*, Vol. 38, No. 1, pp. 117–139.

[33] Taylor, R. G. and Howard, K.W.F. (1996) Groundwater recharge in the Victoria Nile basin of East Africa: Support for the soil-moisture balance method using stable isotope and flow modeling studies. Journal of Hydrology, 180, 31-53

[34] Hulme, M. (1992) Rainfall Changes in Africa: 1931-1960 to 1961-1990, Int. J. Climatology 12:685-699.

[35] Nicholson, S. E. (1993) An overview of African rainfall fluctuations of the last decades, Journal of climate, vol. 6 1463-1466

[36] Gbuyiro, S. O. & Aisiokuebo, N (2003) Climate change in Nigeria—Its reality, expectations and impacts. In: Proc. International Symposium on Climate Change (March–April, 2003, China), 238–240. WMO/TD No. 1172.

[37] Okpara J.N., Akeh, L.E. and Anuforom, A. C. (2006): Possible impacts of Climate Variability/ change and urbanization on water resources of availability and quality in the Benin-Owena River Basin IAHS Publication 308, ISSN 0144 – 7815, pg 394 – 400.

[38] Okpara, J. N. (2007): Monitoring the impacts of Climate Change/ Variability and Anthropogenic Activities on Lake Chad, Northeastern Nigeria, Using remote Sensing Satellite, Proceedings of IUGG2007 International Conference, Perugia, Italy.

[39] FAO (Food and Agriculture Organization) (1997). *Irrigation potential in Africa: A basin approach*, Land and Water Bulletin 4. Rome: Food and Agriculture Organization of the United Nations. Available at
http://www. fao.org/docrep/W4347E/w4347e00.htm#Contents.

[40] Andersen, I., Dione, O., Jarosewich-Holder, M. and Olivry, J-C. (2005). *The Niger River Basin: A vision for sustainable management*. Washington DC, US: The World Bank.

[41] MahooH.F., Young, M.D.B., and Mzirai, O. B. (1999)Rainfall variability and its implications for the transferability of experimental results in semi-arid areas of Tanzania. Tanzania Journal of Agricultural Science 2(2) 127-140

[42] Keay, R.W.J., 1959. An Outline of Nigerian Vegetation. Lagos

[43] White, F. (1983) The vegetation of Africa, a descriptive memoir to accompany the UNESCO / AETFAT /UNSO vegetation map of Africa. UNESCO, Natural Resour. Res. 20: 1-356

[44] WMO, (1988): Analyzing long time series of hydrological data with respect to climate Variability and change, WCAP-3, WMO/TD 224

[45] WMO, (2000) Detecting trends and other changes in hydrological data/ WCDMP 45, WMO /TD 1013.

[46] Thornthwaite, C. W. (1948): "An Approach towards a Rational Classification of climate" Geophysical review, 38, 55-94

[47] Rosenzweig, C., J. Phillips, R. Goldberg, J. Carroll, and T. Hodges, 1996: Potential impacts of climate change on citrus and potato production in the US. Agr. Systems, 52, 455-479, doi: 10.1016 /0308-521X (95)00059-E.
http://www.ipcc.ch/ipccreports/tar/wg2/index.php?idp=563

[48] Rosenthal, Donald H and Howard K. Gruenspecht 1995 "Effects of global warming on energy use for space heating and cooling in the United States", *Energy Journal*, Volume 16 Issue 2, pp. 77-96.

[49] Mearns, L.O., C. Rosenzweig, and R. Goldberg, 1992: Effect of changes in interannual climatic variability on CERES-Wheat yields: sensitivity and 2×CO2 general circulation model studies. *Agric. For. Meteorol.*, 62, 159-189.

[50] Mearns, L. O., C. Rosenzweig, and R. Goldberg, 1996: The effect of changes in daily and interannual climatic variability on CERES wheat: a sensitivity study. *Clim. Change*, 32, 257-292.

[51] Yue S. Hashino, M. (2003): Long term trends of annual and monthly precipitation in Japan. Journal of American Water Resources Association, Volume 39 Issue 3 Page 587-596, June 2003.

Chapter 4

Climate Change Impact on Tree Architectural Development and Leaf Area

Michel Vennetier, François Girard, Olivier Taugourdeau,
Maxime Cailleret, Yves Caraglio, Sylvie-Annabel Sabatier,
Samira Ouarmim, Cody Didier and Ali Thabeet

Additional information is available at the end of the chapter

1. Introduction

1.1. Context

The response of forests to the forecasted increase in climate stress occurrence is considered a key issue in climate change scenarios [1]. Although forest productivity increased in most ecosystems during the 20th century [2,3], a review by Allen *et al.* [4] underlined an emerging trend of heat and drought induced forest decline and dieback at global scale. Several and generally combined physical and biological causes contribute to observed tree decline or die-off [4-7]. Apart extensive insect outbreaks [8], understanding the respective role of hydraulic failure and carbon starvation due to excessive or long lasting water stress is one of the major research goal in order to predict forest response to climate change [9].

The consequences of climatic events on forest health can be immediate but are often delayed up to 5 to 10 years [5,10], and may be significant for decades and sometimes irreversible on tree growth [11]. Recent studies on tree architectural development and primary growth suggested that the long lasting impact of repeated droughts on tree crown development could be one of the causes of these delayed effects [12-14].

Primary growth corresponds to the creation of new tissues outside existing organs, and includes bole, branch and root length growth, branching (birth of new branches or roots), creation and growth of leaves or needles and rootlets, flowering and fruiting [15,16]. It is therefore fully linked to plant architectural development patterns and processes. In contrast, secondary growth for trees corresponds to the radial growth of existing branches, bole and roots. In single trees and forests, although secondary growth usually exceeds primary growth and leaf production [17], the total amount of biomass allocated to primary growth may be very important [18]. As an example, leaf and fruit production measured through litter bags in French adult broadleaved stands (mainly beech and oak) reached 20 to 40% of wood production by stems [19]. The relative allocation to new shoots was not correctly assessed up to now but it can be inferred from leaf production and the leaf mass fraction (LMF), ratio between leaves and whole twig mass. In a recent study, twig wood/leaf biomass ratio was found to reach from 25 to 50% and from 50 to 75% during respectively dry and humid years for *Pinus halepensis* [20] which means low LMF values (25 to 75%). However, LMF was generally higher on dry sites in south eastern France: 40-70% for *Abies alba* [21], 70-80% for *Quercus ilex* [22]. It was found to vary between 80 and 90% for 107 Chinese species at various elevations [23]. As a whole, primary growth may represent between 25 and 70% of secondary growth. But except for tree height growth, scant literature exists on the relationship between tree primary growth processes and climate change or accidents, whatever the species, and plant architecture is commonly neglected in climate change impact studies.

The mediterranean climate is characterized by high temperatures associated with low rainfall in summer, drought being the main environmental constraint for vegetation growth [24]. For the 21st century, climatic models forecast that the Mediterranean basin will be prone to a faster warming than most other continental areas over the world, associated with a reduction of rainfall during the growth season [1,25]. Therefore, this area is a good place to detect and model any climate change impact on vegetation, all the more since a rapid decline in precipitation and higher temperatures were already noticeable in parts of this basin [26,27]. In Southeastern France, the period 1998-2007 was characterized by mean annual temperature and mean summer temperature 0.9°C and 1.3°C above the 30-year average (*figure 1*) Moreover, eight to ten of the twelve hottest years since 1850 were recorded during this time lapse [28] which give a foretaste of the climate forecasted for the next decades. In addition strong climatic events also recently occurred such as the 2003 heat

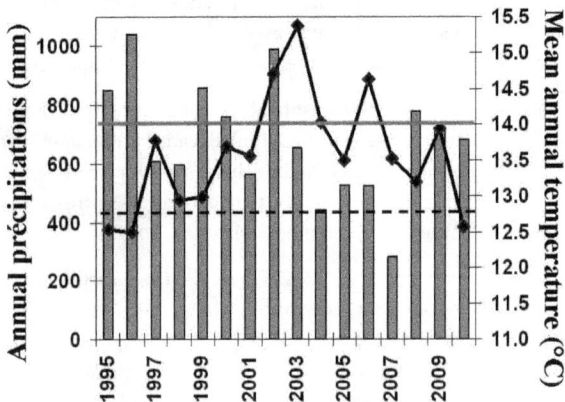

Figure 1. Average annual precipitations (grey bars) and temperatures (black line) in Font-Blanche since 1995. Horizontal lines are 1961–2010 average, grey line = mean rainfall, dotted black line = mean temperature.

wave which significantly impacted French Mediterranean forests as well as most of Europe [29]. Such extreme climatic events are likely to become frequent with global warming [1,30]. Scorching heat had direct and delayed negative effects on tree growth, especially on pine species [31]. The resulting increase in summer and spring water stress may reduce tree growth in Mediterranean areas [32-35]. Raising temperature may also lead to phenological lags [36], particularly in the beginning and end of the growth season, with direct consequences on some primary growth processes and architectural development such as polycyclisme and branching rates [37,38].

1.2. Goals

This study aimed at quantifying the influence of recent climatic trend and events, particularly intense heat and drought, on the primary growth and architectural development of six conifers and one broadleaved species growing in Mediterranean plains (*Pinus halepensis, P. pinea, P. pinaster, Quercus ilex*) and mountains (*P. nigra, P .silvestris, Abies alba*). The three last ones are at the lower limit of their distribution area in southern France. One of the final goals was to model the direct and delayed effects of climatic accidents on tree leaf area in order to help assessing the resulting risk in terms of decline and die-back.

2. Material and methods

2.1. Study area and species

The study area included 8 sites distributed between 80 and 1400 m of elevation, from the coast to hinterland mountains of Provence-Alpes-Côte d'Azur region in south-eastern France (*figure 2*).

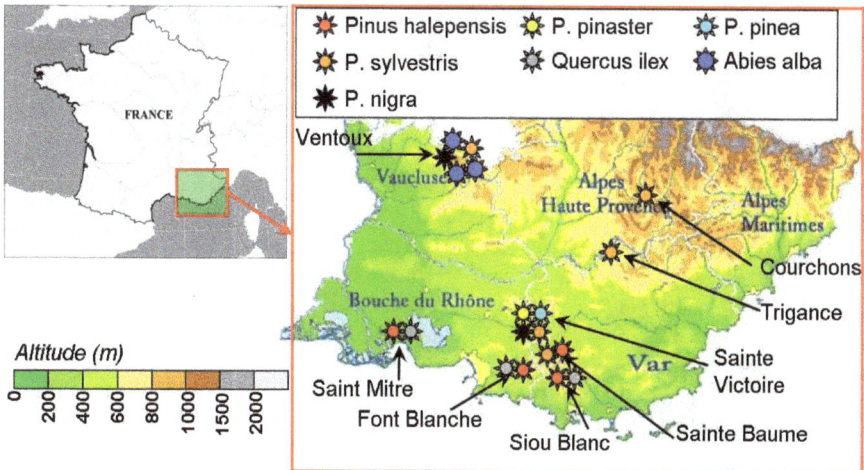

Figure 2. Study area and studied sites.

Nearly 5150 twigs from 1050 branches of 210 trees and 7 species were sampled between 2005 and 2011 (*Table 1*). Their architectural development was retrospectively measured from morphological markers (*figure 3*) over a period of 15 to 25 years. For each tree, the sampling design considered separately three thirds of the crown (top, middle and base), branch orientation (north and south), but also branch hierarchy (principal or secondary axis – *figure 4*) and branch vigor. Secondary axes were chosen according to their relative vigor (strong vs weak axes) within the branch they belong to (*figure 4*). Twig absolute vigor for each species was later split in three groups of equal number (vigorous, medium, frail) according to their total length growth in the last three years before sampling date.

Site name	Species (number of trees)	Altitude (m)	exposition	Dates of mesures
Saint Mitre	PH (11) QI (6)	80	Flat	2008-2009
Font-Blanche	PH (58) QI (34)	420	Flat	2005-2011
Siou Blanc	PH (11) QI (6)	650	Flat	2008-2009
Sainte Baume	PS (5+5+5), PM (5)	950	North	2005-2006
Sainte Victoire 1	PS (5); PN (5)	650	North	2009-2010
Sainte Victoire 2	PM (5); PP (5)	500	Flat	2010
Trigance	PS (5)	1000	Flat	2005-2006
Courchons	PS (5)	1350	North-east	2005-2006
Ventoux	PS (5), PN (10), AA (19)	1100 to 1400m	North	2009-2010

PH = Pinus halepensis Mill, PS = Pinus sylvestris L., PM = Pinus pinaster Aiton, PN = Pinus nigra ssp nigra Arn, PP = Pinus pinea L., AA = Abies alba Mill

Table 1. Study sites

2.2. Growth and architectural parameters

When present, 5 to 10 needles were randomly chosen from the base to the end of each twig and all around it to measure their length and width. Needle number per twig was counted on a subsample of twigs (1/3) with consideration of missing needles which were counted using their scars on the twig. As needles are lined up in three to five lines or spirals along the shoot, counting needles along one or two of these lines or spirals and then bulking up the count proved to be a very reliable assessment (error < 5%)

In order to bridge primary and secondary growth, ring width was measured for all studied trees. Two cores per tree were collected perpendicularly at 1.3 m height. Ring widths of each core were measured using a micrometer (±0.002 mm, Velmex Inc., Bloomfield, NY). Some trees were also logged for stem analysis and rings were counted along 4 perpendicular directions. Ring width series were firstly cross-dated and standardized with the classical methods of dendroecology, to remove age-related tendencies from the growth curve and to obtain a homogenised variance. Then elementary raw and detrended series were respectively averaged for each tree and master chronologies were constructed for each species and for each plot by averaging tree series.

For *Pinus halepensis* and *Quercus ilex*, a follow-up of phenology and architectural development was performed twice a month from 2008 to 2011 on one site (Font-Blanche) to understand their relationship with climate and their interrelations. Branches and twigs were chosen following the same protocols than for architectural studies.

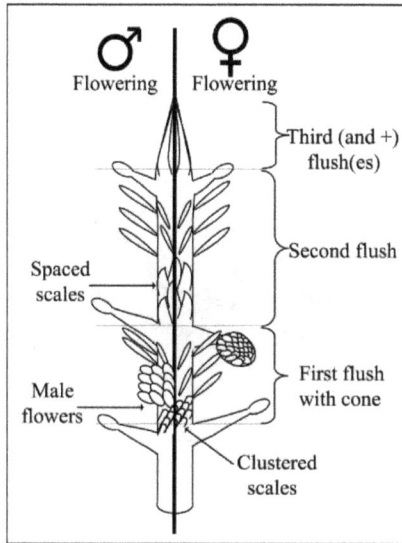

Figure 3. Morphological markers used for twig growth reconstruction in conifers. All growth units start by sterile scales, small and clustered on the first one, larger and spaced out on the following ones in case of polycyclisme (mainly Pinus halepensis and P. pinaster). Male flowering (left): male flowers appear at the base of the first annual growth unit over sterile scales for pines. The scars they left on the twig are generally different from needle scars. Female flowering (right): cones (with very few exceptions) appear at the top of the first annual growth unit. For pines, cones or their peduncles remain a long time on the branch or leave a specific scar. The presence of at least one branch or of a whorl indicates the limit of a growth unit for pines, but some growth units may be branchless. For firs and oaks, intermediate branches may appear, during all the branch life. A given pine twig never bears male and female flowers the same year. For Quercus ilex, a pseudo-whorl of branches indicates the limit between two annual shoots. The retrospective analysis of branch growth is far more difficult than for conifers.

2.3. Branch modeling

The observed variations of architectural parameters, needle number and needle size were integrated into a 2D-model of pine branch architectural development to simulate the impact of climate on pine total leaf area (*figure 5c*). We designed the model for the more complex of studied conifers: Aleppo pine (*Pinus halepensis*). Aleppo pine gives potentially the higher number of growth units per year, so that the model can easily be simplified for other species. In a first attempt we considered medium vigor branches located in the middle part of the crown as representative of the average branch of a tree. Parameters for each twig category (principal and secondary axes, vigor, medium and frail axes) were implemented according to scenarios of successive years considered as "normal" or "bad" (defined afterwards; *figure 5 a -b, table 2*). As pine architectural development for a given year is partly to fully pre-determined by the climate of the previous year, the impact of bad years on polycyclism and the number of needles are delayed compared to the impact on leaf size.

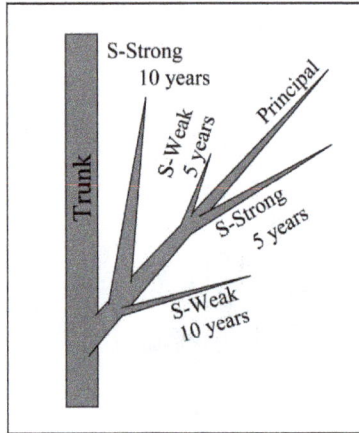

Figure 4. Twig sampling. Five axes are measured on each branch: the principal axis and two pairs of secondary axes one weak and one strong by pair five years-old and 10 years-old. The classes "weak" and "strong" are relative to each other in the concerned pair and their absolute vigour depends on branch vigour.

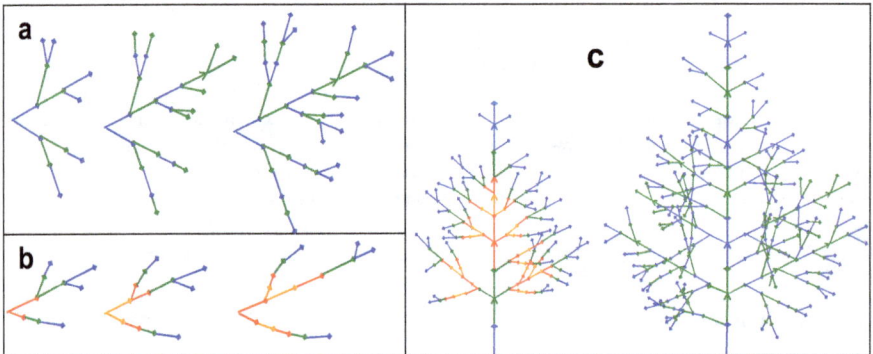

Figure 5. Example of the development of a medium vigor secondary axis of Pinus halepensis without (a) or with (b) 1 to 3 bad years. Example of a middle crown / medium vigor branch (principal and secondary axes) and development after 7 years with 4 successive bad years. The model includes strong and weak secondary axes and 3 classes of twig vigor. Color changes each year along each axis: blue and green for good years, orange and red for bad years.

At each step of branch development and at the end of each year, active twigs (with needles) could be counted and sorted by vigor. The number and size of needles per active twig was set according to *table 2*. Total needle surface (length*width*needle number) was calculated for each twig and bulked up for the whole branch.

This model was used to simulate branch growth for 10 years, as all parameters for secondary axes were obtained for this time span. For longer periods, the interaction and competition with neighboring branches, twig self pruning, branch aging and accidents may significantly change these parameters, so that a 3D model taking these interactions into account is necessary.

For *Abies alba*, some specific sampling and analyses were made to compare trunk, lateral axes and ring width responses to climate. Trees were sampled at 3 different elevations (1150, 1250 and 1350m) [26]. For 14 trees, annual shoot length was measured on the trunk. Laterals axis sampling and measurements followed the same protocol as other species.

2.4. Statistical analyses

As most architectural and growth parameters slowly evolve with branch aging, it was necessary to remove this natural trend. This was systematically done for each parameter using the difference measured at equal cambial age for branches of respectively top vs middle and middle vs base of the crown, for the period 1995-2000 considered as accident-free (*figure 8*, method in references [12,37]).

To quantify inter-annual variability between traits, an individual detrended coefficient of variation (dCV), for the period 1990-2010, was computed, for each trait of each tree, as follow: (i) individual trend was removed by taking the residuals of the linear or non-linear model with time as explanatory variable and (ii) the standard deviation of the detrended sequence was divided by the raw sequence mean (i.e. the mean trait value for each tree).

Thus, the detrended coefficient of variation of trait j for tree i can be written:

$$dCV_{j,i} = \frac{SD\left(X_{j,i} - \beta 0_{j,i} - \beta 1_{j,i} * Time\right)}{\overline{X_{j,i}}}$$ with $X_{j,i}$ the sequence of trait j values for tree i ; $\beta 0_{j,i}$ and

$\beta 1_{j,i}$ the corresponding estimates of the model used for detrending and mean ($\overline{X_{j,i}}$) the mean value of the trait j for tree i.

For a global assessment of the relationship between growth, architectural parameters and climate, a Principal Component Analysis (PCA) was performed considering years as observations and all detrended architectural and growth parameters as variables, species by species. PCA was also used to help sorting good years (favorable climate for tree growth) and bad years (*figure 7*). As some parameters were not common to all species (polycyclism, male flowering, needle number), each PCA was performed with and without these variables to check the stability of years and variables in PCA planes. Needle length was not always available for the same period than other factors. Thus each PCA was also performed with needle length for available years and without needle length on the whole studied period (1995-2010).

All growth and architectural variables were averaged per species for bad and normal years. Bad years were defined as the four worst years in the 2003-2008 period for each individual variable. All other years were merged to compute the data for "normal years": as exceptionally low values due to repeated severe drought were excluded, we considered other data as normally good, mean or bad, representative of the normal interannual variability.

Partial least square (PLS) regressions were used to investigate relationships between architectural or growth parameters and climate. This method was chosen because it handles many variables with relatively few observations [39] and deals with correlated variables [40]. The number of significant PLS components was chosen by a permutation test [41] with a 5% threshold for the explained variance. Variables were tested with a 1000-step cross-validation [42]: they were retained only when the confidence interval (p<5%) for their partial correlation coefficient excluded zero. According to the phenology of the species in South-eastern France, climatic monthly parameters tested in each PLS were rainfall (P), maximum temperature (MaxT), minimum temperatures (MinT) and mean temperatures (MT) from January of previous year (n-1) to November of current year (n) over the period 1995–2010 [43,44]. The low number of observations (16) and high inter-annual variability of climate made grouping monthly climatic parameters necessary to obtain significant variables. We grouped the successive months having same signs for their individual partial correlation coefficients (sum of precipitations, average temperature). To compare exposition, position, status and vigour classes for each growth variable, normality was checked using a standard Shapiro-Wilks test. When the distribution was normal, a variance analysis and a multiple comparison test were performed to look for significant differences globally and further compare the different classes two by two. When the distribution was non-normal, these comparisons were performed respectively by a Kruskall-Wallis test and a Nemenyi test [45].

3. Results

For all species, most growth and architectural variables showed decreasing values in the last 15 or 20 years, and also from the top to the base of the crown (*figure 6.a*). After detrending with the comparison between branch position in the 1995-2000 period (*figure 6.b*), their

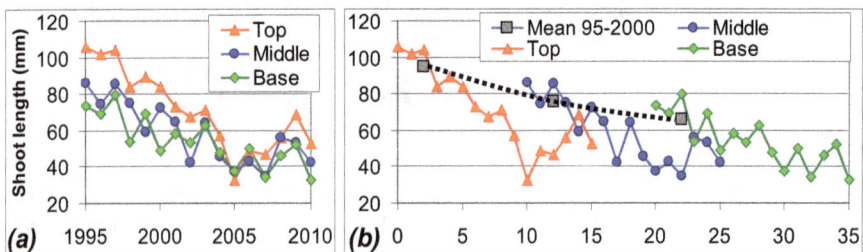

Figure 6. Pinus sylvestris annual branch length growth at Sainte Victoire for the top, middle and base of the crown (respective branch age 15, 25 and 35 years), for the same period (6.a) and with a 10-year shift (6.b) giving the natural trend of growth slowdown with age. Trees are approximately 60 year-old and measured branches are not competing with neighbouring trees.

decreasing rate from 2000 was always faster than the natural trend and showed a deep trough lasting 2 to 5 years after 2003 or 2004 (*figures 8-12 and 14*).

Whatever the species and variables used, PCA axis 1 and 2 were dominant with respectively 50-70% and 17-32% of explained variance, far over Axis 3 (4-12%). PCA results were highly coherent between species for the distribution of years in 3 groups (*figure 7*): bad years (2004 to 2007) and good years (1995 to 1998) were stable across all analyses, other years being more variable, generally in intermediate situation but sometimes good or bad (2003 and 2008). Some variables were stable for most species: shoot length, polycyclism and needle number were correlated with Axis 1 and between each other. Fruiting rate and needle length were better correlated to axis 2. Male flowering was available for *Pinus halepensis* and *P. sylvestris* only and was linked to the occurrence of bad years. Ring width occupied a variable position between axis 1 and 2 according to species. No noticeable difference was observed in year ranking or inter-variable correlations for a given species when PCA was performed with a variable number of years to include needle length.

The period 2004-2007 was characterized by a reduction of all growth and architectural parameters: shoot length, ring widths, needle length and number, and branching rate (*figures 6 and 8 to 14*). Production of female reproductive organs was also disfavored during this period while male flowering twigs were more numerous, particularly in the middle and top crown. This PCA approach indicated that most growth and architectural parameters were inter-correlated, but *figure 8* shows time lags in the response to bad years and recovery. For example, the original position of 2007 in the PCA plane for *Abies alba* is due to the high

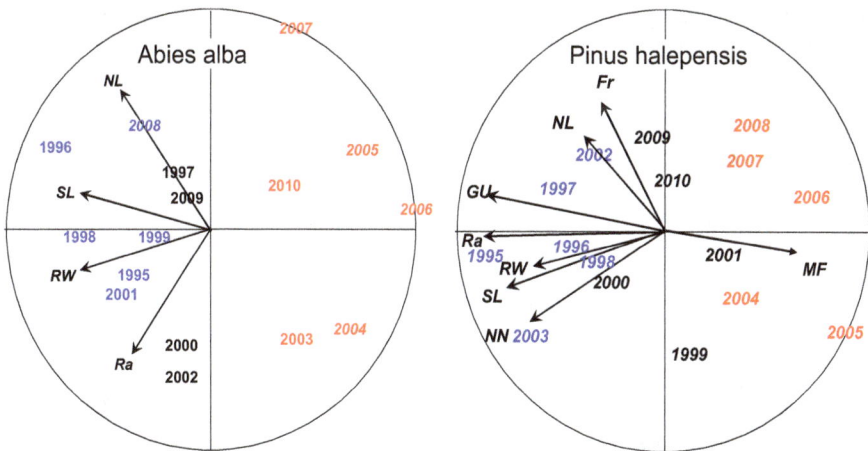

Figure 7. PCA plane for Abies alba (left) and Pinus halepensis (right) with years and variables: SL = shoot length, NL = needle length, Ra = ramification (branching rate), Rw = ring width, GU = number of growth units (polycyclism), NN = needle number, Fr = fructification, MF = male flowers. In red: bad years, in blue, good years.

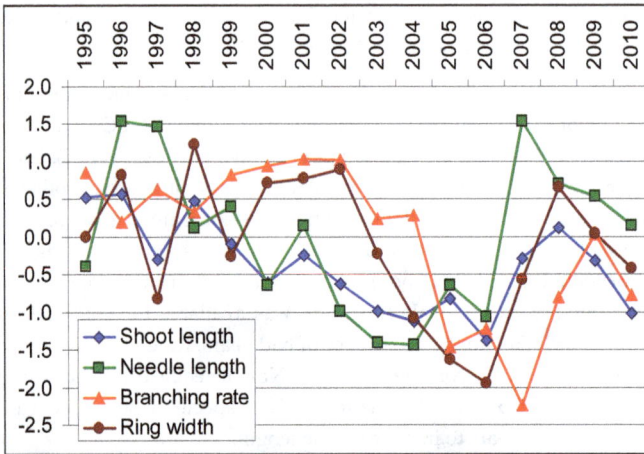

Figure 8. Evolution of detrended and standardized growth traits of Abies alba showing the consistent fall in the heart of the 2003-2007 climatic accident but time lags in their response to climatic variations and recovery after 2006.

value of needle length in 2007 (*figure 8*) while all other variables did not recover so fast, particularly branching rate.

Branching rate, one of the important architectural development indicator, is a good example of the common pattern between all species (*figure 9*). Differences were observed in the response to individual years and in the speed of recovery after the lower values, but the global trend was similar.

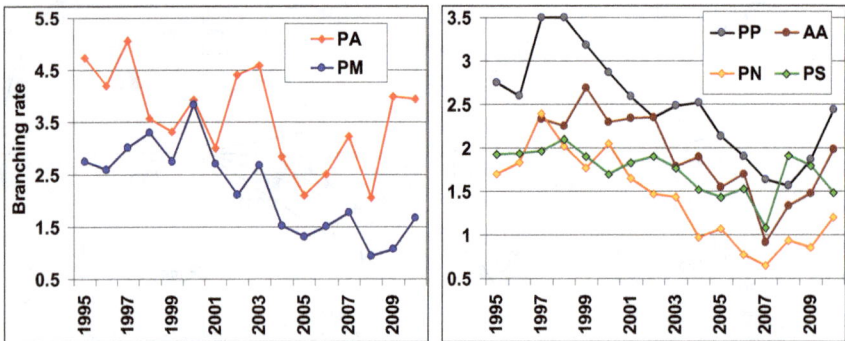

Figure 9. Branching rate (undetrended) on principal axes of the 6 studied conifers. PA = Pinus halepensis, PM = P. pinaster, PP = P. pinea, PN = P. nigra, PS = P. sylvestris, AA = Abies alba

Polycyclism, a fundamental growth trait for some pines species, confirmed this pattern (*figure 10*). It influences branching rate as each growth unit may give birth to new branches. This is why these two parameters are highly correlated (*figures 9 and 10*).

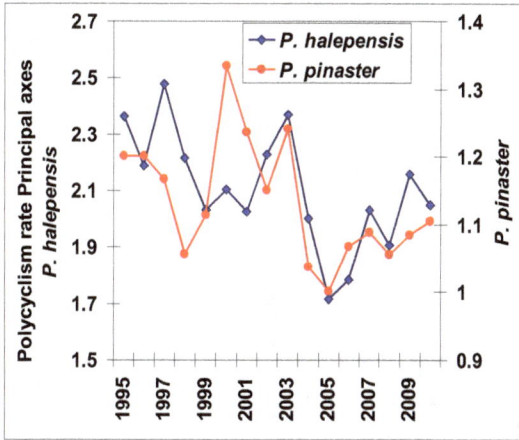

Figure 10. Polycyclism rate of branch principal axes for the two polycyclic pines, P. halepensis (left scale) and P. pinaster (right scale)

All architectural variables were positively correlated to branch vigour. *Figure 11* shows the example of needle number for *P. halepensis* according to branch vigour. The difference between the 3 classes of vigour was always significant all years in the aggregate, and also for most individual years.

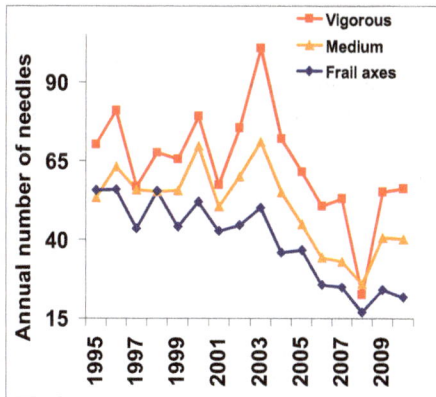

Figure 11. Number of needles per annual shoot for Pinus halepensis according to twig vigour.

The relative fall of growth and architectural variables after 2003 was generally more severe for vigorous and principal axes and at the top of the crown than on weaker and secondary axes and in the middle and bottom of the crown (*figures 6, 11 and 12*). Between 2004 and 2007, during 1 to 4 years according to species and variables, there were no longer significant differences according to branch vigour and hierarchy, and sometimes between positions in the crown. This higher relative sensitivity was also visible in a faster or better recovery: for most variables and all species, low and weak axes started recovering one or two years after top and vigorous ones and sometimes showed no significant increase up to 2010 (*figure 11*). However, the detrended coefficient of variation was not always significantly different between trunk and branch principal or secondary axes length growth, e.g. for *Abies alba* (*figure 12*). Ring width proved to be less variable than shoot length in *Abies alba*, but not in Pines, particularly because of their sharp decrease after 2003 and their high sensitivity to some accidents due to heavy snow falls.

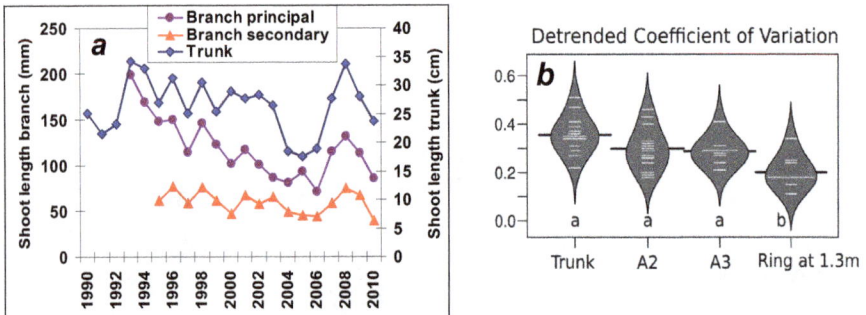

Figure 12. a - Abies alba trunk and branch detrended length growth (LG). Each series starts with its first real value. b - Detrended coefficient of variation (dCV) "beanplot" of the mean value (black lines), the kernel density function (in grey) and the raw values (white lines with size proportional to the number of measurements stacked. Letters summarize the results of a pairwise comparisons using Wilcoxon rank sum tests (P=0.05).

Although branch vigour was correlated to branch hierarchy and position within the crown, each of these factors significantly influenced branch architectural development (*figure 13*). For the same vigour (same mean length growth during the last 3 years), a branch had higher values on branch principal axes than on secondary axes and on axes of following orders, and decreasing values from the top to the base of the crown. Vigour, hierarchy and position were always highly significant (P<0.01%). South-exposed branches had sometimes higher values than north-exposed ones, but the difference was rarely significant (P<5%) and always at the very limit.

Needle length was highly variable from one year to the other (*figures 7 and 14*). It was severely affected by 2003 heat wave for all species, loosing from 25 to 45% on previous years average and recovered only slowly. According to species and sites, 4 to 8 years were necessary to regain normal values (mean of 1995-2002 values when available, or values from literature and herbariums: see reference [12]).

For *Pinus nigra* and *Abies alba*, 2006 and 2007 respectively were the first years after 2003 to reach normal values, but needle length further decreased again to significantly lower values. *Pinus halepensis* and *Pinus sylvestris* needles remained under normal values up to 2010.

Although we only have short series of data for *Quercus ilex*, it seems to follow the same trends as conifers (*figure 15*). Individual leaf area remained very low in 2009 compared to normal values for the French Mediterranean area, but inter-individual variability is very high and prevents assessing the normal values without older references on the same trees.

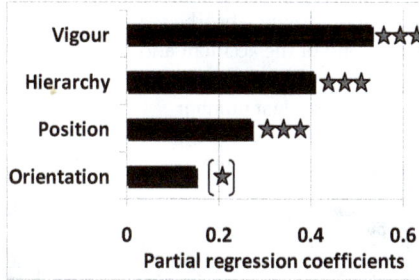

Figure 13. Mean relative weight of branch vigour, hierarchy (principal vs secondary), position (low, middle, top of the crown) and orientation (north vs south) in the determination of branching rate, polycyclism when relevant and needle number and length for the 6 studied conifers. The bars corresponds to the mean partial correlation coefficient of PLS regressions for each growth and architecture parameter and each species. Stars indicate the level of confidence: *** $P<0.01\%$, * $P<0.5\%$.

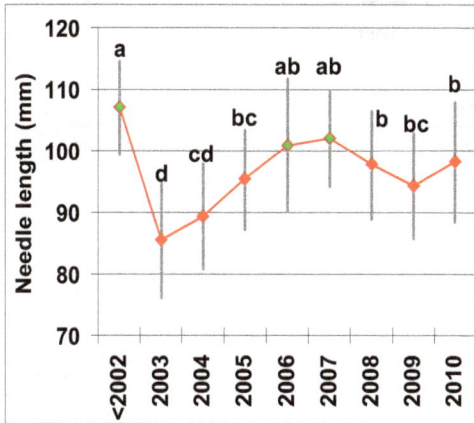

Figure 14. P. nigra needle length for branch principal axis in Mount Ventoux. Vertical bars indicate 2 standard deviation. Letters summarize the results of a Nemenyi test (P=0.05). Years sharing one letter are not significantly different. <2002 = mean values from all previous years (few samples) and from literature and herbariums for the study area.

Table 2 presents *Pinus halepensis* needle mean size and number according to axis vigour for good and bad years, and the resulting leaf area per twig. Leaf area for a given twig decreases with time due to accidents, parasites and aging. Survival rates measured in our plots (at the bottom of the table) varied according to twig vigour: needles remained longer on frail twigs than on vigorous ones.

These values were used to compute the total leaf area of a branch during 10 years with and without a 4-year accident, corresponding to the four worst years observed for many variables in this study (*figure 6*). According to the branch model, the deficit of active twigs (carrying needles) and of leaf area amounted respectively to 59 % and 78 % for *Pinus halepensis* compared to the values simulated with only normal years and average parameters (*figure 16*). This deficit was only slowly absorbed: it remained close to 30% and 40 % respectively two years after the end of the accident and disappeared only after 5 years.

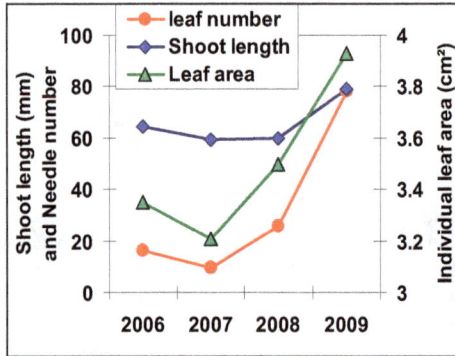

Figure 15. Quercus ilex growth at Font-Blanche.

Axis vigor	vigorous	medium	frail
needles number per twig			
normal year	69	59	49
bad year	52	38	28
needles length (mm)			
normal year	81	71	65
bad year	67	54	42
Needle width (mm)			
normal year	0.9	0.8	0.6
bad year	0.75	0.65	0.55
Needle surface (mm²)			
normal year	72.9	56.8	39
bad year	50.25	35.1	23.1
Leaf area per twig (mm²)			
normal year	4995	3357	1919
bad year	2601	1351	645

Needle survival rates: year 1 = 100%, year 2 = 85%, year 3 = 5%, 15% and 40% respectively on vigorous, medium and frail twigs.

Table 2. Needle number and size and leaf area according to twig vigor and climate for Pinus halepensis.

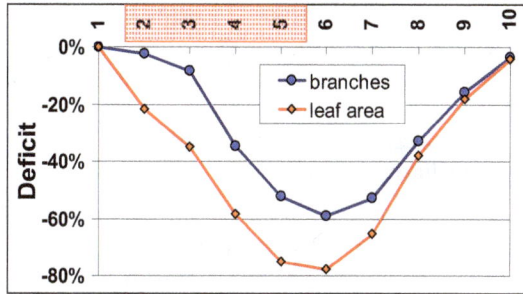

Figure 16. Evolution of the relative deficit of active twigs and total leaf area for a branch of Pinus halepensis submitted to a 4-year long climatic accident (2nd to 5th year), compared to a branch developing with mean climate parameters.

4. Discussion

4.1. Endogenous and climatic effects on crown architecture components

Ontogeny and axis morphogenetical gradients

Most of the studied morphological traits present a progressive decreasing trend along time (*figures 6, 9, 10, 11 and 12*). This decrease corresponds to a morphogenetical gradient named axis drift [15,46,47]. Morphological traits values are also driven by axis vigor, hierarchy and position within the crown (*figure 13*).

Theses endogenous constraints can be viewed as a variational module sensu Wagner *et al.* [48] that constraints any independent trait variations and thus partly explain the strong covariations found between morphological traits (*figures 7 and 8*).

Pre-induction of the primary growth characteristics (and when relevant of several flushes) in buds induces a strong dependence of primary growth on the global health status of trees and branch vigour at the time of bud formation, i.e. in previous year. Except in case of successive intense stress or extreme events, tree vigour and health rarely sharply changes between two years as trees can use non-structural carbohydrates stored in the stem and branches for growth and respiration. Such an autocorrelation between growth units is well known in ring width series [5,49,50] and was logically found for primary growth in this study.

Climatic effect on primary and secondary growth

Climatic effect on tree growth are generally analyzed using classical dendroclimatic analyses based on tree-ring data [51]. They usually conclude that tree radial growth is affected by continuous changes in climatic conditions (trends) and by strong climatic events [3,26,52]. Height growth was more occasionally used as indicator due to time-consuming measurements in the field, but gave reliable results [53]. This study revealed that a temporal survey of branch elongation and architectural parameters is of interest to this aim.

The occurrence of successive drought years between 2003 and 2007 led to a clear fall in all growth and architectural parameters: branch length and tree height growth, branching rate and polycyclism, needle number and needle length reached very low values during two to five years. These reductions could be clearly attributed to climate as the natural trends induced by branch or tree aging were slower, and all parameters finally recovered, at least partly, with more favorable climatic conditions.

As an extreme event with deleterious effects on forest health and productivity throughout Europe [54] and on studied species in Southeastern France [13], the 2003 summer heat-wave was supposed to have deeply impacted tree architectural development and growth. The direct impact of 2003 was mainly visible on needle length and individual leaf area (*figures 8 and 14*), on ring width and to a lesser extent on some parameters of *Abies alba* (*figure 8*). Leaf development and ring width were stopped very early at the end of spring due to extreme water stress. But for most other parameters and species (*figures 8 to 12*), 2003 was a normal or even a good year. The predetermination of most architectural parameters in the buds during 2002 which was a rainy year, and for polycyclic species the ability to add a new growth unit at the end of the year explain that 2003 impact is mainly visible from 2004. But our study cannot accurately assess this impact as 2003 was followed by several very dry years. However, the fact that some of the traits (*Pinus halepensis* and *Pinus pinaster* polycyclism and branch length growth, needle length for many species) started or completed their recovery in 2006 or 2007 (*figures 8 to 10, 12 and 14*) although they were extremely dry years, could be interpreted as the end of the delayed effects of 2003. This is consistent with the time necessary for all studied pine species to built up a full set of normal needles again after the loss of a large amount of old needles in summer 2003 [31], and to get read of the short needles of 2003. This time could have been also necessary to overcome the disorders caused by hydraulic failure in tree sapwood in summer 2003 [9,55,56].

Intra-specific variability in traits response to climate

All the traits did not respond similarly to climate variability and accidents (*figures 8, 12, 15*) for a species on a given site. Each of them is driven by many factors, related to climate or to functional relations between organs within a tree.

Needle length is mainly determined by the climate of the year of their development. For polycyclic species, needles of the later cycle can be very short when they lack time to complete their lengthening before the end of the growth season. Conversely, needle width and thickness mainly depend on twig diameter and vigour [57] and therefore, at tree level, on the climate of previous years and more generally on branch and tree health status. Although they are slightly sensitive to climate conditions of their growing year, they do not follow the rapid changes of needle length. Consequently, needle area is a compromise between its length and width which do not vary synchronously. Needle number and branching rate on a given twig are predetermined in the terminal bud of the twig. For monocyclic species, they are fully controlled by the climate of the previous year and by twig vigour. For polycyclic species, several growth units can be predetermined by previous year conditions but additional cycles may be formed later in the growing year according to climate and other constraints [15]. These new growth units can give birth to both needles and new branches, or only needles or branches. Branches are, however, also controlled by the climate of their first growth season: some of the preformed buds remain dormant and young shoots abort in case of unfavourable conditions. The total leaf area at branch level is thus complex to model due to the many factors at stake.

The variable response of traits in time and intensity may also be linked with phenology: secondary growth occurs longer than shoot extension in monocyclic species [58], but the opposite is observed in some polycyclic species when shoot growth occurs late in autumn and even in winter without cambial activity [20]. Early or late climate-related stresses may not have the same impact.

Ring width seemed to present a reduced plasticity compared with annual shoot lengths for *Abies alba* (*figure 12b*). This result, poorly investigated in the literature, highlight the critical need of multi-traits approaches to assess the effect of climate change on trees and open questions about tree carbon allocation under stress conditions. Although ring width is known to present autocorrelated series, it was generally mainly driven by the climate of its growth season and the previous winter for studied sites and species [13,21,59].

Recently, Girard *et al.* [12] found a significant expanded influence of the previous year for *Pinus halepensis* ring width: rainfall from February to June and temperature from April to September. Moreover, Sarris et al. [60] showed the significant influence of cumulated rainfalls expanding from one to five previous years from the more humid to the driest area of this species distribution range. Thus, climate change should increase the autocorrelation in tree response to climate and make it far more complex in integrating medium term climate variations in each annual growth pattern. The slow response of many growth variables after the 2003-2007 droughts may be a normal adaptation to this arid period and globally to an exceptionally hot and dry climate during one decade in the study area [61].

The inertia of branching patterns, driving leaf number and total leaf area, may explain the increasing length of the integration period with climate aridity.

Inter-specific response variability

Although some differences can be observed in the response of species to inter-annual climate variability, this study showed globally consistent trends in time for all traits and all species (*figures 7, 9 and 10*). Discrepancies consist mainly in time lags of one or two years for the lower values in the 2003-2007 periods, and in the speed of recovery after 2007. In some cases, a significant decreasing trend was observed from 1998 or 2000 for some species, while in the same sites the other species showed no fall before 2003. Inter-specific differences seamed relatively higher before 2003 than during and after the crisis. This proves that species and traits varied resistance to drought or other stresses, which allowed these initial differences, was smoothed by the intensity of the crisis. This conclusion should be tempered as we only present measures for living trees: *P. sylvestris* and *Abies alba* mainly, *P. nigra* to a lesser extent, experienced high rates of die-back in the hinterland. Dead trees could have shown different response patterns.

As already stated, within the six studied conifers, only two (*P. halepensis and P. pinaster*) present frequent polycyclism that should contribute to a different response pattern to climate change. As illustrated by *figure 10,* the number of growth cycles within a year present huge inter-annual variations that are directly linked with climatic conditions [37]. Polycyclism can be viewed as a fast way to improve foraging abilities when growth conditions are favourable. On the other hand, late cycles induced by hot autumns and recently observed even in winter in the study area are detrimental to tree health and growth due to frost damages [38,62].

4.2. Effect of crown architecture components on tree health

Leaf area deficit vs branching rate

Tree leaf area is the product of numerous architectural components. It is the result of the number and size of leaves per growth unit and the number of growth units per annual shoot and the branching rate. At tree level, it also depends on competition between branches and between trees, and on crown shape related to age and tree history. At all scales, it depends on twig, branch and tree vigour and health status, not to mention external factors as defoliators and diseases. Up to now, leaf area deficit and tree growth (radial, height, or volume) were the main factors used to quote tree health [63,64]. If leaf area deficit can initially occur as an avoidance mechanism to maintain a favorable water balance by reducing transpiration, it also induces a reduction in carbon assimilation [65]. Consequently leaf area deficit may be the early warning of a sequence leading to tree death [63,66], and could be used to predict tree mortality [67]. But this deficit was always assessed globally without desentangling its distinct causes. In this study, we quantified with our branch model an extreme leaf area real deficit, reaching nearly 80% after 4 very bad years (*figure 16*). This is far over the "leaf deficit" usually quoted by crown transparency, reaching such an extreme only for dying trees which is far from being the case in our plots. This discrepancy can be partly explained by the reduction of branch length growth and of the distance between leaves or needles along the twigs, which concentrates them in a smaller crown volume.

According to the branch model, during the two first bad years, branch deficit remained under 10% and could not explain the leaf area reduction which reached 35%. Thus, leaf number and size were the main factors at stake. With the lengthening of the drought period, the deficit in branch number became dominant, explaining most of the gap in total leaf area. Finally, during the recovery period after the end of bad years, branching shortage entirely matched leaf area deficit. This is consistent with *figures 8 and 9* showing that branching rate reached its minimum in 2007 or 2008, after the other variables, and particularly after needle length. Its recovery rate was slower than needle number (*figure 11*), which also showed a minimum in 2008 for *Pinus halepensis*. Finally, branching deficit during bad years is probably one of the key issues in leaf area long lasting deficit leading to forest decline.

Functional equilibrium

According to the concept of functional equilibrium [68,69] plants allocate biomass in priority to organs concerned by the most limiting factors [70]: roots if case of nutrients or water shortage, shoots and leaves when their environment is deficient in CO_2 or light. Accordingly, plants show remarkable resilience when part of their leaves, branches or roots are destroyed or artificially removed. They generally recover a normal leaf area or root length and biomass in a few years [71]. This may be true when repeated or long lasting climatic stresses reduce first their aerial growth (shoot and needle length) (*figures 8 and 16*), in order to maintain and develop the root system while limiting evapotranspiration. After the release of the stress, tree favors leaf production to reach the balanced level (*figures 8 and 14*) as well as shoot length. This priority in resource allocation to primary growth seams to the expense of secondary growth as indicated by a slower recovery in ring width, as already described by Girard et al. [12] for *Pinus halepensis*.

5. Conclusion and prospects

Modeling tree responses to climate change and particularly dieback hazard is a key issue since strong changes in tree productivity, survival and recruitment were observed recently [4,63]. A main concern is the assessment of tree vulnerability to increasing drought periods. Empirical models based on statistical relationships are not reliable as they cannot accurately take thresholds and extreme events into account. In contrast, mechanistic models explicitly represent the processes by splitting them into different blocks, which describe the response of the process to some input variables. However the variability of architectural components is poorly represented up to now in process-based models of individual tree growth. Most of them ignore their spatial variations and differentiated temporal response according to axis position in the crown, hierarchy in the branch and vigour. Their improvement with these new findings is urgently needed.

Our analysis made on many sites and species in Southeastern France revealed common patterns of response of tree architectural development to climate change and accidents. The role of long lasting delayed consequences of climate accidents on branching rate, holding back the potential leaf area for years, is one of the key issues to be tackled. Low leaf area, through carbon shortage, may contribute to forest decline and die-back.

This study highlights the necessity of more thorough investigations, in terms of field work and modeling. Our preliminary results must be confirmed for new species and climates and with longer data series to disentangle the multiple and contradictory effects of climate change on tree architectural development.

Author details

Michel Vennetier*, François Girard, Samira Ouarmim and Cody Didier
Irstea - EMAX, Aix en Provence, France

Michel Vennetier
Fédération de Recherche ECCOREV, FR 3098, Aix en Provence, France

* Corresponding Author

François Girard
Direction de la Recherche Forestière, Ministère des Ressources Naturelles et de la Faune, Québec, Canada

Olivier Taugourdeau
UMR AMAP – Université Montpellier 2, Montpellier, France

Yves Caraglio and Sylvie-Annabel Sabatier
UMR AMAP - CIRAD, Montpellier, France

Maxime Cailleret
INRA - URFM, Avignon, France

Samira Ouarmim
Université du Québec en Abitibi-Témiscamingue, Rouyn-Norenda, Canada

Ali Thabeet
Aleppo University Faculty of Agronomy, Aleppo, Syria

Acknowledgement

We would like to thank Christian Ripert, Roland Estève, Willy Martin, Aminata N'Diaye Boucabar, Frédéric Faure-Brac, Maël Grauer, Hendrik Davi, Nicolas Mariotte, William Brunetto and Florence Courdier for their assistance in the field and laboratory work. This research was funded by the French National Research Agency (DROUGHT+ project N° ANR-06-VULN-003-04, and DRYADE project n° ANR-06-VULN-004), the French Ministry for Ecology, Energy and Sustainable Development (GICC–REFORME project, No MEED D4ECV05000007), the Conseil Général des Bouches-du-Rhône (CG13), ECCOREV Research Federation (FR3098), the "F-ORE-T" LTER network and Cemagref.

6. References

[1] Hesselbjerg-Christiansen J, Hewitson B (2007) Regional Climate Projection. In: Solomon S, Qin D, et al., editors. Climate Change 2007: The Physical Science Basis Contribution of Working Group I to the Fourth Assessment Report of the Intergovernmental Panel on Climate Change. Cambridge University Press, Cambridge (U.K.) and New York (U.S.A), pp. 847-940.

[2] Boisvenue C, Running S.W (2006) Impacts of climate change on natural forest productivity - evidence since the middle of the 20th century. Glob. Change Biol. 12: 1-21.

[3] Spiecker H, Mielikäinen K, Köhl M, Skovsgaard J.P (1996) Growth trends in European forests : studies from 12 countries. Springer-Verlag, Heidelberg, 372 p.

[4] Allen C.D, Macalady A.K, Chenchouni H, Bachelet D, McDowell N, Vennetier M, Kitzberger T, Rigling A, Breshears D.D, Hogg E.H, Gonzalez P, Fensham R, Zhang Z, Castro J, Demidova N, Lim J.H, Allard G, Running S.W, Semerci A, Cobb N (2010) A global overview of drought and heat-induced tree mortality reveals emerging climate change risks for forests. For. Eco. Manage. 259: 660-684.

[5] Bigler C, Braker O.U, Bugmann H, Dobbertin M, Rigling A (2006) Drought as an inciting mortality factor in Scots pine stands of the Valais, Switzerland. Ecosystems 9: 330-343.

[6] Galiano L, Martínez-Vilalta J, Lloret F (2010) Drought-induced multifactor decline of scots pine in the pyrenees and potential vegetation change by the expansion of co-occurring oak species. Ecosystems 13: 978-991.

[7] Manion P.D (1981) Tree Disease Concepts. Prentice-Hall, Englewood Cliffs, NJ. 416p.

[8] Kurz W.A, Dymond C.C, Stinson G, Rampley G.J, Neilson E.T, Carroll A.L, Ebata T, Safranik L (2008) Mountain pine beetle and forest carbon feedback to climate change. Nature 452: 987-990.

[9] McDowell N.G, Beerling D.J, Breshears D.D, Fisher R.A, Raffa K.F, Stitt M (2011) The interdependence of mechanisms underlying climate-driven vegetation mortality. Trends in Ecol. Evol. 26: 523-532.

[10] Becker M,Lévy G (1982) Le dépérissement du chêne en forêt de Tronçais. Les causes écologiques. Ann. For. Sci. 39: 439-444.

[11] Becker M (1987) Bilan de santé actuel et retrospectif du sapin (Abies alba Mill.) dans les Vosges. Etude écologique et dendroécologique. Ann. For. Sci. 44: 379-402.

[12] Girard F, Vennetier M, Guibal F, Corona C, Ouarmim S, Asier H (2012) Pinus halepensis Mill. crown development and fruiting declined with repeated drought in Mediterranean France. Eur. J. For. Res., 131 (4): 919-931.

[13] Thabeet A, Vennetier M, Gadbin-Henry C, Denelle N, Roux M, Caraglio Y, Vila B (2009) Response of Pinus sylvestris L. to recent climate change in the French Mediterranean region. Trees 28: 843-853.

[14] Vennetier M, Girard F, Ouarmim S, Thabeet A, Ripert C, Cailleret M, Caraglio Y (2010) Climate change impact on tree architecture may contribute to forest decline and dieback. Int. For. Review 12 (5), p. 45.

[15] Barthelemy D, Caraglio Y (2007) Plant architecture: A dynamic, multilevel and comprehensive approach to plant form, structure and ontogeny. Ann. Bot. 99: 375-407.

[16] Halle F, Oldeman R.A.A, Tomlinson P.B (1978) Tropical trees and forests: an architectural analysis. Springer-Verlag, New-York. 441p.

[17] Grier C.C, Vogt K.A, Keyes M.R, Edmonds R.L (1981) Biomass distribution and above- and below-ground production in young and mature Abies amabilis zone ecosystems of the Washington Cascades. Can. J. For.Res. 11: 155-167.

[18] Cannell M.G.R (1982) World forest biomass and primary production data. Academic Press, Londres. 391p.

[19] Saint-André L, Vallet P, Pignard G, Dupouey J.-L, Colin A, Loustau D, Le Bas C, Meredieu C, Caraglio Y, Porté A, Hamza N, Cazin A, Nouvellon Y, Dhôte J.-F (2010) Estimating carbon stocks in forest stands: Methodological developments. In: Loustau D. editor. Forests, carbon cycle and climate change. QUAE edition, Versailles. pp. 79-100.

[20] Rambal S, Misson L, Huc R, Vennetier M, Guibal F, Girard F, Martin N, Joffre R, Limousin J.-M, Mouillot F, Ourcival J.-M, Ratte J.-M, Rocheteau A, Lavoir A.-V, Brewer S, Thomas A, Myklebust M, Gounelle D, Jouineau A, Simioni G, Ripert C, Prevosto B, Curt T, Martin W, Estève R, Ndiaye-Boubacar A (2011) Mediterranean ecosystems face increasing droughts: vulnerability assessments. Drought+ Project, Final report. CNRS - INRA - Cemagref, Montpellier - Avignon - Aix en provence. 27p.

[21] Taugourdeau O (2011) Le sapin pectiné (Abies alba Mill.,Pinaceae) en contexte méditerranéen : développement architectural et plasticité phénotypique. PhD thesis, Montpellier, Montpellier University. 255p.

[22] Didier C (2010) The impacts of repeated drought on the aboveground primary growth of Pinus halepensis Mill. and Quercus ilex L. in Mediterranean France. Master 2R, GR EMAX, Cemagref Aix en Provence, Université de Poitiers, 50 p.

[23] Yang D.M, Li G.Y, Sun S.C (2008) The generality of leaf size versus number trade-off in temperate woody species. Ann. Bot. 102: 623-629.

[24] Le Houerou H.N (2005) The Isoclimatic Mediterranean Biomes: Bioclimatology, Diversity and Phytogeography Vol 1 and 2. Copymania Publication, Montpellier (France). 760p.

[25] Gibelin A.L, Deque M (2003) Anthropogenic climate change over the Mediterranean region simulated by a global variable resolution model. Clim. Dyn. 20: 327-339.

[26] Cailleret M, Davi H (2011) Effects of climate on diameter growth of co-occurring Fagus sylvatica and Abies alba along an altitudinal gradient. Trees 25: 265-276.

[27] Sarris D, Christodoulakis D, Körner C (2007) Recent decline in precipitation and tree growth in the eastern Mediterranean. Glob. Change Biol. 13: 1187-1200.

[28] Météofrance (2009) Données météorologiques des stations d'Aubagne, Gémenos, Cuges-les-pins (data from the National Meteorological Survey Network).

[29] Zaitchik B.F, Macalady A.K, Bonneau L.R, Smith R.B (2006) Europe's 2003 heat wave: A satellite view of impacts and land-atmosphere feedbacks. Int. J. Clim. 26: 743-769.

[30] Deque M (2007) Frequency of precipitation and temperature extremes over France in an anthropogenic scenario: Model results and statistical correction according to observed values. Glob. Planetary Change 57: 16-26.

[31] Vennetier M, Vila B, Liang E.Y, Guibal F, Thabeet A, Gadbin-Henry C (2007) Impact of climate change on pine forest productivity and on the shift of a bioclimatic limit in a Mediterranean area. Options Méditerranéennes, Série A, 75: 189-197.

[32] Borghetti M, Magnani F, Fabrizio A, Saracino A (2004) Facing drought in a Mediterranean post-fire community: tissue water relations in species with different life traits. Acta Oecologica 25: 67-72.

[33] Ogaya R, Peñuelas J, Martínez-Vilalta J, Mangirón M (2003) Effect of drought on diameter increment of Quercus ilex, Phillyrea latifolia, and Arbutus unedo in a holm oak forest of NE Spain. For. Eco. Manage. 180: 175-184.

[34] Rathgeber C, Nicault A, Guiot J, Keller T, Guibal F, Roche P (2000) Simulated responses of Pinus halepensis forest productivity to climatic change and CO2 increase using a statistical model. Glob. Planetary Change 26: 405-421.

[35] Sardans J, Peñuelas J (2007) Drought changes the dynamics of trace element accumulation in a Mediterranean Quercus ilex forest. Environmental Pollution147: 567-583.

[36] Kramer K, Leinonen I, Loustau D (2000) The importance of phenology for the evaluation of impact of climate change on growth of boreal, temperate and Mediterranean forests ecosystems: an overview. Int. J. Biomet. 44: 67-75.

[37] Girard F, Vennetier M, Ouarmim S, Caraglio Y, Misson L (2011) Polycyclism, a fundamental tree growth process, decline with recent climate change. The example of Pinus halepensis Mill. in Mediterranean France. Trees 25: 311-322.

[38] Vennetier M, Girard F,Didier C, Ouarmim S, Ripert C, Estève R, Martin W, N'diaye A, Misson L (2011) Adaptation phénologique du pin d'Alep au changement climatique. Forêt méditerranéenne 32: 151-166.

[39] Cramer R.D, Bunce J.D, Patterson D.E, Frank I.E (1988) Cross-validation, bootstrapping, and partial least-squares compared with multiple-regression in conventional qsar studies. Quantitative Structure-Activity Relationships 7: 18-25.

[40] Wold S (1995) PLS for multivariate linear modelling. In: Waterbeemd V.D editor. Chemometric methods in molecular design, Weinheim (Germany), pp. 195-218.

[41] Good P (1994) Permutation tests. Springer-Verlag, New-York, 228 p.

[42] Amato S, Vinzi V.E (2003) Bootstrap-based Q²kh for the selection of components and variables in PLS regression. Chemometrics and Intelligent Laboratory Systems 68: 5-16.

[43] Orshan G, Le Floc'h E, Le Roux A, Montenegro G (1988) Plant phenomorphology as related to summer drought mediterranean type ecosystems. In: di Castri F, Floret C, et al. editors. Time scales and water stress ; 5th Int. Conf. on mediterranean ecosystems, Paris. pp. 111-123.

[44] Serre F (1976) Les rapports de la croissance et du climat chez le pin d'Alep (*Pinus halepensis* (Mill)) II L'allongement des pousses et des aiguilles, et le climat Discussion Générale. Oecologia Plantarum 11: 201-224.

[45] Nemenyi P.B (1963) Distribution-free multiple comparisons. PhD thesis. Princeton University, New Jersey.

[46] Nicolini E (1998) Architecture and morphogenetic gradients in young forest-grown beeches (Fagus sylvatica L. Fagaceae). Can. J. Bot. 76: 1232-1244.

[47] Taugourdeau O, Dauzat J, Griffon S, Sabatier S, Caraglio Y, Barthelemy D (2012) Retrospective analysis of tree architecture in silver fir (Abies alba Mill.): ontogenetic trends and responses to environmental variability. Ann. For. Sci., vol. in press.

[48] Wagner G, Pavlicev M, Cheverud J (2007) The road to modularity. Nature Reviews Genetics 8: 921-931.

[49] Dekort I, Loeffen V, Baas P (1991) Ring Width, Density and Wood Anatomy of Douglas-Fir with Different Crown Vitality. Iawa Bulletin 12: 453-465.

[50] Schweingruber F.H (1996) Tree rings and environment: Dendroecology. Paul Haupt Publishers, Bern (Swizerland), 609p.

[51] Fritts H.C (1976) Tree ring and climate. Academic Press, New York, 567 p.

[52] Vila B, Vennetier M, Ripert C, Chandioux O, Liang E, Guibal F, Torre F (2008) Has global change induced opposite trends in radial growth of Pinus sylvestris and Pinus halepensis at their bioclimatic limit? The example of the Sainte-Baume forest (south-east France). Ann. For. Sci. 65: 709.

[53] Mäkinen H, Nöjd P, Isomäki A (2002) Radial, Height and Volume Increment Variation in Picea abies (L.) Karst. Stands with varying thinning intensities. Scand. J.For. Res. 17: 304-316.

[54] Ciais P, Reichstein M, Viovy N, Granier A, Ogee J, Allard V, Aubinet M, Buchmann N, Bernhofer C, Carrara A, Chevallier F, De Noblet N, Friend A.D, Friedlingstein P, Grunwald T, Heinesch B, Keronen P, Knohl A, Krinner G, Loustau D, Manca G, Matteucci G, Miglietta F, Ourcival J.M, Papale D, Pilegaard K, Rambal S, Seufert G, Soussana J.F, Sanz M.J, Schulze E.D, Vesala T, Valentini R (2005) Europe-wide reduction in primary productivity caused by the heat and drought in 2003. Nature 437: 529-533.

[55] Anderegg W.R.L, Berry J.A, Smith D.D, Sperry J.S, Anderegg L.D.L, Field C.B (2012) The roles of hydraulic and carbon stress in a widespread climate-induced forest die-off. PNAS 109: 233-237.

[56] Zeppel M.J.B, Adams H.D, Anderegg W.R.L (2011) Mechanistic causes of tree drought mortality: recent results, unresolved questions and future research needs. New Phytol. 192: 800-803.

[57] Ouarmim S (2008) Impact des sècheresses extrêmes sur la croissance du Pin d'Alep Pinus halepensis Mill. Master 2R, GR EMAX, Cemagref, Université Paul Cézanne, Aix en Provence, 43 p.

[58] Rossi S, Simard S, Rathgeber C.B.K, Deslauriers A, Dezan C (2009) Effects of a 20-day-long dry period on cambial and apical meristem growth in Abies balsamea seedlings. Trees 23: 85-93.

[59] Vila B, Vennetier M (2003) Impact du changement climatique sur le déplacement d'une limite bioclimatique en région méditerranéenne. Cemagref, IMEP, ECOFOR, Aix en Provence, 141 p.

[60] Sarris D, Guibal F, Vennetier M, Christodoulakis D, Arianoutsou M, Körner C (2012) From one month to multiple years: which drought length matters for Pinus halepensis s.l. along its latitudinal distribution. Ann. For. Sci.; vol. accepted.

[61] Vennetier M, Ripert C (2010) Climate change impact on vegetation: lessons from an exceptionally hot and dry decade in South-eastern France. In: Simlard S.W, Austin M.E. editors. Climate Change and variability, Sciyo, Rijeka, Croatia, pp. 225-241.

[62] Rigby J.R, Porporato A (2008) Spring frost risk in a changing climate. Geophy. Res. Lett. 35: L12703.

[63] Carnicer J, Coll M, Ninyerola M, Pons X, Sanchez G, Penuelas J (2011) Widespread crown condition decline, food web disruption, and amplified tree mortality with increased climate change-type drought. PNAS 108:1474-1478.

[64] Dobbertin M (2005) Tree growth as indicator of tree vitality and of tree reaction to environmental stress: a review. Eur. J. For. Res. 124: 319-333.

[65] Breda N, Huc R, Granier A, Dreyer E (2006) Temperate forest trees and stands under severe drought: a review of ecophysiological responses, adaptation processes and long-term consequences. Ann. For. Sci. 63: 625-644.

[66] Galiano L, Martínez-Vilalta J, Lloret F (2011) Carbon reserves and canopy defoliation determine the recovery of Scots pine 4 yr after a drought episode. New Phytol. 190: 750-759.

[67] Dobbertin M, Brang P (2001) Crown defoliation improves tree mortality models. For. Eco. Manage. 141: 271-284.

[68] Brouwer R (1963) Some aspects of the equilibrium between overground and underground plant parts. Jaarboek van het Instituut voor Biologisch en Scheikundig onderzoek aan Landbouwgewassen, n° 1963, pp. 31-39.

[69] Iwasa Y, Roughgarden J (1984) Shoot / root balance of plants: optimal growth of a system with many vegetative organs. Theor. Pop. Biol. 25: 78-105.

[70] Poorter H, Niklas K.J, Reich P.B, Oleksyn J, Poot P, Mommer L (2012) Biomass allocation to leaves, stems and roots: meta-analyses of interspecific variation and environmental control. New Phytol. 193: 30-50.

[71] Poorter H, Nagel O (2000) The role of biomass allocation in the growth response of plants to different levels of light, CO2, nutrients and water: a quantitative review. Aust. J. Plant Physiol. 27: 595-607.

Climate Change and Its Hydrological Effects

Shifts and Modification of the Hydrological Regime Under Climate Change in Hungary

Béla Nováky and Gábor Bálint

Additional information is available at the end of the chapter

1. Introduction

Hydrological regime of water bodies is highly dependent on climatic factors. The runoff is mainly defined by seasonal distribution of precipitation and intensive rainfall events on one side and potential evapo-transpiration on the other. Near surface air temperatures (and other factors of heat budget) regulate the phase of precipitation and consequently snow accumulation, ablation and snowmelt induced runoff. Change of climate evidently would lead to changes in the hydrological regime. Nevertheless, hydrological regime can also be modified by different human activities towards water bodies directly (river training, flood control, flow regulation, water abstractions and inlets) or indirectly to catchments (urbanization, land use changes, deforestation). The tasks of water management and water related policies may change with climate fluctuations, as it is attested by the historical past. The long lasting wet period in the second half of 19th century resulted the framing the law on water regulation in 1871, and frequent droughts in 1930s led to the law on irrigation in 1937.

The given chapter gives an overview of changes in the hydrological regime of water bodies observed in the historical and more recent past and also projected for the future under observed climate change and future climate change scenarios in central regions of Danube River Basin, i.e. in Hungary with limited outlook beyond the national borders, the adaptation possibilities for certain water management sectors to cope with adverse effects of the climate change are also discussed.

2. Mean features of hydrological conditions in Hungary

The country is located in the central part of the Carpathian Basin; the territory is surrounded from all sides by the mountain ranges, in particular by the Carpathians, foothills of East-Alps and Dinaric-Alps. The inner part of the basin, the Pannonian Depression consists mostly of lowlands, including the Hungarian Plain, and in a smaller extent hilly and undulating regions.

River Danube forms the axis of the drainage network (Fig. 1). It originates from Schwarzwald (Black Forest Hills) of Germany, passes the Bavarian and Vienna Basins, and enters into the Carpathian Basin through the Hungarian Gates ('Porta Hungarica' or Devín Gates), and leaves it after a 930-km route at the Iron Gate. Only few smaller streams crosses the Carpathians consequently the Danube collects almost the all runoff of the Basin. The climate plays an important role in shaping the drainage network. Aridity has an increasing tendency from the mountains towards the inner parts of the Basin. Considerable climatic water surplus characterises mountain regions where precipitation exceeds potential evaporation. Inside of the basin the precipitation less than the potential evaporation, so here, Especially the Hungarian Plain shows a considerable climatic water deficit as in internal regions the amount of precipitation remains less than that of the potential evaporation. Following spatial distribution of climatic water surpluses and deficits and also the topography both surface and subsurface runoff are directed from the mountains to the inner parts of the Basin. In mountainous and hilly regions dense drainage network is formed, while in inner parts of the Basin only sparse, ephemeral and artificial stream network exist, while all major rivers are dominated by transit flow.

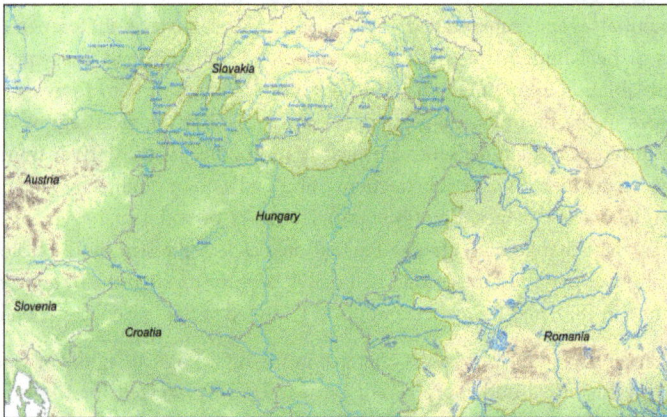

Figure 1. The drainage network of Carpathian Basin

Hungary falls into the moderate climatic zone with four seasons. The hydrological regime of rivers is well expressed with high stream-flow during spring and low stream-flow in late summer and early autumn. In winter the precipitation falls in the form of snow during winter in headwaters, the water is stored in the snow cover and enters into the water cycle only in occasional thaw and time of spring snowmelt. Moving from the inside of the basin towards the mountains the duration of snow cover is increasing from inner parts of the Basin 1.5-2 months up to 5-6 months towards higher mountains, consequently the period of snowmelt is shifted to later months. The earliest occurrence of mostly snowmelt related high flow is in February-March in the lowlands and lower hills, and latest, in April-May on rivers originating in high mountains. The Danube has Alpine regime, with high flow period usually in June generated by the coincidence of lasting snowmelt and rainy period. Glacier melt contributes to high waters and compensates late summer low water. Streams with headwaters in hills and inner Carpathian ranges have earlier low flow periods occurring in late summer or early autumn. Low flow period typical in November on the Danube.

3. Changes in mean hydrologic features through-out historical times

Reconstruction of climate and hydrologic conditions of Hungary in historical times is mostly based on data reported in written sources (chronicles, annals, local and religious histories, municipal documents, letters), firstly collected by Réthly (1962, 1970). The last decades have produced other methods (data collected from sediment layers, pollens, tree-rings) particularly for reconstruction of local climate (Kiss, 2009; Kern et al., 2009). The written sources without doubt contain useful information, but those are often subjective, incidental, sporadically distributed in space and time, non-contemporary, and incomplete especially prior to the 17th century. Written sources contain relatively large amount of information on the extreme climatic events, like floods and droughts, which allows to follow their decadal frequency from the 10th century on (Fig. 2).

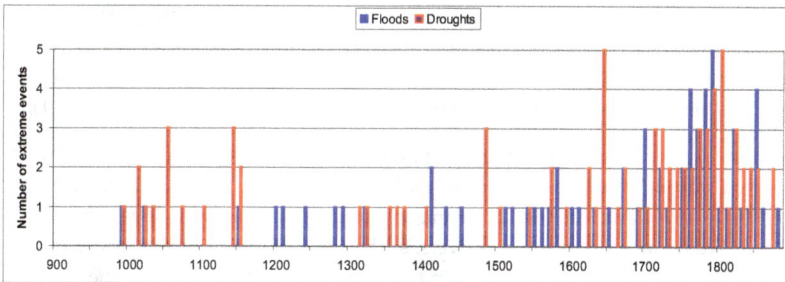

Figure 2. Decadal change of number of extreme floods and droughts from historical sources in Hungary

The increase of frequency of extreme events is evident, but mainly it can be explained by steadily increasing number of written sources, and also by growing climate sensitivity of developing economy, rather than by climate fluctuation. That is why these and similar time series can only cautiously be used for climate reconstruction. Nevertheless, some relevant conclusions can be obtained.

After the last ice-age in the Holocene the climate of the Carpathian Basin stabilized, and during the historical times changed little, only moderate fluctuations occurred (Rácz, 1999). The first millennium A.D. started with the so called *Roman climatic optimum* when climate even warmer than today prevailed. In the middle of 4th century A.D. the climate turned cooler and dryer and was not favourable for pastures. The dry period ended in late 8th century, when the climate became warmer and wet, with that a new period, the *medieval climatic optimum* started. The climate stipulated agricultural and demographic growth and as such was favourable for Hungarians to change semi-nomadic economy, to settle down and organize a state in the Carpathian Basin. Following the new climatic optimum the *Little Ice Age* started in Hungary in the middle of 16th century. Cool and wet climate followed. In the middle of 17th century a milder period started, but the last decades of the century were characterised by strong cooling down which continued until mid-19th century. In the middle of 18th century the climate turned to somewhat milder but the cool period ended up only in the 1860s. During the little ice-age the climate was cooler and wetter than today: temperatures were by 1-1,5°C lower, precipitation was by 10% higher, relative to the second half of 20th century. The magnitude of this difference compared to today's climate is similar to the one expected for the near future due to the projected global warming for the mid-21st century.

Although the *Medieval climatic optimum* was in general favourable for the productivity of pastures and arable lands, written sources report many extreme drought events. In 11-12th and 14th century extreme droughts occurred in 1015, 1022, 1142, 1147, and 1363. In the second half of 15th century long atmospheric droughts resulted in extreme low flows in rivers. In 1473 even the Danube levels dropped so that the usually deep river could be waded, and in 1478-1479 navigation stopped on all major rivers (Danube, Drava, Sava). The water level of the Neusiedler/Fertő Lake highly depleted in years 1466, 1479 and 1494. Less is reported about extreme floods, which is explained by the fact that the grazing agriculture adapted to temporarily inundated pastures in lowland and was less sensitive to floods. The extreme (sometimes ice jam induced) floods of the Danube are mentioned from years 1012, 1210 and 1267-68, the most extreme flood of the Danube occurred in 1501 (Fejér, 2001).

The *Little Ice Age* increased the frequency of floods. An increasing number of extreme floods are reported starting from the last decades of 16th century. Extreme floods occurred in some years in the entire Carpathian Basin or major parts of it (1598, 1691, 1694), in other years only floods are reported in individual rivers, such as on the Drava and the Mura in 1594, on the Danube in 1619. Also several ice jam induced flooding is mentioned for the 17th century. The 18th and 19th centuries were plenty of extreme flood events including many severe ice jam induced floods occurred in the 18th and 19th centuries, when the population was forced to shift their settlements into higher grounds or more protected areas. Extreme floods happened on the Danube and the Tisza in 1712. The cool and wet climate of the Little Ice Age caused the rise of water levels in lakes. The water level of Lake Balaton was several metres higher than the present one (Bendefy & V. Nagy, 1969), the extent of Lake Neusiedler/Fertő was the largest ever known. Despite of cold and wet climate during the Little Ice Age extreme droughts occurred also frequently. Droughts were reported in 1540, 1585, 1638, 1718, and 1720. Even the large rivers depleted in some years (1718), small streams dried up totally (1720), in 1779, 1790 and 1794 the Tisza, and its tributaries the Körös and Maros, could be waded (Fejér, 2001). The Neusiedler/Fertő Lake times dried up in 18th century and in 1790 many small lakes of the Hungarian Plain dried up completely and was turned to arable land. The Little Ice Age ended up in the middle of 19th century when climate turned to warmer and drier. During the 19th century 26 droughts were observed, the especially severe one occurred in 1860s. The most severe drought of the last two centuries caused drying out of lakes Neusiedler/Fertő and Velencei in 1863, the water level of Balaton was the lowest for the period of instrumental hydrological observations.

Under climate fluctuations during the historical times the extent of floodplains also varied in lowlands, particularly in the Hungarian Plain. Floods added to the water supply of lowlands and resulted higher forest rate than justified by local climate. Water of inundated areas was transpired by forests, and a climatic equilibrium was formed between forest ratio and inundated areas (Orlóci & Szesztay, 1994). Land use changes, growing deforestation from 17th century induced expansion of floodplains, and the evaporation of water bodies replaced the transpiration of forest patches. Climate of the Little Ice Age contributed to this tendency. At the turn of 18-19th centuries around 20 000 km² of land was regularly inundated and on 5000 km² persistent water cover remained due to floods of Tisza and its tributaries (Lászlóffy, 1982). Flooding was the obstacle to expand agricultural land and production, much needed by increasing population and growing demand for commercial grain during the Napoleonic Wars in Europe. To increase the extent of the arable lands required protection against floods.

Flood protection works were systematically constructed starting from the mid-19th century. A levee system was built along the lowland rivers, which prevents inundation on major part of floodplains. As a consequence of the construction of the dikes flood levels raised and - due to increasing specific energy of water movement – scouring of the riverbed started, these processes are active also today. The change in the levels of flood waters has a jumping character, while the decrease of the low water levels was a steady process (Fig. 3.). Learning from the consequences of severe droughts major lakes were also regulated. Gates, sluices were built to regulate outflow from these water bodies. The regulation of outflow resulted decreased range of water level fluctuations in Lake Balaton from 2.5 m to 1.0 m. The start of major hydraulic construction works (river training, flood embankments, protection of banks and lake shores) coincided with the beginning instrumental observations.

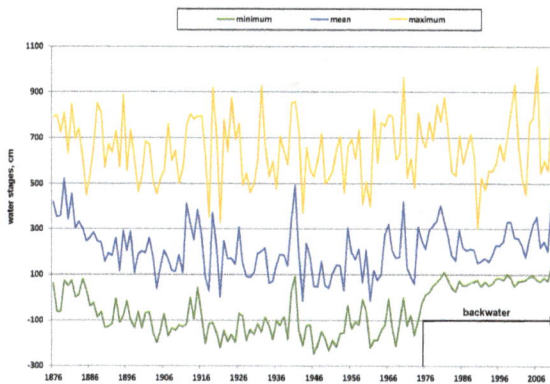

Figure 3. The change of annual flood and low water stage of the River Tisza at Szeged

4. Climate and water trends in the period of instrumental measurements

4.1. Climate tendencies

The average annual temperature increased in Hungary during the 20th century by 0.86 °C, half of that has been observed in the last 50 years. Higher rate of increase is associated with western and lower ones with eastern regions of the country. The temperature increased in all of the seasons, to the less in winter, and more in summer, exceeding 1 °C. Daily maxima and minima also raised, heat waves occurred more frequently, and the number of extremely cold days decreased (Bartholy et al., 2005; Szalai, 2011). Annual precipitation averaged for the whole country decreased during 20th century by about 7%, being equal to one average monthly precipitation. Much less precipitation is measured in last 50 years in western regions of the country, while certain areas in eastern regions experienced some increase (Szalai, 2011). Precipitation decreased in every season with the exception of summer, when it showed a small, non-significant trend towards increase. The number of days with precipitation decreased, while one-day precipitation and the duration of days without precipitation increased, particularly in summer (Bartholy & Pongrácz, 2005). The rate of precipitation falling in the form of snow, the number of days with snow cover showed a small decreasing trend (Szalai, 2011). The tendencies observed in temperature and precipitation over 20th century, more strongly for last decades can be explained to some extent by changes in atmospheric circulation patterns, and by increasing variability of NAO-index highly influencing weather in Europe (Pongrácz & Bartholy, 2000). The climate of Hungary became warmer and drier; aridity increased demonstrated by trends in annual maxima and minima of Palmer-drought index since 1901 (Fig. 4, Szalai, 2011).

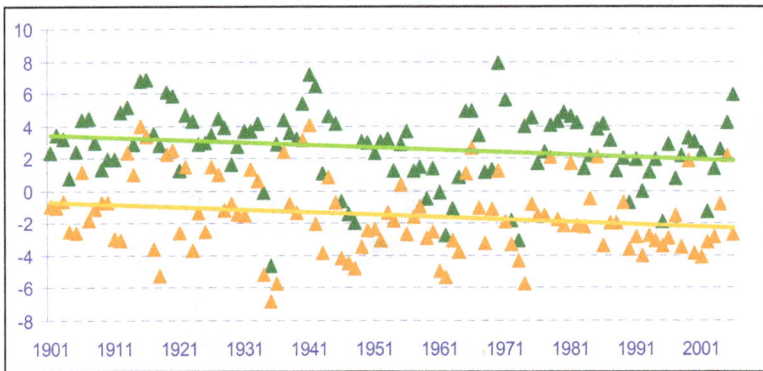

Figure 4. Tendency in annual maximum (green) and minimum (orange) of Palmer droughts-index in 1901-2006 in Debrecen (Szalai, 2011)

4.2. Trends in characteristics of the water regime

Increasing aridity and extreme weather events during 20th century had some affect on water bodies. Based on earlier studies this section gives an overview of detected tendencies in hydrological characteristics: in annual flow, floods and low flows, water balance of lakes, and temperature and ice regime of rivers for certain river sections and lakes. Investigated river sections are located in Hungary and along the Slovak Danube reach at Bratislava. Headwaters and larger part of catchments generating runoff of these streams cover territories of the Carpathian Basin outside the national borders of Hungary; consequently main features of detected tendencies are valid for major parts of the region.

Trend analyses was mainly based on estimated daily discharge series available for the Danube at Bratislava starting from 1876, and 1883 at Nagymaros, for other major rivers, Tisza and tributaries since the beginning of 20th century, and for smaller rivers usually only from the mid-20th century. Tools, applied techniques of flow rate measurement and methods of calculating of daily discharges (instrumentation, measurement rules, coverage of high flows, floods, rating curves) have changed in time, nevertheless there most of the series did not undergo comprehensive checks of data accuracy. Data checks were limited standard statistical tests.

4.2.1. Annual flow and its seasonal distribution

Annual flow is one of the most important characteristics of rivers, it can be considered as an integrated index of climatic and all non-climatic factors within the given catchments. Annual flow of near-natural catchments is controlled mainly by climatic elements, especially by precipitations and temperature (Nováky, 1991, Fig. 5), so it can be a good index of climatic variability and/or climate change. Decreasing but non-significant trends in annual flow are detected on some rivers of the country, particularly in the Upper Tisza originating from the Eastern Carpathians and on the Raba coming from the foothills of Eastern-Alps, on small streams (Zala, Zagyva) the last two catchments of those are located completely inside of Hungary (Nováky, 2002; Kravinszkaja et al, 2010). Decreasing tendencies of annual flow of small rivers are coherent with tendencies indicated for south-eastern regions of Europe given for the flow of near-natural small rivers from 15 European countries including some Slovakian rivers, catchment of which belong to the Carpathian Basin (Stahl et al., 2010).

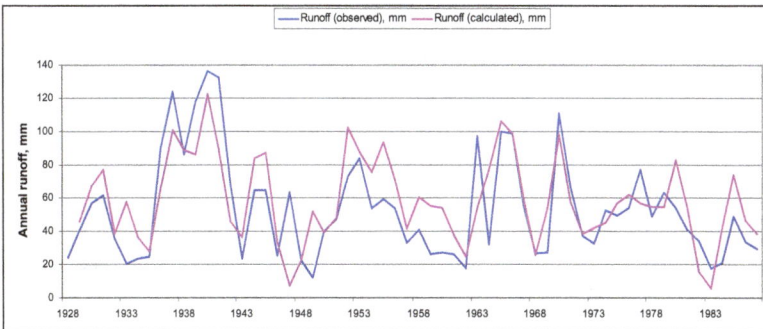

Figure 5. Time series of observed and climatically determined annual flow (Zagyva - Jásztelek) (Nováky, 1991)

Time series of annual flow are analysed in more details at two stations, on the Danube at Bratislava and Nagymaros using different observation periods. At the Nagymaros all examination detected a decreasing trends of annual flow since 1883, and all consecutive examination confirmed the results of previous ones (Lovász, 1985; Gillyénné Hofer, 1994). Differently results were received for the upstream station Bratislava where no trend is proved in annual flow for the period 1876-2008 (Pekarová et al., 2008a), also downstream at Turnu-Severin on the Lower Danube located outside of the Carpathian Basin for the period 1840-2000 (Pekarová & Pekar, 2005). Regarding seasonal distribution of flow at Nagymaros decreasing tendency is observed for August-October and some increase in November-December since 1883 (Lovász, 1986; Gilyénné Hofer, 1994). Similarly, at Bratislava a decrease in flow is detected in summer (May-August), an increase in winter and spring (November-April) for the period 1876-2007 (Fig. 6). The tendencies can be explained mainly by increasing temperature in winter resulting in early snowmelt in upper parts of the Danube Basin and the snowmelt induced runoff does not coincide with runoff from monsoon type rainfall season in early summer months. The opposite trends in summer and winter months led to considerable changes in seasonal distribution of the flow (Pekarová et al., 2008a).

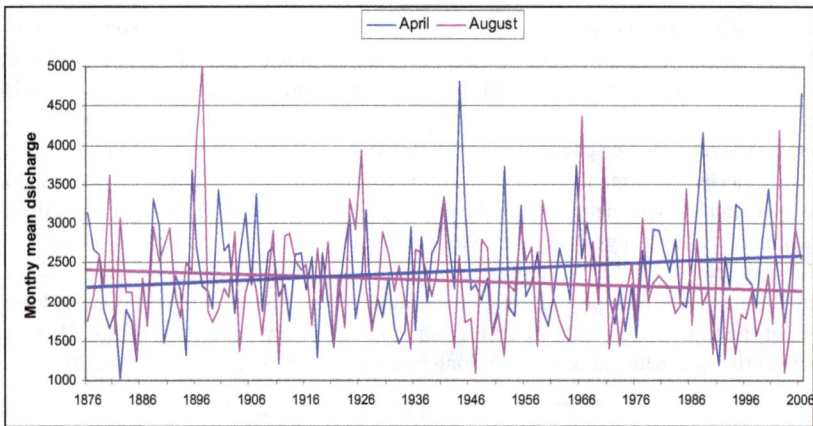

Figure 6. Trends of monthly flows of River Danube at Bratislava (after Pekarová et al., 2008a)

4.2.2. Floods

Different types of floods are observed in the Carpathian Basin. Floods originating from snowmelt accompanied with rainfall are typical for major rivers with headwaters in high mountains occur usually in late winter or early spring months, in February-April. The largest rivers, Danube and Drava have Alpine regime and snowmelt-dominated floods occur later, usually in May-June. Rainstorm generated flash floods on small streams may appear any time in warmer half year. Floods caused by ice jams became extremely rare in the last 40-50 years, which can be explained by both anthropogenic (river training, barrages, reservoirs, cooling and waste water inlets) and climate impact, in particular by increasing winter temperatures (Takács et al., 2008; Takács, 2011). Floods of medium and large rivers (Danube, Drava, Tisza and their tributaries) propagate on floodplains constrained by flood embankments, i.e. in the main channel and on the so called floodberm. Valleys and floodplains of small streams are seldom protected from inundation. The floods can be characterised by frequency of flood crests and peak discharges.

Frequency of extreme floods of Danube and Tisza rivers was examined taking into account only ice free floods, with flood crests exceeding 700 cm for Danube at Nagymaros and 800 cm gauge readings for Tisza at Szolnok. During 1901-2010 extreme flood occurred on the Danube 10 times and on the Tisza 14 times, but frequencies changed considerably in time. Extreme floods occurred only twice in the first half of the period on the Danube and four times on the Tisza, while more than half of 24 extreme flood events on the two rivers were observed during the last two decades between 1991-2010. Flood crests on Tisza in 1997, 1998 and 2000, and on the Danube in 2002 and 2006 exceeded the earlier observed peaks (Fig. 7). Similarly to rivers in Hungary the frequency of extreme floods on some big rivers of Central Europe also increased in the last two decades (EEA, 2008), as on the Vltava (Brázdil et al., 2006), Vistula (Cyberski et al., 2006), Oder and Elbe (Mudelsee et al., 2003).

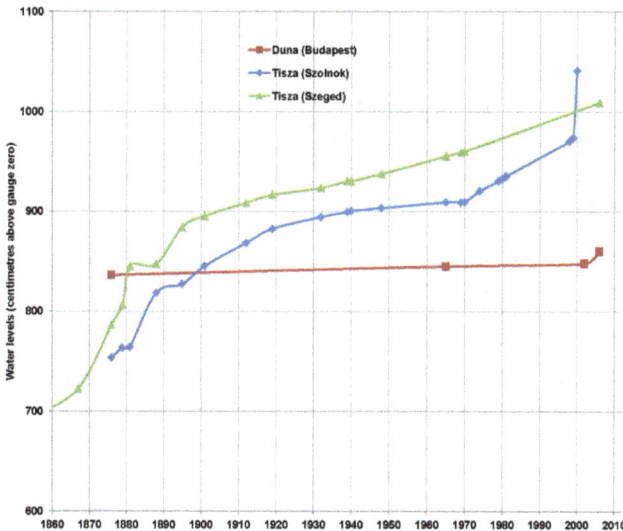

Figure 7. Record breaking extreme floods on the Danube and Tisza before2010

Extreme floods became more frequent on some tributaries of Tisza (Körös, Hernád), but no significant change in frequency of extreme floods is observed for other rivers (Szamos, Bódva, Zagyva, Rába) (Bárdossy et al., 2003; Somlyódy et al., 2010). The frequency of floods seemingly increased in some smaller rivers, especially those originating from the northern parts of the Carpathian Range, or Matra Hills inside of Hungary, however limited flood frequency analyses does not allow to make a final conclusion.

Only few studies examined peak discharges during floods. On the Slovakian Danube reach at Bratislava a slight but no significant upward tendency of annual maxima discharge was detected for the period 1876-2006 (Halmová et al., 2008, Pekarová et al., 2008a, Fig. 8). Annual maximum discharges of the Danube at Nagymaros show a decreasing tendency during 1883-1980 (Lovász, 1986), which is explained by increasing role of rainfall induced flood waves against to snowmelt. The non-significant decreasing tendency at Nagymaros is confirmed for periods 1883-1985 (Gillyénné Hofer, 1994), 1883-2003 (Bálint & Konecsny, 2004), 1883-2006 (Bálint, 2009) and for 1901-1990 on the Hungarian lower reaches of the Danube at Mohács (Keve, 1994).

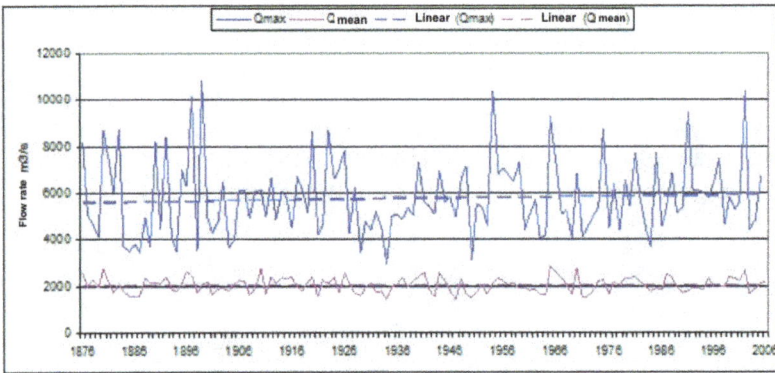

Figure 8. Trends of annual mean and annual maximum discharges of Danube at Bratislava (after Pekarová et al., 2008)

The frequency of extreme floods, flood crests of major rivers (Danube, Tisza and their main tributaries) has been growing quite unambiguously during the last 110 years, however flow peaks being directly connected to climate do not show any significant tendency. The increase of flood crests, peak stages can be explained more by non-climatic factors, like land use changes (Szlávik, 2002), worsening conditions of flood propagations on floodberms during the last decades due to the growing sedimentation, changing canopy cover, and some geomorphologic processes like formation of bank-side bars (Koncsos & Kozma, 2007). The increase of flood crests caused by sediment accumulation may lead to more frequent of extreme flood crests exceeding threshold stages. Nevertheless, the increasing frequency of extreme floods to some extent can be also explained by climatic factors, such as the growing North Atlantic Oscillation, which may led to occurrence of extreme floods in Europe (Kron & Bertz, 2007). More frequent and intensive flash floods probably may be connected to increasing frequency and intensity of short-time rainfall events during the last decades.

4.2.3. Low flow

Water resources management is sensitive to low flow conditions, it may constrain water consumption, primarily for agriculture use, limit self-purification capacity of rivers, make difficult to maintain good ecological conditions, adversely affect navigation on major rivers. Low flow season usually occurs in late summer or early autumn on most of Hungarian rivers, with the exception of some transit flow dominated major streams originating from high Alpine regions, where low flow is typical in late autumn or early winter months.

River training and construction of flood embankments resulted decreasing tendency of low water levels is detected starting from the end of 19th century as a consequence of the scouring of the low flow riverbed. The decrease of low water level is not linked to the decrease of low flow discharges. Moreover, examining the times series of annual minima, and other low flow parameters (duration of low flow period, the deficit during flow periods) for several major and medium size streams (Tisza, Kraszna, Szamos, Maros, Körös, and Berettyó, Hernád, Zagyva) show increasing tendency (Konecsny, 2010; Konecsny, 2011; Konecsny & Bálint, 2009; Konecsny & Bálint, 2011; Konecsny & Nováky, 2011). The increasing tendency is explained mainly by water management measurement (runoff regulation, transfer from other catchments and groundwater abstraction). For example, the operation of a reservoir for flow regulation on river Kraszna increased the low flow from 0.29 m³/ up to 1.15 m³/s (Konecsny & Sorochovski, 1996). Explanation of any detected trends in low flow time series requires a correct separation of those into climate induced and other affects. In the example shown below the separation revealed that the 'natural', climate induced low flows show an increasing tendency for the period 1951-2009 explained by the increasing summer precipitation and increasing rate of short-time intensive rainfall events (Fig. 9). Wider application of the given approach could contribute much to the analysis of low flow tendencies (Konecsny & Nováky, 2011).

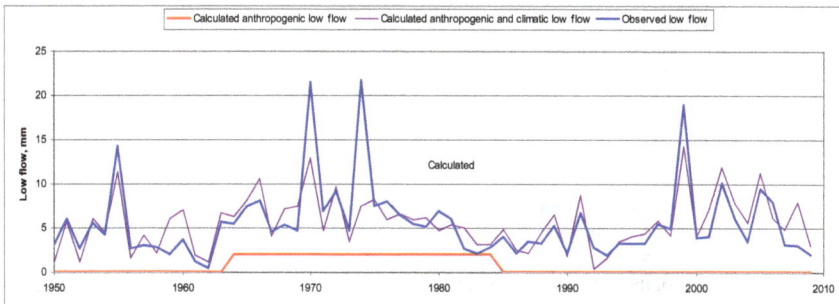

Figure 9. Separation of observed low flow time series into anthropogenic and climatically determined components (Zagyva - Jásztelek) (Konecsny & Nováky, 2011)

The annual minimum discharges of the Danube at Nagymaros also increased as it was detected for the period 1883-1983 (Lovász, 1986), later it is confirmed for an extended period 1883-2009 (Konecsny, 2011). Increasing tendency is detected for the Danube at Bratislava during 1876-2005, the statistical tests resulted that these increases are not significant (Pekarová et al., 2008b).

4.2.4. Lakes

Major shallow lakes in Hungary, Balaton, Neusidler/Fertő and Velence are very sensitive to climate fluctuations. The water balance and water level of lakes are regulated by structures controlling outflow released from the lakes. The aim of regulation is to maintain the water level within the prescribed interval, the defined range or water levels to avoid the inundation of the coastal zone on the one hand and to store enough water to maintain sufficient depth for recreation purposes (Varga, 2005). In case of long drought periods the water levels may fall below the thresholds and cause a critical situation for recreation, hitting tourist industry. Such critical situations in lakes occurred during 20th century. In 1990s the water level of Lake Velence was below the critical limit for a long period. Water level of Balaton was low in 1952-1954, and in 2001-2004. In 2003 depths along the southern cost were remained less than 1.0 m even 0.5-1.0 km distance from the shore.

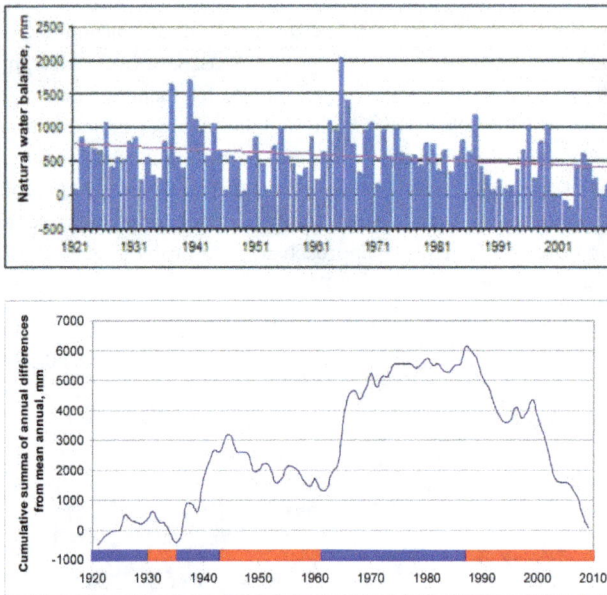

Figure 10. The change of the natural water budget of the Balaton Lake during 1921-2010, annual values and cumulative anomalies are expressed in lake millimetres of water layer (Dry periods are indicated with red, wet periods with blue colour)

Although the water level is regulated the climate fluctuation remains the important factor in fluctuation of water level and water balance of lakes. The role of climate may be followed especially well through the fluctuation of natural water budget (NWB). NWB is the difference between the total inflow (that is the sum of precipitation and inflow to the lake) and the evaporation from lake surface for a given time interval (Szesztay 1959). Annual NWB is available since 1921 for the Balaton Lake show a decreasing tendency (Varga, 2005;

Nováky, 2008; Kravinszkaja et al., 2010) with 30 mm (or 6%) depletion for 10 years (Fig. 10). The comparison of probability distribution functions (which is adequately described by gamma-type distribution function) calculated for two periods proves that the change is significant: the probability distribution function calculated for 1980-2009 is outside of the 95% confidence interval calculated for 1921-1990 (Fig.11). Time series of annual NWBs for Lake Balaton and cumulated annual anomalies demonstrate the fluctuation and length of altering dry and wet spills (Fig. 10).

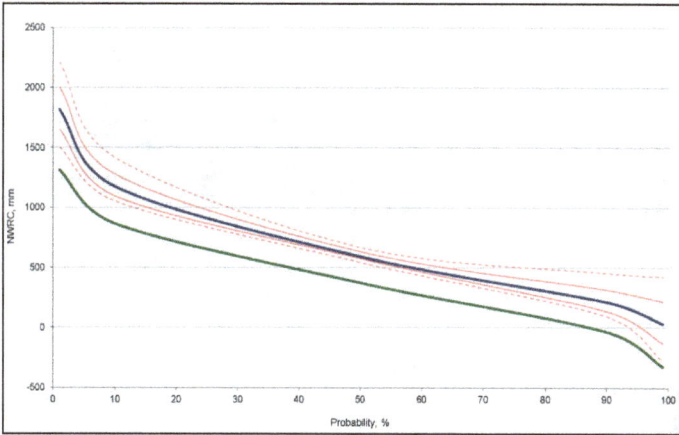

Figure 11. The probability distribution functions of NWRC of Balaton for 1921-1990 with 70% and 95% confidence intervals (blue) and 1980-2009 (green)

4.2.5. Water temperature and ice regime

Water temperature is an essential physical characteristic of natural water bodies influencing directly the aquatic ecosystems. Analyses of water temperatures are availably mostly for the Danube, where rising trends have been detected for almost all of the investigated river reaches, including those where such rise of air temperatures could not be proved (Stancikova, 1993). The contradiction can be explained by that there is no unambiguous connection between local air and water temperature, the later can be influenced by the water temperature (or heat content) of inflow along a given section. The longest analysed time series are available for Bratislava where the mean annual water temperatures increased by 0.6 °C during the period 1926-2006., while there was no rise until 1970s the increase is assigned to the last decades. The rate of water temperature increases is somewhat less than that for air temperatures (Pekarová et al., 2008c). No trends were detected in time series of mean annual water temperatures weighted by daily discharges.

Mean annual and summer temperatures, and maximum daily temperatures show a considerable increase for the last decades (1974-2009) along the lower Hungarian Danube reach, however no increase is detected in winter temperatures. The number of days with water temperature exceeding 20°C and 24 °C thresholds show an upward trend with the rate of 10 days/decade for the 20 °C and 1,5 day/decade for 24°C thresholds (Fig. 12, Nováky 2011). Although the rise in the water temperatures can be related to some extent to the impact of cooling water inflow from thermal and nuclear power stations upstream of Baja, it is believed that the main reason can be in the increase of air temperatures having the same rate of increase during the period analysed as the water temperatures (Nováky, 2011).

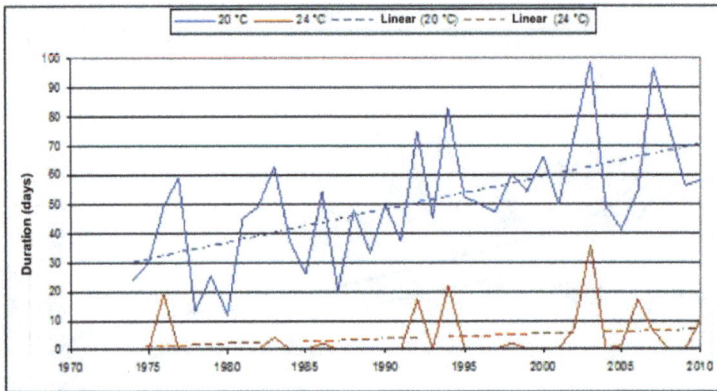

Figure 12. The number of days with temperatures exceeding threshold values 20° and 24°C, Danube - Baja (Nováky, 2011)

Some early studies already pointed out that there was a decreasing tendency in the frequency of ice phenomena (Déri, 1985). The number of days with ice phenomena present decreased during the period 1880-1980 on the Danube and Drava rivers by more than 80%, on the Tisza and Szamos by 30-40%. The number of days with ice cover on the Balaton Lake also decreased considerably. The analyses of the central reaches of the Danube by Stanciková (1993) indicated an almost overall decrease. The observed change in the ice regime was mainly attributed to anthropogenic impact, particularly to river training, construction of barrages, sewage and industrial cooling water inlets, which delay river ice formation (Déri, 1985; Starosolszky, 1989; Stanciková 1993).

The same tendencies were confirmed for River Danube and Drava using longer time series (Takács et al, 2008; Takács, 2011). The dates of the beginning of ice formation and freeze-up were analysed, earlier break-up and start of ice free conditions were observed at all investigated river reaches on the Danube (Fig. 13). The duration of ice-cover shortened on average by about 15 days in 100 years, while the total length of river ice season decreased by 32 days in 100 years. Changes in the duration of ice-cover and the total length of the ice season grow while moving downstream along the river. All the recorded trends are significant. The reason of the changes in river ice regime could be the warming winter weather, however it should not be ignored that many other factors outside of changing climate could also influence the river ice regime changes (Takács et al., 2008).

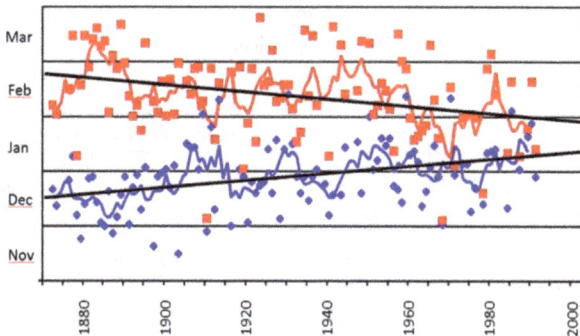

Figure 13. Changes in dates of first ice formation and final disappearance on the Danube River, at Nagymaros, 1874-2004 (Takács, 2011)

5. Impacts of climate change on hydrology and water management

The instrumental observations during the last more than 100 years prove increasing aridity of climate, decreasing rate of snow in winter precipitation, increasing frequency and intensity of extreme events mostly affecting the runoff regime. The recorded tendencies of hydrological characteristics (annual flow, extreme flows, regime of lakes, water temperature, ice regime) are in harmony with climatic tendencies, however statistically those are not significant, and no clear evidence can be given that the hydrological changes already took place. Learning from the past proves that projected climate change would considerable affect the hydrological regimes of rivers and water management in Hungary.

5.1. Climate change in Hungary (and in the Carpathian Basin)

Most of climate projections predict further increase of temperature and climatic aridity for the 21th century in Hungary and in the Carpathian Basin. The rate of warming depends on the emission and climate scenarios and can be in the interval 2-5°C. Temperature rise is expected in all of the seasons. The annual precipitation is expected to decrease with considerable seasonal shifts, i. e. mostly increase in winter and decrease in summer. Studies related to the National Climate Change Strategy foresee that extreme weather events (heat waves, heavy rains) will be more frequent and intensive. Regional climate scenarios for the period 2020-2040 based on medium emission scenarios (A1B) and three different global circulation models (ECHAM, NCAR, ARPEGE) have been projected using regional climate models (REMO, RegCM, ALADIN) (Bartholy et al., 2009, Bozó, 2010). Generally all regional scenarios outline increase of annual and seasonal temperatures with substantial spatial variability, while the highest rate of rise is expected in eastern regions, especially on the Hungarian Plain. The change of precipitation is largely uncertain, predicted rates in the regional scenarios differ not only in the magnitude but also in the direction of the change. Precipitation projections indicate both increase and decrease depending on the given scenario (Table 1). Most likely that annual and summer precipitation will decrease and winter precipitation increase, the frequency and intensity of intensive rainfall and dry spells will grow, but the predicted change is not significant, and might be the consequence of natural climate variability.

	Annual	Spring	Summer	Autumn	Winter
Temperature , °C	0.8 – 1.8	1.0 – 1.6	0.5 – 2.4	0.8 – 1.9	0.8 – 1.2
Precipitation, mm	(-40.8; +2.4)	(-15.9; +6.0)	(-15.0; +3.0)	(-4.8; +5.1)	(-22.8; +10.8)

Table 1. Projected annual and seasonal changes of temperatures and precipitation for the period 2021-2040 based on the three regional climate scenarios as compared to 1961-1990

5.2. Impact of climate change on the hydrological regime

5.2.1. Annual flow and its seasonal distribution

Climate impact assessment studies in hydrology started some decades ago in Hungary. The early impacts assessments can be characterized as follows: (i) they are based on assumed change of climate and without any climate scenario based projections; (ii) those use simple empirical-statistical approaches, and address averaged hydrological parameters, (iii) some simple physical based models are also applied in, the input time series of precipitation and temperature were based on weather generators. Some important results of these early impact assessments are:

• the decrease of mean annual precipitation coupled by the increase of mean temperature would lead to the decrease of mean annual flow with higher rate than in the mean annual precipitation, and the regions with arid climate are more sensitive to change (Nováky, 1991),

• the increase of temperatures by up to 3 °C in catchments of the upper Danube would significantly affect mean annual flow, and even stronger its seasonal distribution, earlier occurrence of snowmelt induced floods (Gauzer, 1994; Bálint & Gauzer, 1994),

• the increase of winter temperature would result in earlier snowmelt, some increase of winter flow would appear on the Danube (Gauzer 1994), and more increase on the Upper Tisza and some its tributaries (Bálint et al., 1995; Bálint & Gauzer, 1998),

• the decrease of summer precipitation would lead to significant decrease of low flow on several rivers, the lowest decrease rate expected on the Danube and the highest one on the Maros River (Nováky, 1994),

• an increase of early spring flood peaks are likely on the other hand later spring floods may decrease. Snowmelt induced floods in Upper Tisza and Zagyva would occur earlier and have a higher peaks. The peak discharges of floods generated by intensive rainfall would increase by a rate of up to 30% in the catchment of Sajó (Bálint et al., 1995; Bálint & Gauzer, 1998),

• sensitivity analysis proved that decrease of precipitation coupled by increase of temperatures would lead to a slight decrease of the (regulated) outflow from Lake Balaton to maintain the present regulated water surface (Nováky, 1994).

Approaches used for climate impact assessment have considerably developed in recent years. Approaches using climate change scenarios generated climate time series feeding physically based daily time step hydrological models are widely used. Models with higher temporal resolution allow to predict extreme hydrological events with higher certainty and especially are important for water management decision making.

Comprehensive climate impact assessment was carried out within the frame of CLAVIER project for the Tisza Basin (Jacob & Horányi, 2009; Bálint et al., 2010). Hydrological projections are based on regional climate model REMO 5.7 under A1B emission scenario and ECHAM5 global climate model. The reference period 1961-1990 was used. The climate change impact on the hydrological regime characteristics of 30 year period 2021-2050 as representative period for the future was estimated. According to the regional climate scenario the annual temperature is expected to rise in all of the catchments by 1,3-1,4 °C for 2021-2050 while the rise is more in autumn and winter and less in spring and summer. Annual precipitation is likely to decrease by 2-5% in most of catchments, and no change or a slight increase is predicted only for the Upper Tisza and some of its tributaries. Changes in seasonal distribution of precipitation is predicted with significant spatial variability especially during summer. The precipitation is likely to increase by 5-15% in winter, and to decrease by 6-8% in spring and summer in most of the catchments while these figures are only 1-2 % of increase and 1-2% decrease consequently in catchments of the Upper Tisza and its tributaries (Szamos, Kraszna). Based on the regional climate models produced meteorological input VITUKI-NHFS and VIDRA conceptual hydrological models were used to produce long term hydrological series. The impact assessment indicates slight decrease of annual mean flow almost throughout the region with significant spatial variability and even some increase (less, than 5%) for high elevation zones in the Upper Tisza. The highest rate of decrease, up to 15% is indicated in the southern regions including the Mures Basin. Simulation results indicate significant change in seasonal distribution of flow. In winter months, especially for February and December an increase of mean monthly flow is indicated with significant spatial variability from 5% for Sajo and up to 40% for Upper Tisza. In others months the mean monthly flow is likely to decrease with the highest rates up to 15-20% in southern catchments.

Climate change impact assessment on the hydrological regime was evaluated for a more distant perspective for 30 years period of 2061-2090 for the entire Tisza Basin represented by Senta cross section (Radvánszky & Jacob, 2008; Radvánszky & Jacob, 2009). The climate scenario predicts further increase of temperature, particularly for winter by 5 °C and for summer by 2°C, further decrease of annual and summer precipitation is expected, and an increase for winter. As a consequence of climate change annual runoff is expected to decrease in most of the catchments, increase is likely in North-eastern Carpathian Mountains. The monthly flow is projected to increase in March-April and to decrease in the other months up to 30%.

Some sporadic studies outside of the Tisza Basin indicate similar tendencies in annual flow. The impact of climate change was evaluated for the end of 21th century for catchments of Lake Balaton within the framework of CLIME Project (Padisak, 2006). The ECHAM4/OPY

(E) and HadAM3p (H) global circulation models under high (A2) and low (B2) emission scenarios were used. The output of a global circulation model was downscaled to the Lake Balaton region by the RCAO regional climate model using the weather generator to produce multiple data sets for future climate scenarios. Simulation of runoff was carried out by hydrological models, the GWFL, and ARES model (Padisák, 2006). All climate scenarios show clear increase in annual temperature ranging from 2,7 °C (H-B2) to 5,8 °C (E-A2), and slight increase in annual precipitation ranging from 0 to 15%. The simulation indicate that increase in temperature in the interval 2,7-5,5 °C would lead to a decrease of annual mean flow by 1-18%, so the increase in precipitation would not able to offset the effect of increasing evapotranspiration on the runoff. Both hydrological models predict reduced monthly flow for the period April-December. January-March monthly flows also decreased in the ARES simulation, while GWFL simulation indicates a slight increase.

Within the framework of the CECILIA (Central and Eastern Europe Climate Change Impact and Vulnerability Assessment) project the potential impact of climate change on river runoff was evaluated in mountainous catchments of the Váh Basin (Macurová et al., 2011). In the impact analysis the ALADIN regional climate model was used. Using a conceptual water balance model possible changes in the mean monthly runoff were estimated for the time horizons of 2021-2050 and 2071-2100. The general conclusion for both time horizons is that monthly flows are likely to increase in winter months from October to April, while to decrease in months from May to September/October.

5.2.2. Floods

The possible impact of climate change on flood conditions was also investigated in the CLAVIER project (Bálint et al., 2010). No clear picture can be drawn about possible changes in flood characteristics. While more frequent winter floods are expected, the decrease of mean flow in some seasons is not followed by the decrease of flood peaks. The simulations indicate for the Upper Tisza a rise of the frequency of smaller and larger floods and a decrease around the median. Torrential type of flood events may occur even more frequently, while the frequency of floods with long duration and large volume may become lower. In Lower Tisza at the location Szeged the frequency of medium size floods, that is the floods with peak discharge between 1800-2300 m^3/s, is likely to decrease by about 20%, while the frequency of extremely high or low floods would increase. The frequency of floods with peak discharge higher than 3000 m^3/s at Szeged could be doubled due to increasing winter precipitation especially to increasing rate of rainfall (Fig. 14). The simulation carried out with the hydrological model CONSUL for the Mures River indicates a likely decrease of maximum flow discharge of 1% probability in Lower Mures/Maros by 40%, and slight increase, by less than 5% only in Middle Mures/Maros reach (Corbus et al., 2011).

5.2.3. Low flow

No comprehensive evaluation of climate change impact on low flow conditions has been made yet for the entire Hungarian river network. Some investigations indicate that low flow

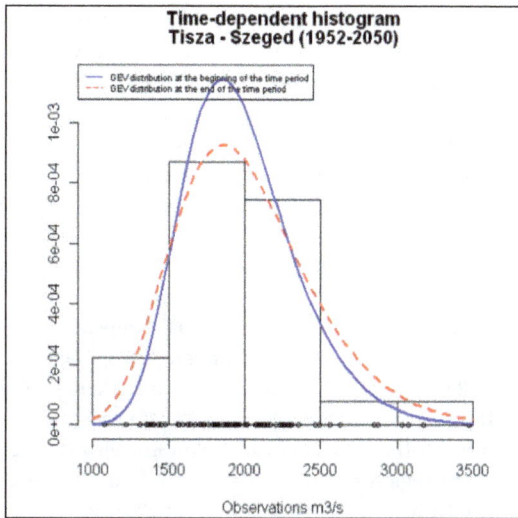

Figure 14. Change in distribution functions of flood peaks under climate change for Lower-Tisza at Szeged

would occur more frequently as compared to present. The study of the Lower Danube low water conditions indicates the possibility of more expressed and longer flow periods (Bálint et al., 2011; Jacob & Horányi, 2009). Conclusion is in harmony with the more detailed climate impact studies evaluated for the Upper Danube in frame of GLOWA-Danube project (Mauser et al., 2008; Prasch &.Mauser, 2011). The Upper Danube watershed with its current water surplus serves the large downstream regions of the Danube. In the framework of project the impact of climate change on the low flow conditions was investigated for Danube in Germany at Achleiten (Mauser et al., 2008). The reference period of 1970-2005 was applied, and the prediction made for the transient period 2011-2060. Climate change impact analyses based on emission scenario A1B and regional climate models REMO, MM5, CLM were used. For the simulation of climate impact on the low flow regime the PROMET hydrological model was applied. To evaluate the uncertainty of climate impact assessment 12 statistically equivalent realization of the same climate scenario were created by a synthetic weather generator, each realizations differing with the storyline of precipitation and temperature during the transient period. All hydrological simulations indicate a significant change in low flow: the reduction of the minimum annual 7-day average discharge range from half of the present discharge by 2030 and to one third by 2060. The 430 m^3/s minimum low flow 7-day discharge is likely to decrease to 210-360 m^3/s. The consequence on the low flow conditions of Hungarian reach of River Danube is evident.

5.2.4. Lakes

Water balance of lakes in the future will depend on projected precipitation and evaporation conditions largely linked to air temperatures. Regional climate scenarios predict a rise in temperature, but its effect on the evaporation is in many uncertain. The expected change in potential evaporation for the Lake Balaton was studied in the CLAVIER project. The impact study based on the regional climate model REMO 5.7 runs, boundary conditions for which were taken from global model ECHAM5/MPI-OM, based on an A1B emission scenario. The potential evaporation was calculated by the modified Morton equation (Kovács & Szilágyi, 2010; Szilágyi & József, 2009). According to the results no significant change in mean annual lake evaporation can be expected, that rise in annual temperature by 1.2-1.4° would lead only to an increase of evaporation from 858 mm/year in reference period 1921-2007 to 888 mm/year for 2001-2050 period.. The small change in the evaporation is explained by the fact that the impact of increasing temperature will be compensated by the increase in humidity of air. Other studies suggest that the increase in annual temperature by 1-2° C would be accompanied by 5-10% increase of potential evaporation (Mika, 1999). Regarding the future water balance of Lake Balaton there is also a great uncertainty in projections of the precipitation change. The regional climate scenarios REMO based on the ECHAM model, and ALADIN based on the ARPEGE global circulation models expect a slight increase in annual precipitation in the region of Lake Balaton, while the regional climate scenario RegCM based on NCAR GCM   predicts a significant decrease in annual precipitation, particularly in the region of Lake Balaton up to 20% (Bartholy et al., 2009). The tendencies observed in the past and expected in the future for various hydrological characteristics are summarized in Table 2.

Hydrological parameter	Tendencies in the past	Expected tendency in the future
Annual flow and seasonal distribution	No or slight decrease in annual flow, change in timing of annual flow	Up to 15% decrease in annual flow, increase in winter flow by up to 20%
Flood	Water stage increasing, water discharge is uncertain or unchanged	Earlier occurrence of spring floods, change in peak flows is uncertain
Low flow	Uncertainty because of anthropogenic impact	Decreasing low flow
Water balance of lakes	Decreasing natural water resources	Decreasing natural water resources, water budget is uncertain
Water temperature and ice regime	Increasing temperatures, decreasing ice phenomena	Increasing temperatures and decreasing ice phenomena

Table 2. Past and projected trends in hydrological regime parameters in Hungary

5.3. Water management implications of climate change in Hungary

Due to decreasing of annual and summer flows climate change would pose additional challenges to water management (Nováky, 2011). It is almost certain that the flow generated inside of the country will decrease; the other more substantial component of the possible decrease of water resources is linked to the decrease of flow generated outside of the country and entering in trans-boundary streams. This reduction may have adverse effects particularly in summer, limiting water uses connected to riverbeds (fishing, navigation) and water abstraction from rivers for agriculture, industry, and drinking water supply. Warming climate is likely to lead to increasing water consumption for irrigation, fish ponds, and power station. Reducing water resources and increasing water demand would result in more frequent conflicts between water uses, particularly in the region of the Hungarian Plain mostly prone to climate change (Simonffy, 2011). Reducing water resources would make it difficult to maintain good ecological state in rivers and lakes as it is prescribed by the Water Framework Directive. Maintaining prescribed regulation water levels in lakes will not be always possible limiting their recreational use. Climate change is likely to bring unfavourable changes for flood management. More frequent floods in winter, earlier occurrence of snow melt induced spring floods with likely increase of their peak discharge, increasing frequency and intensity of short-time rainfall induced flash floods would be superimposed on the adverse effects from non-climatic factors, and increase the risk of floods.

To reduce adverse effects resulting from climate change adaptation measures are required. Options to maintain the balance between water resources and water uses remain the same as used in the past, such as flow regulation by reservoirs, especially seasonal flow regulation, water transfer from areas with surplus water in water resources, forced use of groundwater resources. Some additional new water sources also in demand: rainfall water retention in lowlands, or rain-water harvesting from the roofs of buildings. Nevertheless, these adaptation options have a lot of constrains. Building of reservoirs is limited by topography, particularly in lowlands, worsening of hydrological conditions (increase in inter-annual variability of flow, increase in evaporation), and not least ecological requirement. Water transfer is limited by investment and energy costs, also by ecological requirements. The forced extraction of groundwater would be limited by decreasing recharge possibilities, and adverse ecological affects. There is a need to enlarge the role of water-demand regulation, such as economic use of water regulated by financial and legal rules, and technology improvement (water circulation, innovative technology in irrigation), using of "dry technologies" in cooling systems. Structural measures will prevail in adaptation to increasing flood risk will be based on traditional structural measures, such as the reservoirs in highland upstream sections and flood embankments in lowlands, however some less conventional structural measures, like lowland flood detention or emergency reservoirs, "space for floods" extension of the floodberms, non-protected floodplains maybe used more (Koncsos, 2011). There is a need to develop non-structural adaptation measurements, regulation of flood generation by land use management, flood zoning, development of flood warning systems and contingency planning and disaster mitigation.

6. Conclusions

Climate of Hungary and that of the Carpathian Basin in historical times underwent cyclic fluctuations. The variability was expressed by extreme events, droughts and floods occurred frequently. Historical documents mention several extreme events and their number had increased in time, what can be explained with the spread of the literacy, and increasing vulnerability of economy to natural disasters. In the Little Ice Age climate became cooler and wetter, the abundance of water became typical: runoff increased, permanently and/or temporarily inundated areas expanded, surface of lakes increased. As the water abundance impeded the development of agriculture being the main economy sector in the country the urgent need to regulate waters aroused. Hydraulic construction works, including flood protection can be considered as a response of the economy to changing climate.

The Little Ice Age ended up in the middle of 19th century. Climate turned again to warmer and drier, this trend is proved by the results of instrumental observations in 20th century. This tendency seemingly accelerated over the last three-four decades. Following the climatic changes some trends in hydrological regimes can also be detected, although their recognition is difficult due to the absence of long term observations, especially regarding flow rates, lack of detailed studies, and not least owing to the fact that rivers lost their natural character due to increasing impacts of anthropogenic impact. High natural variability especially that of extreme hydrological events can offset general trends of change. Although some tendencies in hydrological regimes, such as the decrease of annual flow in some rivers, seasonal shifts of flow, and significant decrease in water budget of Lake Balaton, increases in water temperature and change in the ice regime are in good harmony with climatic trends, still their significance can be disputed at many places.

Climate change scenarios predict unequivocally the warming and drying of the climate for Hungary in the 21th century, and predict the change in overall precipitation amounts and extremes with high uncertainty. Presently it can be stated that if these climate trends remain as it was observed in 20th century, especially during the last decades the changes in hydrological regimes will also continue even at accelerated rate. Changes in hydrological regimes would pose two major challenges for water management: reduction of water resources is likely to lead to more conflicts among water consumers, and higher risk of floods. Generally, adaptation options to cope with these challenges are well-known; nevertheless there is a high uncertainty if these options would be sufficient under changing conditions. The uncertainty is high how to implement climate policy to everyday water management decision making. Reduction or at least handling of this uncertainty in the nearest future is one of the most important tasks for the policy in the water field.

Author details

Béla Nováky
Department of Hydraulic and Water Resources Engineering,
Budapest University of Technology and Economics, Hungary

Gábor Bálint
VITUKI Environmental Protection and Water Management Research Institute, Budapest,
Hungary

Acknowledgements

The present work was supported by the Hungarian Academy of Sciences in the framework of the MTA-BME Water Management Research Group (Programme leader: Prof. János Józsa). Part of this work has been supported by the ECCONET (Effects of Climate Change on the Inland Waterway Networks) FP7 project, contract number 233886 – FP7, funded by the European Commission, DG TREN and conducted by an interdisciplinary consortium of ten partners, under coordination of Transport & Mobility Leuven.

7. References

Bálint, G. (2009). Közepes és nagy vízfolyásaink lefolyási sajátossága. MTA Meteorológiai Tudományos Bizottsága Légköri Energiák Munkabizottsága előadói ülés, 2009, január 15.

Bálint, G. & Gauzer, B. (1994). A rainfall runoff model as a tool investigate the impact of climate change. *XVIIth Conference of the Danube Countries on Hydrological Forecasting and Hydrological Bases of Water Management,* , Budapest, 5-9 September, 1994.

Bálint, G., Gauzer, B., Dobi, I. & Mika, J. (1995). The use of global warming scenarios for low flow simulation, In: *Proceedings of the International Workshop on Drought in the Carpathians Region,* Budapest-Alsógöd, 3-5 May, 1995.

Bálint, G & Gauzer, B. (1998). Az éghajlatváltozás hatása a lefolyás alakulására a Tisza vízgyűjtőjében. (alapozó tanulmány). In: *A hazai vízgazdálkodás stratégiai kérdései. Strategic Issues of the Hungarian Water resources Management (in Hungarian with English Summary),* Ed. Somlyódy, L. Stratégiai Kutatások a Magyar Tudományos Akadémián. Budapest

Bálint, G. & Konecsny, K. (2004). Folyóink vízjárása 2003-ban. Kisvizek – árvíz után. Magyar Meteorológiai Társaság és Magyar Hidrológiai Társaság előadói ülése 2004. január 22.

Bálint, G., Gauzer, B., Gnandt, B., Lipták, G. & Mattányi, Zs (2011). Projected effects of climate change for the low water periods of river Danube (Project ECCONET). Proceedings of XXVth Conference of the Danubian Countries, ISBN 978-963-511-151-0, Budapest, 16-17 June, 2011

Bálint, G., Zsolt Mattányi, András Csík, Boglárka Gnandt, Adrienn Hunyady (2010). Regional Climate Change Model Based Assessment of Hydrological Regime Changes and Other Regional Characteristics Geophysical Research Abstracts Vol. 12, EGU2010-14472, 2010 EGU General Assembly 2010

Bárdossy, A.; I. Kontur; J. Stehlik; G. Bálint, 2003: Could the global warming cause the last floods of the Tisza River? In: *G. Blöschl (Ed.) Water Resources Systems—Hydrological risk, management and development (Proceedings of symposium HS02b held during IUGG2003 at Sapporo, July 2003).* IAHS Publ. no. 281,

Bartholy J. & Pongrácz R. (2005). Tendencies of extreme climate indices based on daily precipitation in the Carpathian Basin for the 20th century. *Időjárás,* Vol. 109, No. 1, January-March, 2005, 1-20. ISSN 0324-6329

Bartholy J., Pongrácz R., Matyasovszky I. & Schlanger V. (2005). Trends of climate having taken place in the 20th century and to be expected in the 21th century on the territory of Hungary (in Hungarian). *"AGRO -21" Brochures. Climate change – impact – responses.* No. 33, 2005, pp. 3-18. ISSN 1218-5329

Bartholy, J., Pongrácz, R. & Torma, Cs. (2009). Regional climate change projected for the Carpathian Basin in 2021-2050 based on RegCM simulations. *"CLIMA- 21" Brochures. Climate change – impact – responses.* Vol. 60, pp. 3-13. ISSN 1789-428X

Bendefy, L. & V. Nagy, I. (1969). A Balaton évszázados partvonalváltozásai (Millennial shoreline changes of Lake Balaton). Műszaki Kiadó, Budapest,

Bozó, L (ed.) (2010). *Environmental vision – safety of environment and climate (Környezeti jövőkép – Környezet- és klímabiztonság,* Academy of Science of Hungary, ISBN 978-963-508-597-2.

Brázdil, R., Kotyza, O. & Dobrovolný, P. (2006). July 1432 and August 2002 – two millennial floods in Bohemia? *Hydrological Sciences Journal,* Vol. 51, No. 5, pp. 848-863, ISSN: 0262-6667

Corbuş, C., Mic, R. & Mătreaţă, M. (2011). Assessment of climate change impact on peak flow regime in the Mureş river basin. Proceedings of XXVth Conference of the Danubian Countries, ISBN 978-963-511-151-0, Budapest, 16-17 June, 2011

Cyberski, J., Grześ, M., Gutry-Korycka, M., Nachlik, E. & Kundzewicz, Z. W. (2006) History of floods on the River Vistula. *Hydrological Sciences Journal,* Vol. 51, No 5, pp. 799–817, ISSN: 0262-6667

Déri, J. (1985). Changes in the ice-regime of the Danube River (In Hungarian). *Vízügyi Közlemények,* Vol. 67. No 4, pp. 613–621, ISSN 0042-7616

EEA (2008). *Impacts of Europe's changing climate -2008 indicator-based assessment,* European Environment Agency, ISBN 978-92-9167-372-8, Copenhagen.

Fejér, L. (2001). The chronicle of our waters (In Hungarian) *Vizeink krónikája,* Vízügyi Múzeum, Levéltár és Könyvgyűjtemény, ISBN 963 00 8967 X, Budapest

Gauzer, B. (1994). Impact of air temperature change on the runoff regime of the Danube at Nagymaros gauging station, In: *Effect of climate change on hydrological and water quality parameters* (in Hungarian), Starosolszky Ö, pp. 37-58, VITUKI, ISBN 963 511 117 7, Budapest

Gilyénné Hofer A. (1994). Detection of consequences of climate change by analysis of hydrological data series, In: *Effect of climate change on hydrological and water quality parameters* (in Hungarian), Starosolszky Ö, pp. 19-36, VITUKI, ISBN 963 511 117 7, Budapest

Halmova, D., Pekarova, P., Pekar, J. & Onderka, M. (2008). Analyzing temporal changes in maximum runoff volume series of the Danube River. XXIVth Conference of the Danubian Countries. IOP Conf. Series: Earth and Environmental Science 4 (2008), 012007

Jacob, D & Horányi, A. (2009). Climate Change and Variability: Impact on Central and Eastern Europe. CLAVIER Newsletter, iusses 4-5-6, August 2009. www.clavier-eu.org

Kern, Z., Grynaeus, A. & Morgós, A. (2009). Reconstructed precipitation for southern Bakony Mountains (Transdanubia, Hungary) back to 1746 AD based on ring widths of oak trees. *Időjárás,* Vol. 113, No. 4, (October–December 2009), 113, pp. 299-314.

Keve, G. (1994). Hollandia egyik rajnai árterületének újraélesztése - a magyar Duna hidrológiai hasonlóságai alapján. *Hidrológiai Közlöny*, Vol.74, No.1, pp. 40-46, ISNN 0018-1323

Kiss, A. (2009). Historical climatology in Hungary: Role of documentary evidence in the study of past climates and hydrometeorological extremes. *Időjárás*, Vol.113, No.4, (October–December 2009), pp. 315–339

Koncsos, L. (2011). Flood management and strategy, In: In: *A hazai vízgazdálkodás stratégiai kérdései. Strategic Issues of the Hungarian Water resources Management (in Hungarian with English Summary)*, Ed. Somlyódy, L, Academy of Science of Hungary, pp. 206-232, ISBN 978-963-508-608-5, Budapest

Koncsos, L. & Kozma, Zs. (2007). Modelling of floodplain sedimentation on the Hungarian Tisza (in Hungarian). A hullámtéri feltöltődés becslése a Tisza magyarországi szakaszán. *Hidrológiai Közlöny*, Vol.87, No.5, pp. 59-63, ISNN 0018-1323

Konecsny, K. (2010). Hydrological-statistical evaluation of low flow on common Romanian-Hungarian lower course of Crisul Negra/Fekete-Körös. *Proceedings of XXVII. Conference of Hungarian Hydrological Society*, ISBN 978-963-8172-25-9, Sopron, July 7-9, 2010

Konecsny, K. (2011). The hydrological-statistical characteristics of the low water stage on the River Danube at Nagymaros (in Hungarian). *Hidrológiai Közlöny*, Vol. 91, No. 5., September-October 2011, 51-58, ISSN 0018-1323

Konecsny, K. & Bálint, G. (2009). Low water related hydrological hazards along the lower Mureş/Maros river. *Riscuri şi catastrofe*, Vol.8, No.7, pp. 202-207, ISSN 1584-5273

Konecsny, K. & Bálint, G (2011). Main hydrological statistical characteristics of low water on the Somes/Szamos River. *Riscuri su Catastrofe*, An IX, Vol.8, No.2, pp. 115-132, ISSN: 1584-5273

Konecsny, K. & Nováky, B. (2011). Climatic and anthropogenic effects on low flow time series of Zagyva (In Hungarian). *Proceedings of XXVII. Conference of Hungarian Hydrological Society*, ISBN 978-963-8172-28-0, Eger, July 6-8, 2010

Konecsny, K. & Sorocovshi, V. (1996): The effect of water reservoirs ont he flow int he Romanian and Hungarian catchments of Rivers Túr and Kraszna (In Hungarian). *Proceedings of Conference on the Water and Water environment in Carpathian Basin*, Eger, October 1996.

Kovács, A & Szilágyi, J. (2010). Future evaporation estimation for Lake Balatom. *Hidrológiai Közlöny*, Vol.90. No.1, January-February, 2010, pp. 15–18, ISSN 0018-1323

Kravinszkaja, G., Pappné Urbán, J. & Varga, Gy. (2010). Evaluation the effects of natural and anthropogenic factors on the long-term variation of water balance of Lake Balaton. (in Hungarian). *Proceedings of XXVII. Conference of Hungarian Hydrological Society*, ISBN 978-963-8172-25-9, Sopron, July 7-9, 2010

Kron, W., Bertz, G., 2007: Flood didasters and climate change: trends and options – a (re)insurer's view. Global Change: Enough Water for All? J.L.Lozán, H.Grassl, P.Hupfer, L.Menzel, C.D.Schönwiese (eds.), University of Hamburg, Hamburg, 268-273

Lászlóffy, W. (1982). *A Tisza*, Akadémiai Kiadó, ISBN 963 05 2681 6, Budapest

Lovász, Gy. (1985). Tendencies of runoff at Nagymaros section of Danube during 1883-1980 (in Hungarian). *Földrajzi Értesítő (Hungarian Geographical Bulletin)*, Vol.34, No.1-2. 47-57, ISSN 0015-5403

Macurová, Z., Výleta, R., Hlavčová, K., Szolgay, J. & Stěpánek, P. (2011). An evaluation of climate change impacts on simulated monthly runoff in Slovakia. *Proceedings of XXVth Conference of the Danubian Countries*, ISBN 978-963-511-151-0, Budapest, 16-17 June, 2011

Mauser, W., Marke, T. & Stoeber, S. (2008). Climate change and water resources: Scenarios of low flow conditions in the Upper Danube River Basin. *Proccedings of XXIVth Conference of the Danube Countries*, ISBN 978-961-91090-3-8,, Bled, Slovenia, 2-4 June, 2008.

Mika J., 1999: A hazai vízgazdálkodási stratégia alakításánál figyelembe vett éghajlati szcenáriók (alapozó tanulmány). *A hazai vízgazdálkodás stratégiai kérdései (szerk.: Somlyódy L.). Stratégiai Kutatások a Magyar Tudományos Akadémián.* Budapest. ISSN 1586-4219

Mudelsee, M., Börngen, M., Tetzlaff, G. & Grünewald, U. (2003). No upward trends in the occurrence of extreme floods in Central Europe. *Nature*, 425, pp. 166–169, ISSN: 0028-0836

Nováky, B. (1991). Climatic effects on runoff conditions in Hungary. Special Issue on the landscape-ecological impact of climatic change. *Earth Surface and Landforms*, Vol. 16, No.7, pp. 593-600, ISSN 0197-9337

Nováky, B. (1994). Expected effects of climate change on extreme phenomena of the water balance, In: *Effect of climate change on hydrological and water quality parameters* (in Hungarian), Starosolszky, Ö, pp. 43-178, VITUKI, ISBN 963 511 117 7, Budapest

Nováky, B. (1994): Climate and Natural Water Resurces Change of Balaton. *Proceedings of XII. Conference of Hungarian Hydrological Society*, ISBN 978-963-8172-28-0, Siófok, 17-19 July, 1994.

Nováky, B. (2002). Impacts of climate change on water management, In: *Strategic Issues of the Hungarian Water resources Management* (in Hungarian with English Summary), Somlyódy L, pp. 75-106, Academy of Science of Hungary, ISBN 963 508 333 5, Budapest

Nováky, B. (2008): Climate change impact on water balance of Lake Balaton. *Water Science and Technology – WST*, 9. Vol.58, No.9, pp. 1865–1869 doi:10.2166/wst.2008.563

Nováky, B. (2011). Climate change and its effects, (Az éghajlatváltozás és hatásai), In: In: *Water management of Hungary: Helyzetkép és stratégiai feladatok (Current review and strategic tasks)*, Ed. Somlyódy, L, pp. 85-102, Academy of Science of Hungary, ISBN 978-963-508-608-5, Budapest

Orlóci I., & Szesztay, K. (1994). Main hydroecological factors of climate change on the Great Hungarian Plain, In: *Effect of climate change on hydrological and water quality parameters* (in Hungarian), Starosolszky Ö, pp. 179-204, VITUKI, ISBN 963 511 117 7, Budapest

Padisák J. [ed]. (2006) Testing of Delivery and Internal Dynamics of P and N Models in „Warm World" Mode. CLIME Project (EVK1-CT-2002-00121), Department of Limnology, University of Pannonia, Veszprém, Hungary.

Pekarová, P. & Pekar, J. (2005). Long-term discharge prediction for the Turnu Severin station (the Danube) using a linear autoregressive model. *Hydrological Processes,* Vol. 20, No. 5, March 2006, pp. 1217-1228, ISSN: 1099-1085

Pekárová, P., Skoda, P., Miklánek, P., Halmová , D. & Pekar, J. (2008a). Detection of changes in flow variability of the upper Danube between 1876-2006. *IOP Conference Series: Earth and Environmental Science,* Vol. 4, No 1, 1-9, doi: 10.1088

Pekárová, P., Halmová, D., Miklánek, P., & Onderka, M. (2008c). Is the Water Temperature of the Danube River at Bratislava, Slovakia, Rising. *Journal of Hydrometeorology,* Vol.9 No5. October 2008, pp.115-1122, ISSN 1525-755X

Pekárová, P., Miklánek, P., Onderka, M., Halmová, D., Bacová Mitková, V., Mészáros, I. & Skoda, P. (2008b). Flood regime of rivers in the Danube River Basin, In: *National report for the IHP UNESCO Regional cooperation of the Danube Countries,* Institute of Hydrology SAS, Slovak Committee for Hydrology, Bratislava. http://147.213.145.2/DanubeFlood/PDF/Report01_Flood.pd

Pongrácz, R. &, Bartholy, J. (2000). Statistical linkages between ENSO, NAO, and regional climate. *Időjárás,* Vol.104, No.1, pp. 1-20.

Prasch, M. & Mauser, W. (2011). Glowa-Danube: integrative techniques, scenarios and strategies for the future of water in the upper Danube basin. *Proceedings of XXVth Conference of the Danubian Countries,* ISBN 978-963-511-151-0, Budapest, 16-17 June, 2011

Rácz, L. (1999) Climate history of Hungary from the 16th century to the present day (in Hungarian). *Magyar Tudomány* Vol.160, No.9, (September 1999), pp. 1127-1139. HU ISSN 0025-0325

Radvánszky, B. & Jacob, D. (2008). Prospective climate changes int he drainage area of the River Tisza and their effects ont he overland flow. Application of the Regional Climate Model (REMO) and the Hydrological Discharge Model (HD). *Hidrológiai Közlöny,* Vol.88, No.3, May-June, 2008, 33-41, ISSN 0018-1323

Radvánszky, B. & Jacob, D. (2009). The Changing Annual Distribution of Rainfall in the Drainage Area of the River Tisza during the Second Half of the 21st Century. *Zeitschrift für Geomorphologie, Supplementary Issues,* Vol.53, No.2, December 2009, 171-195, ISSN 1864-1687

Réthly, A. (1962). *Időjárási események és elemi csapások Magyarországon 1700-ig (Weather events and natural disasters in Hungary until 1700).* Akadémiai Kiadó, Budapest

Réthly, A. (1970*): Időjárási események és elemi csapások Magyarországon 1701-1800 (Weather events and natural disasters in Hungary 1701-1800).* Akadémiai Kiadó, Budapest

Simonffy, Z (2011). Vízkészletek és igények, (Water management of Hungary) In: *Helyzetkép és stratégiai feladatok (Current review and strategic tasks), Ed. Somlyódy, L* pp. 121-168, Academy of Science of Hungary, ISBN 978-963-508-608-5, Budapest

Somlyódy L., Nováky B. & Simonffy Z. (2010). Climate change, extremes and water management (in Hungarian with English abstract). *"CLIMA- 21" Brochures. Climate change – impact – responses.* Vol.61, No.? pp. 15-32. ISSN 1789-428X

Stahl, K., Hisdal, H., Hannaford, J., Tallaksen, M., van Lanen, H.A.J., Sauquet, E., Demuth, S., Fedekova, M. & Jódar, J. (2010). Streamflow trends in Europe: evidence from a dataset of near-natural catchments. *Hydrol. Earth Syst. Sci.,* Discuss. 7, pp. 5769-5804, ISSN 1027-5606

Stanciková, A. (1993). *Temperatur- und eisregime der Donau und ihrer wichtigeren Zubringer/Termicheskyi I ledovyi rezhim Dunaya i ego osnovnyh pritokov (in Russian),* Regional Cooperation of the Danube Countries, ISBN 80-0700622-2, Bratislava

Starosolszky, Ö. (1989). The effect of river barrages on the ice regime (In Hungarian). *Vízügyi Közlemények (Hydraulic Engineering),* Vol.71, No.3, pp. 345-386, ISSN 0042-7616

Szalai, S. (2011). The hydro-climatic characteristics of Hungary (in Hungarian). *"CLIMA -21" Brochures. Climate change – impact – responses.* No.65, 2011, pp. 17-28. ISSN 1789-428X

Szesztay, K. (1959). Water Balance Survey of Lakes and River Basins in Hungary. In: Commission of Surface Waters, Eds., pp. 579-593, IASH Publication No,51, ISBN, Helsinki

Szilagyi, J. & Jozsa, J. (2009). Estimating spatially distributed monthly evapotranspiration rates by linear transformations of MODIS daytime land surface temperature data *Hydrol. Earth Syst. Sci.,* 13, pp. 629-637

Szlávik, L. (2002). Floods: risk and safety. In: *A hazai vízgazdálkodás stratégiai kérdései. Strategic Issues of the Hungarian Water resources Management (in Hungarian with English Summary), Ed. Somlyódy, L,* pp. 205-244, Academy of Science of Hungary, ISBN 963 508 333 5, Budapest

Takács, K. (2011). Changes in river ice regime of the river Danube. *Proceedings of XXVth Conference of the Danubian Countries,* ISBN 978-963-511-151-0, Budapest, 16-17 June, 2011

Takács, K., Nagy, B. & Kern, Z. (2008). Anthropogenic effects on river ice regime – river regulation, reservoir and water pollution. – 1st International Geographical Scientific Colloquium, Mostar-Zagreb-Budapest

Varga, Gy. (2005). Survey of water balence condition of Lake Balaton (in Hungarian). *Vízügyi Közlemények (Hydraulic Engineering),* Special issue on Balaton, pp. 93-104. ISSN 0042-7616

Chapter 6

Effects of Climate Change on Hydric Resources: Some Implications and Solutions

Jesús Efren Ospina-Noreña, Carlos Gay García, Ana Elisa Peña del Valle and Matt Hare

Additional information is available at the end of the chapter

1. Introduction

Among the issues that humanity currently faces and will face in the future, the scarcity of water and the onset of large-scale events linked to it – such as the increasingly frequent periods of protracted drought and heavy flooding in different regions of the world – are undoubtedly some of the most pressing. Accordingly, these conditions will have to be considered in the modeling and analysis of water supply and demand in the coming years. As the distribution and growth of the world population runs parallel both to the increasing demand for water for different uses and the potential reduction of natural resources, these are important factors that hereafter will affect the availability of the liquid for human consumption.

Section 1, in describing the impact of climate change on the availability and distribution of water resources, analyzes this phenomenon under different models and scenarios of greenhouse gases emissions; section 2 refers to and documents the implications of water availability in different regions, taking into account the various production and development sectors; lastly, in section 3 we discuss some cases of adjustment policies implemented through the adaptive management approach (implemented in recent years in different parts of the world).

The first section of the present chapter yields some relevant results from the research that has been advanced at the Centro de Ciencias de la Atmósfera (CCA) [Center for Atmospheric Sciences] and the Programa de Investigación en Cambio Climático (PINCC) [Climate Change Research Program] of the Universidad Nacional Autónoma de México (UNAM) [National Autonomous University of Mexico], based on the modeling of climate change in different regions and covering various aspects, notably the situation of productive and development sectors vis-à-vis water resources.

The chapter provides, among other items, the findings derived from modeling the potential impact of climate change on the availability of water and the degree of pressure exerted on water resources in four hydrological-administrative regions of the Comisión Nacional del Agua (CNA) [National Commission of Water] in the Gulf of Mexico Basin, with the corresponding documentation. In depicting the vulnerability of water resources in the Guayalejo-Tamesí River Basin due to climate change, it highlights the effects of this phenomenon on irrigation districts (e.g., the downturn it has brought about in the supply/demand rate). The scenarios contemplate the coverage of demand, the supply requirements, and the unmet requests in the region. Likewise, the consequences of capturing rainwater for irrigation purposes in a pilot project based in San Miguel de Allende, Mexico, are discussed in the context of strategies for adapting to climate change.

The chapter in question also seeks to address important issues regarding the potential impact of climate change on the hydropower sector by focusing on the relationship of water supply/demand in the Sinú-Caribe River Basin in Colombia. In doing so, it analyzes briefly the eventuality that water resources embody a limiting factor in the course of time for different activities in the region, and provides at the end a series of observations that can help to plan and establish guidelines for the different production and development sectors.

Finally, the chapter points to the importance of adaptive schemes for the sustainable management of water resources under scenarios of high uncertainty and complexity. Moreover, by using several examples from the international realm on the adaptive management of water, this section shall show how the participatory-laden management of public policies, in conjunction with the measures referred to as "flexible," can effectively address several of the dilemmas as regards to water resources and the natural hazards that accompany climate change.

2. Models and scenarios of climate change

In order to observe the effects of climate change on hydric resources, it is necessary to work out future scenarios of those variables that become more relevant or influential as far as the availability of water is concerned, such as: temperature (T), precipitation (PCP), and evaporation (Ev), among others

Nowadays, there are several joint models (Atmosphere/Ocean General Circulation Model [AOGCM]) that are run under different scenarios of greenhouse gases emissions, which result in a wide range of future scenarios on a global and regional scale with respect to climate variables. This allows us to pose different projections whereby multiple analyses are facilitated and solid tools are generated for decision makers. At the same time, however, there appears a high degree of uncertainty and complexity – something that one must take into account when studying the impact of climate change on water availability, as the present chapter intends to do.

The afore mentioned studies are relying on some models and have considered different periods of analysis; this information is given in Table 1, which includes the approximate location of the projects. Table 2 briefly describes the patterns of the various scenarios that have been used.

3. Methodological aspects

In each one of these studies, the projections of climate change's effects on the availability of hydric resources were estimated mainly by adjusting the averages for the mean temperature (meaT), the highest temperature (maxT), and precipitation (PCP) (base lines) taking into account the most representative weather stations that can be found within and/or close to the corresponding areas under scrutiny, while the projections about the anomalies to be ascribed to each region were provided by the program known as MAGICC/SCENGEN[1] v.5.3, and the outlets of the experiments carried out by the Canadian Institute for Climate Studies (CICS) for the models presented in Table 1.

For the analysis of hydric resources, each one of the projects had to use a different and, indeed, numerous series of variables and relations, namely the current availability (or natural offer) of water; evaporation; flow or expenses; the P/T (Precipitation/ Temperature) relationship or Lang Index (I_L); the supply/demand relation; the projections made for hydric resources, for its demand, and for the population; the index of pressure on the resource, etc.

More details about the information on models, scenarios, tools, back-up software, and the methodologies used can be found in: (Sánchez-Torres., *et al*, 2011; Ospina-Noreña., *et al*, 2009a; Ospina-Noreña., *et al*, 2009b; Ospina-Noreña., *et al*, 2010; Ospina-Noreña., *et al*, 2011a; Ospina-Noreña., *et al*, 2011b).

Furthermore, for the information concerning greenhouse gases (GHG) emissions, global climate models or general circulation models (GCMs), programs known as climate scenario generators, and relevant conceptualizations of climate change, we refer to Wigley (1994), Wigley (2003), Hulme, *et al* (2000), Conde (2003); as for the examination of vulnerability and the effects on different sectors, see Gay (2000).

4. Results

4.1. Trends and future scenarios

As for the regions on which the studies referred to in this chapter are concentrated, the climate change scenarios show a tendency to the rise of the mean temperature and the highest temperature in each case. Regarding precipitation, the projections indicate slight to substantial increases or decreases in a given region (Tables 3, 4, 5, 6), which implies a high degree of uncertainty in the results – something that will have to be assessed.

---

[1] Authors such as Wigley (1994), Wigley (2003), Hulme., *et al* (2000), Conde (2003) point out that there are simple climate models which incorporate the gamut of emissions scenarios to the studies of climate change. According to them, these models can simulate the response of global climate to changes in the concentrations of greenhouse gases (GHG) in terms of an increment in temperature and the rise of the sea level. One of them is the model for the evaluation of greenhouse gases effects that is designated as the Model for the Assessment of Greenhouse-Gas Induced Climate (MAGICC). However, for the results of MAGICC to be combined with the outlets of general circulation models (GCMs), it is necessary to use the climate scenario generator called SCENGEN (Regional Climate SCENarioGENerator).

Study	Model*	Scenario				Period					Location	
		A2	B2	B1	A1B	2010-2039	2040-2069	2030	2060	2080	Latitude	Longitude
"Impacts of Climate Change on the Hydric Regions of the Gulf of Mexico" (Ospina-Noreña., et al, 2010).	GFDLCM 2.0	X	X					X		X	17-23° N	89-99° W
	MPIECH-5	X	X					X		X		
"Vulnerability of Water Resources to Climate Change Scenarios. Impacts on the Irrigation Districts in the Guayalejo-Tamesí River Basin, Tamaulipas, Mexico" (Sánchez-Torres., et al, 2011).	GFDLCM 2.0	X	X	X	X	X	X				22°47'39"	98°42'58"
	MPIECH-5	X	X	X	X	X	X					
	UKDADCM3	X	X	X	X	X	X					
"Scenarios of Climate Change for Collecting Rainwater" (Advance of the work in progress by Ingenieros Sin Fronteras México, A.C., and the Instituto Tierra y Cal, A.C.).	MPIECH-5	X	X					X	X		21° N	101° W
	UKHADCM3	X	X					X	X			
	UKHADGEM	X	X					X	X			
	GFDLCM21	X	X					X	X			
	MIROCMED	X	X					X	X			
	CSIRO-30	X	X					X	X			
	CCCMA-31	X	X					X	X			
	Weighted Average	X							X			
"Examination of the Sinú-Caribe River Basin, Colombia" (Ospina, 2009, Ospina-Noreña., et al, 2009a, Ospina-Noreña., et al, 2009b, Ospina-Noreña., et al, 2011a, Ospina-Noreña., et al, 2011b).	CCSRNIES-A21	X				X	X				7-10° N	75-77° W
	CSIROMK2B-A21	X				X	X					
	CGCM2-A21	X				X	X					
	CGCM2-A22	X				X	X					
	CGCM2-A23	X				X	X					
	HadCM3-A21	X				X	X					
	HadCM3-A22	X				X	X					
	HadCM3-A23	X				X	X					
	Weighted Average	X				X						

*Generally, the designation of the models is based on the root or the initials of the institute in charge of the climate modeling; e.g., Geophysical Fluid Dynamics Laboratory (GFDL), Canadian Climatic Center Model (CCCM), National Center for Atmospheric Research (NCAR).

Table 1. Models and Scenarios Used in the Studies.

Scenario Families	Description
A1	It describes a future world with fast economic growth, a world population that attains its highest value by the midcentury to decrease subsequently, and the rapid dissemination of new, more efficient technologies. Some of the most important characteristics are a convergence between regions, the building-up of capacities together with the increase of cultural and social interactions, and the reduction of regional differences as far as income per inhabitant is concerned. It contemplates three groups that are different in their technological orientation: intensive utilization of fossil energy sources (A1FI), intensive utilization of non-fossil energy sources (A1T), or a well-balanced employment of all kinds of sources (A1B).
A2	It presupposes a very heterogeneous world whose distinguishing traits are self-sufficiency and the preservation of local identities, as well as the continuous growth of the world population. The economic development is basically oriented to the regions, whereas the economic growth per inhabitant and technological change are rather fragmentary and slower than in the case of other scenario families.
B1	It describes a world in convergence with a world population that reaches a maximum by the midpoint of the present century and decreases subsequently, as is the case in the evolutionary line A1; also, it presupposes sudden changes of the economic structures leading to an information and service economy, a decreasing utilization of materials, and the introduction of clean technologies whereby it becomes possible to profit efficiently from resources. Notably, its focus is on economic, social, and environmental sustainability.
B2	This scenario family presupposes a world in which local solutions to the need for economic, social, and environmental sustainability predominate; the population increases progressively at a slower pace than in A2, whereas economic growth occupies an intermediate position and technological change is less fast, though more diversified, than in the evolutionary lines B1 and A1. It is focused principally on the local and regional levels (this scenario has already been superseded).

Source: IPCC (2000)

Table 2. Characteristics of the Scenario Families.

Although the estimates used in the projections of climate change's effects on water resources availability for each one of the CNA's hydrological-administrative regions (see, Attached Document I) and in the study on the collection of water were applied every five years, herein we will present only some relevant results regarding the periods mentioned in Table 1. Table 3 registers a slight increase in region XII (the Yucatán Peninsula) that was provided by model GFDLCM2.0 for scenarios A2 and B2, as well as small decreases for the rest of the regions, region IX (Northern Gulf) being the most affected, as it reaches a decrease of 10.8%

in scenario A2 and 4.3% in scenario B2 for 2080. On the other hand, model MPIECH-5 for the A2 and B2 scenarios projects important reductions of precipitation in regions XII (-21.1% and -14.7% for scenarios A2 and B2, respectively, by 2080) and XI Frontera Sur (-26.4% and -18.5 for scenarios A2 and B2, respectively, by 2080), together with slightly minor diminutions in regions X (Central Gulf) and IX.

We can deduce from Table 4 that in the Guayalejo-Tamesí River Basin precipitation displays a tendency downwards, whereas temperature shows a tendency upwards, which in turn provokes a decrease in the P/T relationship or Lang Index (I$_L$) , used to determine the kind of year (very humid, humid, normal, dry or very dry); thus allowing the remark that for the period of 2010-2069 virtually no humid or very humid years are expected at the Guayalejo-Tamesí River Basin, the temperature sticking to normal and with trends towards dry and very dry days as time goes by. Again, though in this study the projections were undertaken year after year and for four emission scenarios (Sánchez-Torres., *et al*, 2011), the only findings that are presented correspond to scenarios A2 and B2 during the span of a decade and for the end of the periods reported in Table 1.

Year/Mod	Region XII[1]				Region XI[2]			
	GFDLCM20-A2		GFDLCM20-B2		GFDLCM20-A2		GFDLCM20-B2	
	% changePrec	Change meaT (°C)	% changePrec	Change meaT (°C)	% changePrec	Change meaT (°C)	% changePrec	Change meaT (°C)
2030	4.7	0.5	4.3	0.6	0.26	0.51	0.5	0.58
2080	11.4	2.0	7.8	1.5	-2.77	2.02	-2.2	1.53
	MPIECH-5 A2		MPIECH-5 B2		MPIECH-5 A2		MPIECH-5 B2	
2030	-6.6	0.7	-4.75	0.8	-7.87	0.85	-6.07	0.86
2080	-21.1	2.7	-14.67	2.0	-26.35	3.02	-18.48	2.22
	Region X[3]				Region IX[4]			
	GFDLCM20-A2		GFDLCM20-B2		GFDLCM20-A2		GFDLCM20-B2	
2030	-1.66	0.49	0.02	0.56	-8.1	0.53	-2.2	0.66
2080	-2.9	1.9	-1.24	1.44	-10.81	2.08	-4.3	1.59
	MPIECH-5 A2		MPIECH-5 B2		MPIECH-5 A2		MPIECH-5 B2	
2030	-2.82	0.84	-0.92	0.84	-6.68	0.79	-1.00	0.9
2080	-6.26	2.9	-3.56	2.12	-6.7	2.85	-1.47	2.1

[1] Current mean precipitation: 1,226.7, current mean temperature: 26.5°C, current P/T: 46.3
[2] Current mean precipitation: 2,105.2, current mean temperature: 26.9°C, current P/T: 79.1
[3] Current mean precipitation: 1,755.9, current mean temperature: 23.6°C, current P/T: 79.4
[4] Current mean precipitation: 1,349.3, current mean temperature: 24.1°C, current P/T: 56.8

Table 3. Anomalies of Precipitation and Mean Temperature for the Four Hydrological-Administrative Regions.

Year	GFDLCM2.0-A2				GFDLCM2.0-B2			
	meaT£	PCP€	P/T	% change*	meaT	PCP	P/T	% change*
2039	25.6	764.1	29.9	-9.6	25.7	810.1	31.6	-4.6
2069	26.4	754.5	28.6	-13.4	26.2	802.8	30.7	-7.1
	MPIECH-5-A2				MPIECH-5-B2			
2039	26.0	751.9	28.9	-12.6	26.0	800.6	30.8	-7.0
2069	27.2	731.9	26.9	-18.6	26.8	786.7	29.4	-11.0

£ Base line: 25°C, € Base line: 826.5 mm, *With respect to the current value of P/T equal to 33.06.

Table 4. Climate Projections for the Study on the Irrigation Districts in the Guayalejo-Tamesí River Basin, Tamaulipas, México.

As is shown by the results of the study "Scenarios of Climate Change for Rainwater Collection" (prepared by Ingenieros Sin Fronteras México, A.C., and the Instituto Tierra y Cal, A.C, 2012), in the case of the A2 scenarios only one model, the GFDLCM21_A2, projects an increase in precipitation, that is, here 85.7% of the models analyzed indicate that a decrease in precipitation is highly likely; nevertheless, all the models are equally likely to occur. Therefore, our proposal is to generate the weighted average scenario, which takes into account the results of all the models and is meant to operate as a planning platform for the collection of rainwater or the availability of this resource, so as to avoid the most adverse effects. The two models that thoroughly undermine the objectives and aims of the project would be the CCCMA-31 and the CSIRO-30 (see Table 5).

As for the B2 scenarios, there are three (42.9%) where an increase is projected and four (57.1%) where a decrease in precipitation is foreseen; however, two of the three models that project an increase are almost insignificant, as can be observed in Table 5 – something that once again highlights the tendency toward a decrease in precipitation in the project's study or influential area.

Regarding the research into the Sinú-Caribe River Basin in Colombia, the findings show a tendency to the rise in the highest temperature, and slight to substantial increases or decreases of the PCP, which leads to the reduction of water availability, the effects projected by model HadCM3 being the most adverse (see Table 6).

4.2. The effect of climate change on water availability and pressure degree in the Gulf of Mexico

Relying on the projections for precipitation and the mean temperature, the Lang Index was calculated for the four scenarios that were obtained from running the models GFDLCM2.0 and MPIECH-5 for each one of the Gulf of Mexico's hydrological-administrative regions. Such index can be interpreted as one measuring the degree of aridity or humidity that predominates in the various regions. Starting from the P/T relationship's current values that are shown in Table 3, we determined the percentage of rise or diminution of such relation (see Table 7) according to the different projections – an amount that in turn is assumed to represent an increase or decrease in the availability of hydric resources.

The findings presented in Table 7 (which follow the classification set in Table 8) allow us to establish a change in trend in the climate zones of each hydrological-administrative region, as is shown also in Table 9. For example, in region XII a humid zone of steppe and savannah (Hzss) would turn into an arid zone (Az), whereas in regions XI and X humid zones of sparse forest (Hzsf) would become humid zones of steppe and savannah (Hzss). These figures were obtained as projected by the model MPIECH-5 for scenarios A2 and B2.

Year	UKHADGEM_A2		UKHADC M3_A2		MPIECH-5_A2		GFDLCM21_A2		MIROCMED_A2		CSIRO-30_A2		CCCMA-31_A2		WA_A2ᵉ
	Change (%)	Change (mm)	Change (%)	Change (mm)	Change (%)	Change (mm)	Change (%)	Change (mm)	Change (%)	Change (mm)	Change (%)	Change (mm)	Change (%)	Change (mm)	Change (mm)
2030	-5.2	479.4	-1.9	496.1	-3.4	488.3	8.4	548.0	-2.6	492.7	-5.9	476.0	-8.7	461.7	482.1
2060	-10.0	455.1	-3.0	490.3	-6.3	473.7	18.5	599.3	-4.5	482.9	-11.4	447.9	-17.3	417.9	460.7
Weighted Factor	3		2		3		1		2		4		5		
Change through 2060ᵉ	-50.5		-15.3		-31.9		93.7		-22.7		-57.7		-87.7		-44.9
	UKHADGEM_B2		UKHADC M3_B2		MPIECH-5_B2		GFDLCM21_B2		MIROCMED_B2		CSIRO-30_B2		CCCMA-31_B2		WA_B2ᵉ
2030	-2.2	494.7	0.7	509.0	-0.7	502.3	9.5	553.4	0.1	506.1	-2.7	491.7	-5.1	479.6	498.7
2060	-3.8	486.6	1.6	513.4	-1.0	500.8	18.0	596.6	0.5	507.9	-4.9	481.1	-9.4	458.2	494.1
Weighted Factor	2		1		2		1		1		2		3		
Change through 2060ᵉ	-19.0		7.8		-4.8		91.0		2.3		-24.5		-47.4		-11.5

ᶠWeighted average scenario, in A2 and B2.
ᵉBased on the current value 505.6 mm

Table 5. Anomalies of Precipitation in the Study "Scenarios of Climate Change for Rainwater Collection."

Model/Variable	maxT.	PCP	Storage Vol.		Flow	Weighted Factor
	Change °C	Change %	Change[a] %	%Change_TMV[b]	Change %	
CCSRNIES_A21	0.5	0.16	-2.3	-12.9	-5.9	1
CSIROMK2B_A21	0.7	13.5	-1.9	-12.6	-2.3	1
CGCM2_A21	0.7	-5.4	-6.9	-17.0	-11.8	2
CGCM2_A22	0.9	-2.6	-7.8	-17.8	-13.3	2
CGCM2_A23	0.8	-3.1	-6.4	-16.6	-11.3	2
HadCM3_A21	1.9	-21.0	-29.7	-37.3	-34.9	4
HadCM3_A22	1.6	-6.2	-20.1	-28.8	-23.8	3
HadCM3_A23	1.4	9.6	-10.5	-20.2	-14.2	2
Weighted Average					-18.8	
Referenced or Current (average values)	37.5 (°C)	2,212.0 (mm)	1,452.8 (MCM) - TMV: 1,630 (MCM)		340.3 m³/s (in the dam) 527.7 m³/s (in all the basin)	

[a]With regard to the scenario in question, equal to 1,452.8 million cubic meters (MCM).
[b]With regard to the Technology Maximum Value (TMV): Technology Maximum Value, equal to 1,630 million cubic meters (MCM).

Table 6. Projections of the Hydrological-Climatic Variables, Period of 2010-2039, Sinú-Caribe River Basin, Colombia.

Year	Region IX				Region X			
	GFDLCM20		MPIECH-5		GFDLCM20		MPIECH-5	
	A2	B2	A2	B2	A2	B2	A2	B2
2030	-11.4	-6.2	-11.0	-5.9	-9.7	-8.4	-12.0	-10.3
2080	-19.1	-11.6	-17.8	-10.8	-15.8	-12.8	-21.7	-17.1
	Region XI				Region XII			
2030	-1.6	-1.6	-10.7	-9.0	2.8	2.1	-9.1	-7.5
2080	-9.6	-7.5	-33.8	-24.7	3.7	1.9	-28.4	-20.7

Table 7. Percentage of Change in the P/T Relationship or Lang's Index.

I_L	Climate zones	Abbreviation
$0 \leq I_L < 20$	Desert	D
$20 \leq I_L < 40$	Aridzone	Az
$40 \leq I_L < 60$	Humid zone of steppe and savannah	Hzss
$60 \leq I_L < 100$	Humid zone of sparse forest	Hzsf
$100 \leq I_L < 160$	Humid zone of dense forest	Hzdf
$I_{la} \geq 160$	Hyperhumid zone of grassland and tundra	Hhzgt

Source: Changed from Urbano-Terrón (1995).

Table 8. Climate Zones according to the Lang Index.

Climate Condition	Region IX				Region X			
	GFDLCM20		MPIECH-5		GFDLCM20		MPIECH-5	
	A2	B2	A2	B2	A2	B2	A2	B2
Current Classification	Hzss	Hzss	Hzss	Hzss	Hzsf	Hzsf	Hzsf	Hzsf
Classification by the Year of 2080	Hzss *	Hzss *	Hzss *	Hzss *	Hzsf *	Hzsf *	Hzss	Hzss

Climate Condition	Region XI				Region XII			
	GFDLCM20		MPIECH-5		GFDLCM20		MPIECH-5	
	A2	B2	A2	B2	A2	B2	A2	B2
Current Classification	Hzsf	Hzsf	Hzsf	Hzsf	Hzss	Hzss	Hzss	Hzss
Classification by the Year of 2080	Hzsf *	Hzsf *	Hzss	Hzss	Hzss *	Hzss *	Az	Az

*They retain the current classification, but it is worth noting that each time they are getting closer to the lower limit of the classification shown in Table 8, by the end of the period.

Source: Ospina-Noreña., et al (2010).

Table 9. Change in the Classification of Climate Zones in the Gulf of Mexico.

As it can be observed, there is a general tendency to change from more humid climate zones to less humid climate zones in the different hydrological-administrative regions, and this could have transcendental implications regarding change in the natural vegetal coverage, with the consequent effects on the various extant biotic-physical elements, namely the floristic, fauna, and ecosystem structures, which might undergo relevant transformations. Likewise, in the future the predominating systems of agricultural production could be affected.

Moreover, we find that in region XII there could be a slight rise in water availability, as projected by model GFDLCM2.0 for scenarios A2 and B2. As for the other scenarios, they display significant reductions nonetheless, as time goes by in all the Gulf of Mexico's regions. Such decreases exacerbate the existing conditions as regards the degree of pressure on hydric resources, especially when considerable increases in the extraction of and demand for the hydric resource are expected in the future (Ospina-Noreña., *et al*, 2010).

On the other hand, Table 10 presents the results obtained via models GFDLCM2.0 and MPIECH-5 with respect to the projections about the degree of pressure on hydric resources in the Gulf of Mexico's hydrological-administrative regions. By looking at the results of model MPIECH-5, scenario A2, we can observe that in region XII the degree of pressure, which in 2010 was of 5.0%, would rise to 19.2% in 2030 and 24.3% in 2080. Such results presuppose that the demand for water projected for 2030 would remain the same until 2080 it goes without saying, the rise in the degree of pressure could be much higher than is reported in this study, and its final value will depend on whether the efficiency of hydraulic systems improves and the right policies are adopted concerning the sustainable management of hydric resources in the hydrological-administrative regions under scrutiny.

It is worth stressing that even though there might be slight increases in water availability, as projected by model GFDLCM2.0 for scenarios A2 and B2 in region XII (see Table 7), when considering the projections about the demand for water, the degree of pressure would go from 4.7% in 2010 to 17% by 2080 for the same scenarios. In each of the cases it is noticeable that region IX is the most affected, attaining a degree of pressure that would go from 29.2% by 2080, as projected by model MPIECH-5, scenario B2, to 32.2%, according to model GFDLCM20, scenario A2.

Year/Mod	MPIECH-5, Scenario A2				MPIECH-5, Scenario B2			
	Region IX	Region X	Region XI	Region XII	Region IX	Region X	Region XI	Region XII
2010	22.6	4.1	1.2	5.0	22.3	4.2	1.2	5.1
2030*	29.3	19.4	13.7	19.2	27.7	19.0	13.4	18.8
2080	31.7	21.8	18.5	24.3	29.2	20.6	16.2	22.0
	GFDLCM2.0, Scenario A2				GFDLCM2.0, Scenario B2			
	Region IX	Region X	Region XI	Region XII	Region IX	Region X	Region XI	Region XII
2010	22.6	4.1	1.1	4.7	22.3	4.1	1.2	4.9
2030*	29.4	18.9	12.4	17.0	27.8	18.6	12.4	17.1
2080	32.2	20.2	13.5	16.8	29.5	19.5	13.2	17.1

*From 2030 on, we took into account the combined effect of climate change and the increase in the demand for water projected for this year; in other words, up until 2025 the only factor to be considered was the decrease in water availability projected by different scenarios of climate change, while the demand for water by 2000 remained the same. After 2030, we considered the decrease in water availability due to the climate change effect, though the constant factor was the demand for water projected by 2030.

Source: adapted from Ospina-Noreña., *et al* (2010).

Table 10. Projections about the Degree of Pressure on the Hydric Resources, according to Models MPIECH-5 and GFDLCM2.0, Scenarios A2 and B2.

The Attached Document II illustrates the evolution undergone by the degree of pressure for model MPIECH-5, scenario A2, in each one of the Gulf of Mexico's regions, keeping in mind that in 2000 regions IX, X, XI, and XII showed a pressure degree of 21.4%, 3.8%, 1.2%, and 4.9%, respectively. For this model in particular, it is clear that in the year 2000 regions X, XI, and XII proved to have a scarce degree of pressure, whereas region IX presented a middle-strong degree of 21.4%. By 2030 the same regions (X, XI, and XII), according to the projections offered in Table 10, will approximately have a moderate degree of pressure of 19%, 14%, and 19%, respectively, and region IX will keep facing a middle-strong degree, equal to 29%. By 2080, regions X and XII will reach a middle-strong degree of pressure; region XI will continue to have a moderate degree, though reaching the upper limit (according to the ranges presented in the Attached Document II), and region IX will retain its category of middle-strong pressure, though it will come ever closer to the upper limit and could well enter the strong degree of pressure (>40%) after 2080.

4.3. The effect of climate change on the irrigation districts in the Guayalejo-Tamesí River Basin

If this study employed the software application known as WEAP (Water Evaluation and Planning), that was because it allows the user to make projections and simulations of the supply/demand relationship in the hydric resource and to undertake the respective analyses; thus, it becomes possible to determine the core facets of such relationship, such as: unmet demand, demand coverage, requirements of the offer, delivered supply, increased demand, and index of pressure on the resource, among others. In this way, it is feasible to generate the elements required for the well-ordered utilization and management of the basin's hydric resources and to develop adaptive strategies vis-à-vis the potential climate changes (Sánchez-Torres., et al, 2011).

Once the projections for the climatic variables and the Lang Index (P/T) were set, we generated climate change scenarios by means of such relationship, thereby determining the type of year, which subsequently was applied in the WEAP model. In doing so, we couldn't overlook that going from the index's present condition to the lower limit would entail getting closer to drier zones each time, and vice versa when being closer to the upper limit (see Tables 4 and 8), so the following categories were established: reduction or increase of the current Lang Index (I_L) up to 5%, normal year; reduction between 5.1 and 10%, dry year; reduction higher than 10%, very dry year; increase between 5.1 and 10%, humid year; increase higher than 10%, very humid year.

Table 11 shows the type of year projected every 10 years for the Guayalejo-Tamesí River Basin during the periods of 2010-2039 and 2040-2069, for models GFDLCM2.0 and MPIECH-5, for scenarios A2 and B2, respectively, taking into account the P/T relation and the aforementioned criteria.

Considering the conditions detailed in Table 11, and regardless of the fact that through the WEAP program it is possible to get a wide variety of results concerning the supply/demand relation of the hydric resource in one region, this chapter makes specific reference to the unmet demands for water in the irrigation districts inside the Guayalejo-Tamesí River Basin.

Year/Mod	GFDLCM2.0-A2					GFDLCM2.0-B2				
	meaT	PCP	P/T	% Change*	Type of Year	meaT	PCP	P/T	% Change*	Type of Year
2010	25.2	811.4	32.2	-2.7	Normal	25.2	819.5	32.5	-1.8	Normal
2020	25.3	790.6	31.3	-5.4	Dry	25.4	817.2	32.2	-2.7	Normal
2030	25.4	773.0	30.4	-8.0	Dry	25.5	813.6	31.9	-3.6	Normal
2039	25.6	764.1	29.9	-9.6	Dry	25.7	810.1	31.6	-4.6	Normal
2040	25.6	763.2	29.8	-9.8	Dry	25.7	809.7	31.5	-4.7	Normal
2050	25.8	755.4	29.3	-11.5	Very dry	25.9	807.7	31.2	-5.5	Dry
2060	26.1	753.7	28.9	-12.6	Very dry	26.0	804.9	31.0	-6.4	Dry
2069	26.4	754.5	28.6	-13.4	Very dry	26.2	802.8	30.7	-7.1	Dry
	MPIECH-5-A2					MPIECH-5-B2				
	meaT	PCP	P/T	% Change	Type of Year	meaT	PCP	P/T	% Change	Type of Year
2010	25.4	807.7	31.9	-3.7	Normal	25.4	816.1	32.2	-2.7	Normal
2020	25.5	784.4	30.7	-7.0	Dry	25.6	811.7	31.7	-4.1	Normal
2030	25.8	763.7	29.7	-10.3	Very dry	25.8	806.0	31.2	-5.6	Dry
2039	26.0	751.9	28.9	-12.6	Very dry	26.0	800.6	30.8	-7.0	Dry
2040	26.1	750.7	28.8	-12.9	Very dry	26.1	799.9	30.7	-7.1	Dry
2050	26.4	739.5	28.0	-15.3	Very dry	26.3	795.8	30.3	-8.5	Dry
2060	26.8	734.3	27.4	-17.1	Very dry	26.5	790.8	29.8	-9.9	Dry
2069	27.2	731.9	26.9	-18.6	Very dry	26.8	786.7	29.4	-11.0	Very dry

*With respect to the current value of 33.06.

Table 11. Type of Year Projected, Models GFDLCM2.0 and MPIECH-5, Periods of 2010-2039 and 2040-2069.

Table 12 summarizes the results for the demand for water that has been unmet annually in the irrigation districts (ID) inside the Guayalejo-Tamesí River Basin; it is expressed in millions of m^3 and measured every ten years for the period of 2010-2069.

ID/Model of Climate Change	Year						
	2010	2020	2030	2040	2050	2060	Total
ID Xicoténcatl 029							
GFDLCM2.0-A2	539.36	546.91	580.08	580.08	592.22	593.83	3,432.48
GFDLCM2.0-B2	539.36	539.36	539.36	539.36	563.63	580.08	3,301.15
MPIECH-5-A2	539.36	561.47	581.43	593.83	593.83	593.83	3,463.76
MPIECH-5-B2	539.36	539.36	555.54	580.08	580.08	580.08	3,374.50

ID/Model of Climate Change	Year						
	2010	2020	2030	2040	2050	2060	Total
ID Río Frío							
GFDLCM2.0-A2	94.45	95.77	101.58	101.58	103.70	103.99	601.06
GFDLCM2.0-B2	94.45	94.45	94.45	94.45	98.70	101.58	578.06
MPIECH-5-A2	94.45	98.32	101.81	103.99	103.99	103.99	606.53
MPIECH-5-B2	94.45	94.45	97.28	101.58	101.58	101.58	590.90
ID San Lorenzo							
GFDLCM2.0-A2	0.00	0.00	0.00	0.00	0.00	0.00	0.00
GFDLCM2.0-B2	0.00	0.00	0.00	0.00	0.00	0.00	0.00
MPIECH-5-A2	0.00	0.00	0.00	0.00	0.00	0.00	0.00
MPIECH-5-B2	0.00	0.00	0.00	0.00	0.00	0.00	0.00
ID 002 Mante M. Izquierda							
GFDLCM2.0-A2	9.82	29.19	15.59	13.68	27.01	32.98	128.26
GFDLCM2.0-B2	9.82	28.79	14.49	12.72	25.71	32.22	123.74
MPIECH-5-A2	9.82	29.97	15.62	14.00	27.08	32.98	129.48
MPIECH-5-B2	9.82	28.79	14.93	13.68	26.46	32.22	125.88
ID 002 Mante M. Derecha							
GFDLCM2.0-A2	15.10	40.96	24.92	21.67	37.03	45.27	184.95
GFDLCM2.0-B2	15.10	40.39	23.17	20.15	35.25	44.22	178.28
MPIECH-5-A2	15.10	42.05	24.98	22.18	37.14	45.27	186.72
MPIECH-5-B2	15.10	40.39	23.87	21.67	36.28	44.22	181.53
ID 1 Las Ánimas							
GFDLCM2.0-A2	0.00	9.69	10.35	10.17	10.45	10.53	51.19
GFDLCM2.0-B2	0.00	9.56	9.62	9.45	9.95	10.29	48.87
MPIECH-5-A2	0.00	9.95	10.37	10.41	10.48	10.53	51.74
MPIECH-5-B2	0.00	9.56	9.91	10.17	10.24	10.29	50.16
ID 2 Las Ánimas							
GFDLCM2.0-A2	0.00	11.61	12.39	12.17	12.52	12.61	61.29
GFDLCM2.0-B2	0.00	11.45	11.52	11.32	11.91	12.31	58.50
MPIECH-5-A2	0.00	11.91	12.41	12.46	12.55	12.61	61.95
MPIECH-5-B2	0.00	11.45	11.86	12.17	12.26	12.31	60.05

Table 12. Annual Demand for Water that Is Unmet in the Irrigation Districts inside the Guayalejo-Tamesí River Basin.

Based on these results, we can conclude that it is under the climate conditions obtained through model GFDLCM2.0-B2 where the coverage of the demand for water in the aforementioned river basin reaches the highest percentages; or, to put it differently, among the projected climate conditions to meet the demand for water in that basin, the least adverse correspond to those obtained by model GFDLCM2.0-B2. On the other hand, the most unfavorable projections about the same conditions correspond to model MPIECH-5-A2.

It is also noticeable that the most adverse conditions to meet that demand belong to the irrigation district Xicoténcatl 029; as a means to illustrate this, Figure 1 shows the projections on the unmet demand for the same district, according to the different models (million cubic meters [MCM] are used).

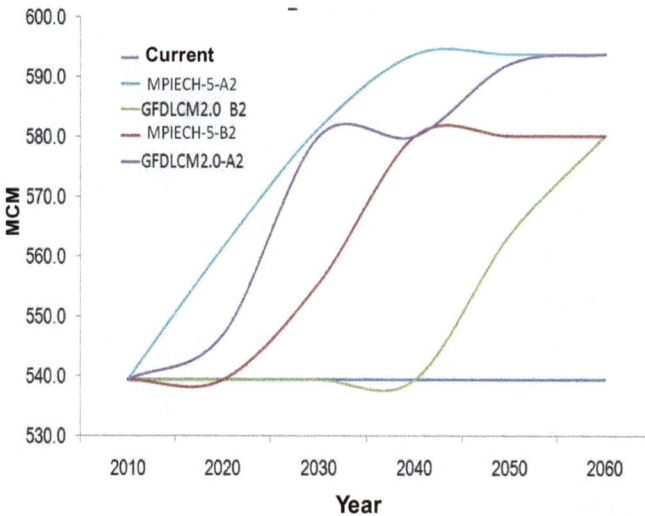

Figure 1. Unmet Demand for Water.

As this figure shows, scenario MPIECH-5-A2 presents the most unfavorable conditions and the fastest changes with regard to the unmet demand for water. On the other hand, scenarios B2 project less drastic changes with less immediate effects.

Given that model WEAP enables the user to incorporate a limitless series of scenarios within the calculus process, we decided to analyze several scenarios in which a number of changes are contemplated as adaptive measures to be possibly taken; the scenarios were:

• Base scenario (BS): It deems it advisable to continue operating with the system of water rights and hydraulic infrastructure as it has been applied so far, overlooking any adaptive measure vis-à-vis climate change;

• Irrigating-technification scenario (ITS): It values the gradual introduction of some kind of irrigating technification (by dripping, by aspersion, etc.) that leads to the optimization of the water volumes in concession for agricultural use;

- Irrigating-technification scenario plus changing of the crops (ITS+CC): It entails the gradual introduction of irrigating technification plus a shift from having highly water-demanding crops to having moderate-demanding ones, such as Sorghum, Soy, Safflower or Grass species;
- Irrigating-technification scenario plus a reduction of the cultivation areas (ITS+RCA): It poses the gradual introduction of the irrigating technification plus the gradual diminution of the cultivation areas to be irrigated.

Table 13 summarizes the results (expressed in percentages) concerning the average demand for water that is met monthly throughout the period in the irrigation districts, according to model MPIECH and scenario A2, and taking into account the adaptive measures that are proposed.

Irrigation Districts	Scenarios			
	BS	ITS	ITS+CC	ITS+RCA
Xicoténcatl 029	5.0	15.7	16.3	35.8
Río Frío	50.9	88.6	90.4	89.3
San Lorenzo	91.0	91.0	91.0	91.0
002 Mante Margen Izquierda	51.9	79.3	81.3	80.4
002 Mante Margen Derecha	41.1	57.1	58.3	63.2
Unidad de Riego 1 Las Ánimas	87.6	91.0	91.0	91.0
Unidad de Riego 2 Las Ánimas	87.6	91.0	91.0	91.0

BS=Base scenario, which includes the model's projections about climate change; ITS=Irrigating-technification scenario; ITS+CC= Irrigating-technification scenario plus changing of the crops; ITS+RCA= Irrigating-technification scenario plus a reduction of the cultivation areas.

Table 13. Results of the Adaptive Measures vis-à-vis Climate Change in the Irrigation Districts, Model MPIECH-5-A2.

It can be inferred from this information that the volumes of water in concession are higher than the natural offer of the hydric resource in the area under study. Such conclusion finds support in the fact that, even when the adaptive measures vis-à-vis climate change are considered, the levels of efficiency attained for the coverage of water demand keep being quite low indeed.

4.4. The effect of climate change on rainwater collection

As mentioned earlier, the two models that are most adverse for the main objective and purpose of capturing rainwater for irrigation would be the CCCMA-31 and the CSIRO-30, which can be observed in Figures 2 and 3, where the projections about precipitation according to different models and under scenarios A2 and B2 are presented.

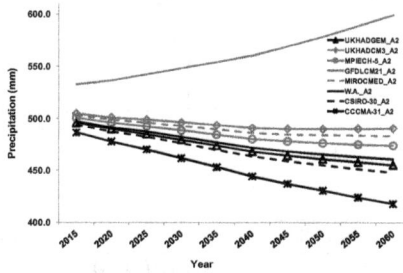

Figure 2. Projections about Precipitation according to Different Models, Scenarios A2.

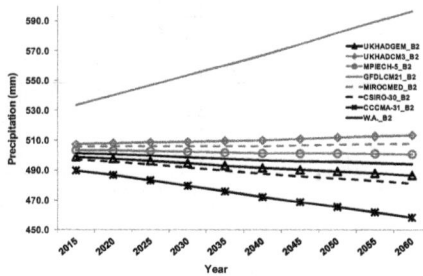

Figure 3. Projections about Precipitation according to Different Models, Scenarios B2.

Table 14 shows the findings of models CSIRO-30, CCCMA-31, UKHADGEM, and MPIECH-5 under scenarios A2, which are the most negative for rainwater collection; the periodicity is of 10 years, with stations December, January, February (DJF), March, April, May (MAM), June, July, August (JJA), September, October, November (SON); likewise, the weighted average scenario is proposed (Table 15).

	CCCMA-31_A2								
	Change (%)				*PCP (mm)£*				
Year/Station	*DJF*	*MAM*	*JJA*	*SON*	*DJF*	*MAM*	*JJA*	*SON*	*Annual*
2020	-3.5	-3.2	-7.9	0.91	28.3	56.4	272.2	123.5	480.5
2030	-4.7	-4.7	-12.6	0.08	28.0	55.5	258.5	122.5	464.5
2040	-6.6	-6.6	-17.5	-0.65	27.5	54.4	243.9	121.6	447.4
2050	-6.3	-7.6	-22.4	-0.24	27.5	53.8	229.3	122.1	432.8
2060	-5.1	-8.4	-27.5	0.57	27.9	53.4	214.3	123.1	418.7
Weighted Factor					1	2	4	3	
	CSIRO-30_A2								
	Change (%)				*PCP (mm)£*				
Year/Station	*DJF*	*MAM*	*JJA*	*SON*	*DJF*	*MAM*	*JJA*	*SON*	*Annual*
2020	-3.9	2.5	-5.0	-1.8	28.2	59.7	280.7	120.2	488.8

2030	-5.1	3.6	-8.47	-3.9	27.9	60.3	270.8	117.6	476.6
2040	-7.2	4.6	-11.9	-6.0	27.3	60.9	260.5	115.1	463.7
2050	-7.0	6.6	-15.2	-7.1	27.3	62.1	250.5	113.8	453.7
2060	-6.1	9.1	-18.7	-7.8	27.6	63.5	240.2	112.8	444.2
Weighted Factor					2	1	3	4	

				UKHADGEM_A2					
	Change (%)				PCP (mm)$^£$				
Year/Station	DJF	MAM	JJA	SON	DJF	MAM	JJA	SON	Annual
2020	-13.7	-3.6	-2.9	1.5	25.4	56.1	287.1	124.3	492.8
2030	-19.5	-5.3	-5.2	1.0	23.7	55.1	280.2	123.6	482.6
2040	-26.5	-7.4	-7.6	0.5	21.6	53.9	273.1	123.1	471.7
2050	-31.6	-8.6	-9.8	1.3	20.1	53.2	266.5	124.0	463.8
2060	-36.2	-9.6	-12.1	2.4	18.8	52.6	259.8	125.4	456.6
Weighted Factor					4	3	2	2	

				MPIECH-5_A2					
	Change (%)				PCP (mm)$^£$				
Year/Station	DJF	MAM	JJA	SON	DJF	MAM	JJA	SON	Annual
2020	-5.7	-7.1	-0.7	4.9	27.7	54.1	293.5	128.4	503.6
2030	-7.8	-10.4	-2.1	5.8	27.1	52.2	289.5	129.6	498.3
2040	-10.7	-14.3	-3.4	7.1	26.2	49.9	285.6	131.1	492.9
2050	-11.5	-17.4	-4.4	9.6	26.0	48.1	282.5	134.2	490.8
2060	-11.6	-20.3	-5.5	12.7	26.0	46.4	279.4	137.9	489.8
Weighted Factor					3	4	1	1	

Table 14. Change of Stationary Front.

	Weighted Average (WA)									
	Precipitation (mm)					Reduction (mm)$^£$				
Year/Station	DJF	MAM	JJA	SON	Anual	DJF	MAM	JJA	SON	Total
2020	26.9	55.7	279.9	122.8	485.4	-2.4	-2.5	-15.7	0.4	-20.3
2030	26.0	54.5	269.6	121.5	471.6	-3.4	-3.7	-25.9	-0.9	-34.0
2040	24.7	53.1	258.9	120.2	457.0	-4.7	-5.1	-36.7	-2.2	-48.7
2050	24.1	52.2	248.5	120.4	445.0	-5.3	-6.1	-47.1	-2.1	-60.6
2060	23.6	51.4	237.7	120.9	433.6	-5.8	-6.8	-57.9	-1.5	-72.0

$^£$With respect to the current values: 29.4 (DJF), 58.2 (MAM), 295.6 (JJA), 122.4 (SON), 505.6 (Annual).

Table 15. Change of Stationary Front, Weighted Average Scenario.

In this case the calculation of the weighted average was based on the station, according to the reduction attained by 2060; thus, for instance, for the DJF station the model that showed the highest reduction was the UKHADGEM-A2, so it was assigned the value of 4 (weighted factor), followed by models MPIECH-5-A2, CSIRO-30-A2, and CCCMA-31-A2, which were assigned the values of 3, 2, and 1, respectively. In this way, the weighted average scenario for the DJF station is built as follows:

$WA_{(DJF)}$ = (UKHADGEM-A2$_{(DJF)}$*4 + MPIECH-5-A2$_{(DFJ)}$*3 + CSIRO-30-A2$_{(DFJ)}$*2 + CCCMA-31-A2$_{(DJF)}$*1)/10

It is noticeable that in the calculus made for the weighted average the biggest reduction happens in summer (JJA), reaching a reduction of 19.6% as it goes from 295.6 mm (the current value) to 237.7 mm by 2060, i.e., 57.9 mm less in this station; by adding the reduction in all of the stations, we would get 72.0 mm, that is, 14.2% annually, which implies going from 505.6 mm at present to 433.6 mm by 2060.

After considering such amounts, it becomes manifest that there would have to be an annual reduction of 720 m³/ha or 7,200 m³ by 2060 in the project's influential area which equaled 10 ha – a loss concentrated mainly in station JJA that would attain the value of 579 m³/ha, which would amount to a decrease in water availability of 5,790m³ or 5,790,000 lt in the project's influential area.

In view of these results, it is recommended to put into effect and develop to a substantial degree programs of infrastructural design, integral management, and efficient use of hydric resources, together with the setting of appropriate calendars for irrigation that truly correspond to the species to be sown; the minimum and the maximum of available water, as well as the maximum reduction as projected by the climate change scenarios, must be taken into account. In other words, great attention should be paid to the results obtained in stations JJA and DJF, aside from assessing especially the scenario designated as the weighted average (Figure 4).

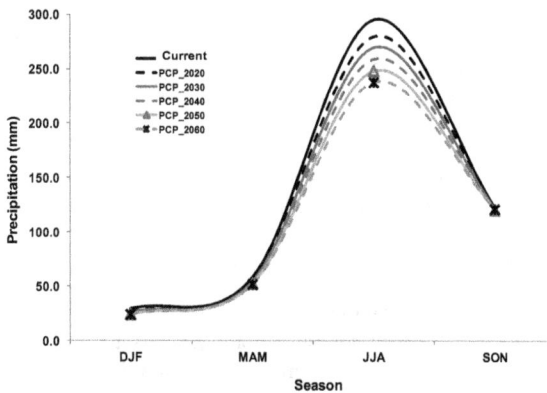

Figure 4. Projection of Stationary Precipitation, Weighted Average Scenario.

Table 16 provides the monthly results found with the models that project more extreme changes; these values can also bring support to the programs of infrastructural design, integral management, and efficient use of hydric resources.

Month	Current Precipitation (mm)	CCCMA-31-A2				CSIRO-30-A2			
		2030		2060		2030		2060	
		Change (%)	mm	Change (%)	mm	Change (%)	mm	Change (%)	mm
January	13.3	-5.2	12.6	-2.2	13.0	19.6	15.9	49.8	20.0
February	6.2	-2.1	6.1	-4.8	5.9	*-6.6*	5.8	*-14.2*	5.4
March	6.9	-3.6	6.6	-11.3	6.1	0.3	6.9	-3.1	6.6
April	15.4	*-24.3*	11.6	*-46.6*	8.2	26.3	19.4	59.7	24.5
May	36.0	2.3	36.8	10.5	39.8	-6.4	33.7	-8.0	33.1
June	91.3	*-14.4*	78.1	*-33.2*	61.0	*-25.8*	67.7	*-57.2*	39.1
July	114.9	-4.1	110.2	*-7.7*	106.1	7.0	123.0	15.6	132.9
August	89.4	*-18.4*	72.9	*-40.5*	53.2	-1.3	88.3	-4.3	85.5
September	76.6	4.5	80.0	7.0	81.9	-5.5	72.4	-13.9	66.0
October	36.4	*-20.8*	28.8	*-36.2*	23.3	*-11.1*	32.4	*-15.8*	30.7
November	9.4	22.9	11.6	43.0	13.5	11.9	10.5	19.8	11.3
December	9.8	-4.7	9.4	-4.2	9.4	*-17.3*	8.1	*-30.6*	6.8
Annual (mm)	505.6		464.9		421.3		484.2		461.9
Change (mm)			-40.7		-84.3		-21.5		-43.8
Change (%)			8.1		16.7		4.2		8.7

Table 16. Change of Monthly Precipitation, Models CCCMA-31 and CSIRO-30, Scenario A2.

We can observe in this table that the months with a higher percentage of reduction are, in descending order, April, August, October, and June, respectively, for model CCCMA-31, scenario A2, both for 2030 and 2060, whereas for model CSIRO-30 the months with a higher reduction are, in descending order, June, December, October, and February, respectively.

As the monthly results show, the decrease in annual precipitation could be of 40.7 mm (8.1%) by 2030, according to model CCCMA-31-A2, and 21.5mm (4.2%) according to model CSIRO-30-A2; by 2060 the decrease would be between 84.3 mm (16.7%) and 43.8 mm (8.7%) for models CCCMA-31-A2 and CSIRO-30-A2, respectively.

It is opportune to mention that obtaining monthly averages or totals makes no sense, as the distribution of precipitation doesn't follow a homogeneous pattern, i.e., the increase or decrease of 100% in a month with scarce precipitation may be insignificant, whereas the increase or decrease of 10% in a month with abounding precipitation can prove to be quite meaningful; thus, for instance, an increase or decrease of 50% in February and June, according to the values of current precipitation presented in Table 16, would correspond to 3.1 mm and 57.45 mm, respectively, the first one of which may be taken it to be insignificant and the second as significant.

Due to the monthly analyses of the different models, it is highly likely that the uncertainty regarding the percentage of monthly variation in precipitation will grow, insofar as there will be few coincidences and a dearth of consistency as far as the deviations' magnitude and direction are concerned; therefore, we can recommend too highly the utilization of the findings obtained through the analyses by station: Winter (DJF), Spring (MAM), Summer (JJA), and Autumn (SON) in the weighted average scenario, embodying as they do, worthy input for the task of planning all the aspects and analyzing all the variables required for the implementation of the project devoted to rainwater collection.

Aside from the changes in the average conditions, it is recommended to keep in mind the effects of extreme events, which seem to be steadily growing while their magnitude rises: such happenings may well foster avenues that run counter to the aims of the project, causing damage in the infrastructure's solidity and endurance as well as in the cultivation areas.

4.5. The effect of climate change on hydroelectric generation and the supply/Demand relation in the Sinú-Caribe River Basin in Colombia

The climatic-hydrological projections presented in Table 6 for the Sinú-Caribe River Basin point to certain changes and adverse effects on the volume that the Urrá1 Dam keeps in storage and, of course, on the generation of electric energy (Table 17, Figure 5a), as well as on the supply/demand relation of the whole basin.

By looking at the values of the first column (changes concerning the reference scenario), it becomes clear that all the models and scenarios indicate there is a reduction in the generation of electric energy that goes from 0.7% to 35.2%, while in the second column (changes concerning the maximum generation capacity) all the scenarios, including the reference scenario, indicate decreases with a range of 15.4% to 45.5%. The diminution in the Sinú River's flow contribution to the Urrá 1 Reservoir is not only directly linked to the generation of electric energy (Figure 5a) but also to the volume kept in storage by the dam (Figure 5b).

On the other hand, taking into account the demand for the hydric resource in the domestic, industrial (agricultural and livestock), and commercial realms in 28 sites located inside the basin, we find that currently the required offer (190.1 MCM) is approximately equivalent to the supply delivered at present. However, as time goes on, all the scenarios project supplies lower than the required offer: at the end of the period of 2039 the delivered supply is to be between 402.3-619.4 MCM, depending on the climate scenario under analysis, with the weighted average pointing to a supply of 563.3 MCM (Figure 6); in the meantime, the requested offer reaches a value of 636.1 MCM, which suggests a strong pressure on the system and underscores the necessity of putting the hydric resources in the basin in order and under a sensible management.

Model/Variable	Electric Generation		Storage Vol.		Flow
	Change£ %	Change∞ %	Changeᵃ %	Change_TMVᵇ %	Change %
CCSRNIES-A21	-0.7	-16.5	-2.3	-12.9	-5.9
CSIROMK2B-A21	-11.3	-25.4	-1.9	-12.6	-2.3
CGCM2-A21	-0.8	-16.5	-6.9	-17.0	-11.8
CGCM2-A22	-13.7	-27.4	-7.8	-17.8	-13.3
CGCM2-A23	-13.4	-27.1	-6.4	-16.6	-11.3
HadCM3-A21	-35.2	-45.5	-29.7	-37.3	-34.9
HadCM3-A22	-25.9	-37.7	-20.1	-28.8	-23.8
HadCM3-A23	-2.9	-18.3	-10.5	-20.2	-14.2
Weighted Average					-18.8
Referenced or Current (Average Values)	0.0	-15.9			340.3 m³/s (in the dam) 527.7 m³/s (in all the basin)

£ Calculation made concerning what is designated as the Reference Scenario (1418.9 GWh/year), which would be the beginning or current scenario.
∞ Calculation made concerning the maximum generation capacity (1687.2 GWh/year).
ᵃWith respect to the Reference Scenario, equal to 1,452.8 million cubic meters (MCM).
ᵇWith respect to the TMV: Technology Maximum Value, equal to 1,630 million cubic meters (MCM).
Source: adapted from Ospina (2009)

Table 17. Changes in the Percentage of Electric Generation, the Storage Volume in the Dam, and the Flow.

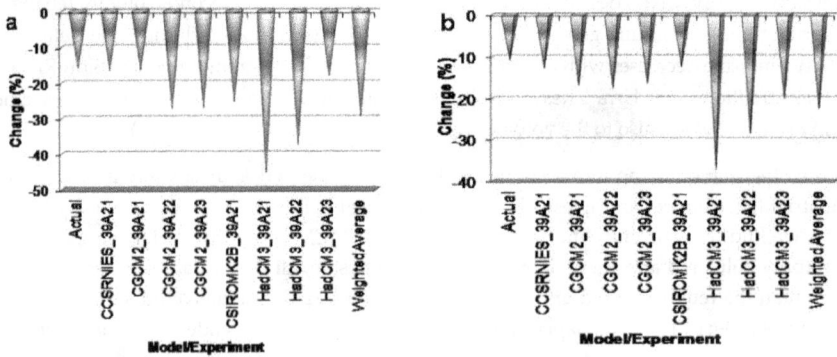

Source: Adapted from Ospina (2009).

Figure 5. Reduction (expressed in %) for the Period of 2010-2039. a) Electric Generation. b) Storage Volume in the Dam.

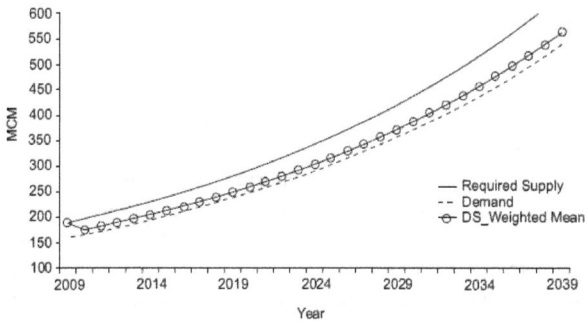

Figure 6. The Demand for the Hydric Resource in the Basin: An Account of the Required Offer and the Delivered Supply.

Although the demand (540.7 to 2039 MCM) lies below the delivered supply (563.3 MCM), it is worth noticing that the requests posed by each one of the sectors and the 28 points in demand are not met in 100%, inasmuch as such appeals do not include losses in the systems, the reutilization of water in the processes or the management in the different petitionary locations, so the required offer (631.1 MCM) is much higher, as can be observed in that figure.

In this scenario the annual reduction of the average flow in the basin, as related to the present one, would be of 18.8%, as Table 17 shows – note that for this calculation there are underlying scenarios which estimate reductions as high as 35%, aside from the fact that for its analysis it only contemplates the period of 2010-2039, the estimates for the period of 2040-2069 being more severe. Such reductions must be analyzed with great care, due to the implications they could have for the region in question, since reports from the Latin American Development Bank (Corporación Andina de Fomento [CAF, 2000]) indicate that in past events of what is designated as "El Niño" (1997-1998) the reductions in the flow of River Sinú have gone up to 33%, thus carrying deeply adverse consequences such as great losses in the agricultural production and other economic sectors, an increase in the demand for water and energy, the costs of resources for consumers, etc.

5. Participatory and adaptive approaches for water management

The effects of climate change on water resources are expected to bring multiple impacts over various sectors, whose consequences would be difficult to prevent completely and therefore to manage effectively. Therefore, from a policy perspective, policymakers are interested in identifying how to increase the capacity of organization and society to integrate uncertainty and complexity of climate change adaptation cross-sectorally. Along with water modeling tools and scenario development, described earlier in this chapter, there are other approaches to the management of water resources in situations of high uncertainty, involving adaptation and participation.

5.1. Participatory water management

Participatory water management is the involvement of stakeholders, who would not normally be involved, in different aspects of the management of their water resources. There are many different participatory methods that can be applied to increase such involvement (see Hare., *et al*, 2006). The participation of a wide range of stakeholders in management activities has been seen as important for supporting the assessment of situations of high uncertainty, from a cross-sectoral, integrated perspective (Rotmans, 1998). It has been promoted as a way of achieving societal acceptance of water management decisions (Mostert, 2003) which is of importance in those increasing cases where enhancing the capacity of local authorities, stakeholders, and interested groups to adopt and support approaches for implementing climate change actions is vital. The latter is becoming a critical policy issue (Brown., *et al*, 2010).

Participation is also being promoted actively for supporting the development of useful models and scenarios (Rotmans, 1998, Pahl-Wostl, 2002, Haag, 2001) used in resource management – for the purposes of increasing their degree of sectoral integration, quality, their validity, acceptance and use. This approach is called participatory modelling (see Hare (2011)) for an overview of the field within the water management sector). Participation is in addition viewed by researchers and policy makers as a means of bridging the policy-science-interface gap, leading to the improved use of research products in the management sector (e.g., Borowski and Hare (2007)). Those advocating a purely model-based approach to supporting adaptation in water resources management should take note of the considerable policy-science interface obstacles that exist to the water-sector adopting and using models for decision-making (see Borowski and Hare (2007), Mysiak., *et al* (2008), Webler., *et al* (2011)).

In Europe, the promotion of participatory water management is primarily driven by academia (see for example Ridder., *et al* (2005)) and a very favourable institutional enabling environment created by the EU Water Framework Directive (2000/60/EC; WFD). Article 14 (see EU (2002)) of the latter, prescribes minimum obligatory levels of stakeholder participation in the development of surface water quality management plans throughout the member countries of the union. It also seeks to encourage quite high levels of participation. The levels of participation (see Mostert (2003)) range from low level participation involving simply informing stakeholders and consulting them on already designed plans, to actively involving them in planning and, at an even higher level, in decision-making. The WFD prescribes the first two and seeks to encourage the decision-makers to actively involve stakeholders in planning. Promoting participation in actual final decision-making is not considered by the WFD, and is not common in European water management.

The WFD and associated EU research projects developing experimental high-level participatory water management processes for it (e.g. EU FP7 NeWater project – (Moellenkamp., *et al, 2010)*; EU FP7 AquaStress project – (Daniell., *et al, 2010)*) have provided a decade-long experiment in the potential uptake of participation as a serious tool of use in the water management sector. To date, the uptake of active forms of participation (i.e. going beyond informing and consulting stakeholders) has been poor outside of research-driven processes, despite the institutional enabling environment of the WFD, and despite the financial and human resources being provided by the EU through the research sector to support this uptake. The reasons for this are manifold (see, for example, the TRUST project – (Krywkow., et al, 2007)). As Hare (2011) reports, there is often a lack of personnel, skills and organizational capacities to implement such active and intensive levels of participation with multiple stakeholder groups. The statutory and reward incentives of water management also often work against managers risking their reputations and perhaps jobs on developing participatory processes that might get out of control or slow down implementation of vital infrastructure projects, in cases where funding is time-dependent (op. cit). Videira., et al (2006) also suggest that the low level of prescribed participation in the WFD itself restricts the incentives for water managers to go beyond what is prescribed – why risk your job etc., when higher levels are not prescribed? Finally, as Borowski and Hare (2007) identified in their research – stakeholders - practitioners, policymakers and decision-makers - often do not have the time to participate in what will normally be time-intensive processes. If the water managers are not inclined to adopt high levels of participatory processes outside a research project, and stakeholders may often not have enough time, active, high levels of participation within the water management sector remains an aspiration.

This remains a critical problem for approaches to managing uncertainty in the face of climate change, such as adaptive management (Mysiak., et al, 2010) and adaptive governance (Huitema., et al, 2009) in which active stakeholder participation is a central component. Fortunately, as Hare (2011) suggests, these approaches, when considered from the perspective of niches, also contain a potential solution to the dilemma, which allows both adaptation and participation to be taken up and implemented. This solution will be returned to at the end of the chapter. First it is important to discuss adaptive management and adaptive governance and their role in managing water resources in times of climate change.

5.2. Adaptive management and adaptive governance

Previous experiences using a participatory water management approach keep being very valuable when trying to mainstreaming climate change adaptation into water resources management. Mainstreaming climate change adaptation is understood as a process of cross-sectoral integration of policy through common policy-making on adaptation which facilitate changes in socio-economic systems and governance regimes, aiming at dealing with uncertainty and capturing the opportunities for synergistic results in terms of increased adaptive capacity and lower vulnerability. However, in order to be able to continually maintain the experience needed to cope with uncertainty and change, the integration of actions and policies for water management needs to increase its own adaptive capacity, too. Various experiences from the literature offer valuable examples on the way in which an adaptive approach has provided a systematic perspective that lead a management system to undertake the necessary adjustments in its structure, function, and performance under varying conditions (Knieper., *et al*, 2010, Pahl-Wostl., *et al*, 2010, Schlüter., *et al*, 2010). Therefore, such a system is better able to deal with complexity and increasing uncertainty.

In a similar vein, Peña del Valle et al, (in submission) have argued that mainstreaming climate change adaptation could succeed in all sectors´ processes if the very practice of mainstreaming climate change adaptation is conceived and undertaken under an adaptive approach. Thus, according to the authors, the use of a conceptual framework based on knowledge from the fields of adaptive water resources management and adaptive governance can help to identify steps to increase the adaptive capacity of managerial and governance regimes. Such increases can be charted along a learning pathway composed of learning cycles, operationalised in terms of 1st, 2nd, and 3rd loop learning activities. These, impact on four specific types of capacities, which are central to a social system to adapt. These are: 1) capacities for multi-stakeholder involvement, 2) integrated analysis, 3) experimentation, and 4) flexible institutions.

Multi-stakeholders involvement: Bringing multi-stakeholders and organizations to work together is central in the process of mainstreaming climate change adaptation through a cross-sectoral perspective (Peña del Valle et al, in submission). By identifying common issues and material inter-linkages between each others' sectors and, by sharing particular concerns and interest, stakeholders are more able to increase their awareness of other sector's issues, along with their constrictions and potentials. This process, however, has to be supported by adequate institutions, which can undertake several adjustments in their practice definition, role's assignment, as well as in their interactions and procedures. It is thought that such actions would result in moving from one decision-making authority to a broader involvement of different stakeholders into the formulation of flexible climate change adaptation policy.

Integrated information and analysis: The extent to which climate change adaptation can match its goals with other policy relevant areas, such as poverty, development, energy security, and health may depend on to what extent other sector's processes and perspectives convey with those of strategic planning and administrative procedures. This requires building an integrated scheme of knowledge and shared information systems, which facilitates the identification of overlapping areas and gaps. Stakeholders can advance together in building up an comprehensive analysis that includes different perspectives, interests, and expected outcomes.

Continuous experimentation in policy-making: Building collective initiatives and common problem-solving solutions may not be sufficient for mainstreaming climate change adaptation all at once given the uncertainty associated with climate change processes. Therefore, it is also necessary to keep up a process of learning. This relates to the notions expressed by various authors, where systems undergo a process of continual reorganization in their goals and methods in order to keep creative, flexible and novel in their approach to problem-solving (Pahl-Wostl and Hare, 2004, Armitage., *et al*, 2008). In such a way, governance and managerial systems can evolve their knowledge on climate change adaptation, so that they can rapidly respond to a range of different situations and needs, but always linked to current socio-economic treats and vulnerabilities, at various levels.

Flexible institutions: A central outcome of a continuous learning process between stakeholders and organizations to address climate change adaptation is the development of customary institutions to the evolution of more open institutional schemes. This implies, for instance, that societies and organizations are more able to modify the sets of rules that regulate social behaviours, actions and motivations. As organization's structures become more flexible, they are also more able to admit modifications on a regular basis; decision-making processes are more inclusive and diverse to accommodate uncertainty and continuous disturbance in problem-solving and decision-making practices. In such a way, mainstreaming climate change adaptation can be undertaken under an enabled environment, where processes of social-networking, multi-level governance, continuous feedback facilitate the delivery of useful policy recommendations to climate change adaptation in a timely and effective manner.

5.3. Learning cycles

Various authors have already shown the benefits of relying on social and societal learning process for improving adaptive water management and, for building adaptive capacity (Pahl-Wostl, 2009, Lebel., *et al*, 2010) in governance regimes to deal with uncertainty associated with climate change.

The use of learning loops to illustrate progressions in social learning processes has been developed in diverse areas (Armitage., *et al*, 2008). The single loop refers to a refinement of current methodological pathways and established practices; the double loop involves a redefinition of assumptions and strategies for institutional participation and shared benefits; in the triple loop there is a transformation of underlying values and common regulative and normative frameworks for thinking and action. Thus, it is expected that the evolution of the learning process in stages or learning-loops would help to identify the sort of capacities and processes that support moving a management regime through the different stages of learning and towards overcoming the barriers to mainstreaming.

As Peña del Valle et al (in submission) commented, multi-stakeholder involvement, a key capacity for adaptive management, can be seen as a key capacity for mainstreaming, since having this capacity facilitates the involvement of cross-sectoral policy makers in integrated policy development for adaptation. In a mainstreaming context therefore, the single loop learning activity for strengthening multi-stakeholder involvement would be to "bring together cross-sectoral policy makers to share their adaptation plans" at regular time intervals. There is no compunction for the policy makers to act upon what they learn at such meetings, but it can strengthen the level of multi-stakeholder involvement in planning, albeit indirectly, if what is learned by each policy maker about others' plans is made use of in their own planning. The double loop learning activity, on the other hand, would "bring the policy makers together to create and identify common goals for adaptation planning". The desired outcome would be a stronger multi-stakeholder involvement in the plans of each policy sector in terms of there now existing shared goals to be the used (voluntarily) as the basis of any planning. The triple loop learning activity, finally, would be for the policy makers to work together to agree on the "creation of new institutions prescribing how each policy sector shall work together to develop joint policy on adaptation".

5.4. Participation and adaptation in support of each other

From the approaches developed above, participatory process of policy integration across sectors can facilitate changes in socio-economic systems and governance regimes. In the field of adaptive water management this would also imply to count with a handy approach to continuously dealing with uncertainty and capturing the opportunities for synergistic results in terms of increased adaptive capacity and lower vulnerability.

Adaptive management and governance can also provide an environment in which high levels of participation can be adopted by water managers to properly support them to manage water resources in the face of climate change uncertainty; but only if one gives up on the idea of placing participation at the heart of formal decision-making and planning, whose current institutions provide few incentives to the adopt active levels of participation. The way forward, as Hare (2011) suggests for participatory modelling, is shown by Moellenkamp., et al (2010) who observed that water managers were more willing to accept higher levels of stakeholder involvement if it took place in niches adjacent to the formal planning cycle (see above). Such an "adjacent" niche might manifest itself as a group of stakeholders who actively participate and collaborate in a series of workshops and other activities focused on developing new ideas for management, with the blessing of the decision-maker but with no formal attachment to the planning cycle (e.g. op cit). In this way the planning cycle stays insulated from the risks of failure associated with participation, and participation can thus be taken up by water managers in this controlled environment with little danger. Those niches, in which stakeholders can actively design and experiment with cutting edge management ideas, can then be connected to the learning cycle as part of an adaptive management process, and participation can support adaptation in return: As and when the formal planning process requires new ideas due to a perceived failure in planning, it can turn to the multi-stakeholder participatory process in the learning cycle niche which can then be used to power the type of reflection and adaptation needed for effective adaptive management and governance (see Figure 1, Hare (2011)). The case study in Moellenkamp., et al (2010) is a demonstration of the potential for such niches to impact on the policy cycle if they are set up properly.

6. Conclusions and recommendations

As the present chapter illustrates, the changes in magnitude and direction of the climate variables can be expressed in various ways according to the different regions of the world, which makes it necessary to provide models and regional as well as local climate scenarios.

In hydrological modeling it is required to produce a scale-reduced design of global replicas and regional copies, to practice studies and analyses at different levels (i.e., those related to regions, localities, and basins), and to undertake the analysis and management of uncertainty.

From what has been discussed here, it can be concluded that climate change will prove to have negative effects on water availability, which in turn will have deep consequences for different sectors, projects, and activities, among which are found not only irrigation, the generation of electricity, and the collection of rainwater (to which we have made reference in this chapter), but also many others (e.g., the economy and development domains). Therefore, it is indispensable to continue examining and analyzing the effects of climate change on hydrological resources and contributing sets of hydrological modeling.

Attached Document I

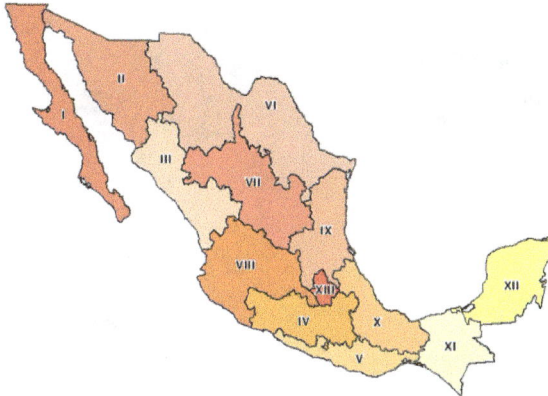

Hydrological-Administrative Regions of the CNA

Attached Document II

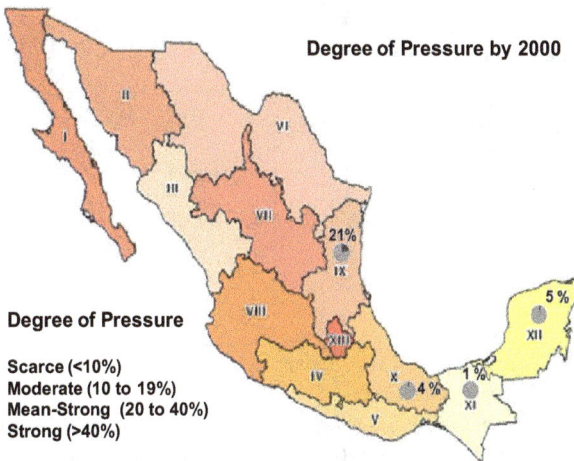

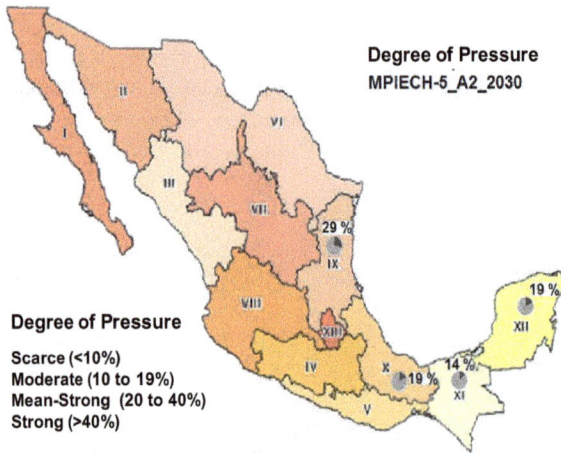

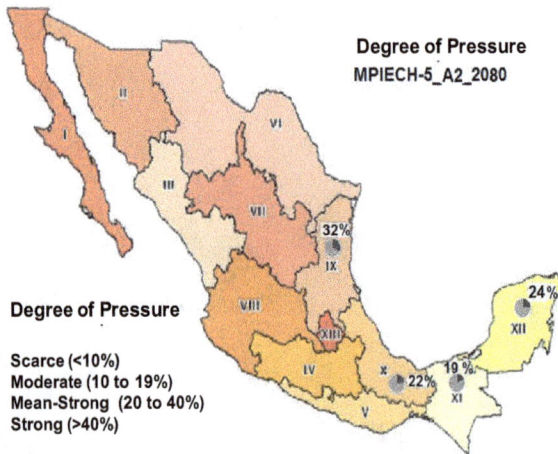

Evolution of the Degree of Pressure, Model MPIECH-5, Scenario A2.
Source: Sánchez-Torres., *et al* (2011).

Author details

Jesús Efren Ospina-Noreña, Carlos Gay García,
Ana Elisa Peña del Valle and Matt Hare
*Programa de Investigación en Cambio Climático (PINCC) [Climate Change Research Program] of the
Universidad Nacional Autónoma de México (UNAM) [National Autonomous University of Mexico]*

Acknowledgement

The authors wish to thank Ingenieros Sin Fronteras México, A.C., and the Instituto Tierra y Cal, A.C., for allowing them to submit the preliminary findings of the project on Rainwater Collection they have been developing.

7. References

Armitage, D., M. Marschke, and R. Plummer. 2008. Adaptive co-management and the paradox of learning. Global Environmental Change 18:86–98.

Borowski I, Hare MP. 2007. Exploring the gap between water managers and researchers: difficulties of model-based tools to support practical water management. Water Resources Management 21: 1049–1074.

Brown, PR, Nelson R, Jacobs B, Kokic P, Tracey J, Ahmed M, DeVoil P (2010). Enabling natural resource managers to self-assess their adaptive capacity. Agricultural Systems 103, 562-568.

CAF (Corporación Andina de Fomento). 2000. El Fenómeno El Niño 1997- 1998 Memoria, Retos y Soluciones Volumen III: COLOMBIA. Caracas, VE. 232pp.

Conde, C. 2003. Vulnerabilidad y Adaptación al Cambio y a la Variabilidad Climática. Conceptos y Métodos Básicos.

Daniell KA, White I, Ferrand N, Ribarova IS, Coad P, Rougier J-E, Hare MP, Jones NA, Popova A, Rollin D, Perez P, Burn S. 2010. Co-engineering participatory water management processes: theory and insights from Australian and Bulgarian interventions. Ecology and Society 15: 11.
http://www.ecologyandsociety.org/vol15/iss4/art11/.

EU. 2002. Guidance on public participation in relation to the Water Framework Directive. Prepared in the framework of the Common Implementation Strategy of the European Commission and the EU Member States.
http://forum.europa.eu.int/public/irc/env/wfd/library.

Gay, C. 2000. México: Una Visión hacia el siglo XXI. El Cambio Climático en México. Resultados de los Estudios de Vulnerabilidad del País Coordinados por el INE con el Apoyo del U.S. Country Studies Program. SEMARNAP, UNAM, USCSP.220 pp.

Haag D, Kaupenjohann M. 2001. Parameters, prediction, post-normal science and the precautionary principle – a roadmap for modelling decisionmaking. Ecological Modelling 144: 45–60.

Hare, M.P. 2011. Forms of Participatory Modelling and its Potential for Widespread Adoption in the Water Sector. Environmental Policy and Governance 21:386–402

Hare, M.P., Barreteau, O. Beck, M.B., et al. 2006. Methods for stakeholder participation in water management. In Giupponi, C. Jakeman, A.J., Karssenberg, D. & Hare, M. (Eds) Sustainable management of water resources: an integrated approach. Edward Elgar: Chichester.

Huitema, D., E. Mostert, W. Egas, S. Moellenkamp, C. Pahl-Wostl, and R. Yalcin. 2009. Adaptive water governance: Assessing adaptive management from a governance perspective. Ecology and Society 4(1): 26.

Hulme, et al. 2000. Using Climate Scenario Generator for Vulnerability and Adaptation Assessment: MAGICC and SCENGEN. Version 2.4 Workbook, Climate Research Unit, Norwich. UK, 52 pp.

IPCC. 2000. Informe Especial del IPCC. Escenarios de Emisiones. Grupo de Trabajo III.

Knieper, C., Holtz, G., Kastens, B., Pahl-Wostl, C., 2010. Analysing water governance in heterogeneous case studies – Experiences with a database approach. Environmental Science and Policy 13 (7), 592–603.

Krywkow J, Rasche K, Moss E, Mitchell T, Vancleemput K, van der Kroef R, Hotting R, Rodenbach A, Noordanus C. 2007. Public and stakeholder participation. Theme Group III Final Report of the TRUST Project, TRUST: Rotterdam.

Lebel, L., Grothmann, T. & Siebenhüner, B. 2010. The role of social learning in adaptiveness: insights from water management. International Environmental Agreements: Politics, Law and Economics, 10(4).

Moellenkamp S, Lamers M, Huesmann C, Rotter S, Pahl-Wostl C, Speil K, Pohl W. 2010. Informal participatory platforms for adaptive management – insights into niche finding, collaborative design and outcomes from a participatory process in the Rhine. Ecology and Society 15: 41.http://www.ecologyandsociety.org/vol15/iss4/art41/.

Mostert E. 2003 The Challenge of public participation. Water Policy 5: 179-197

Mysiak J, Giupponi C, Depietri Y, Colombini G. 2008. A note on attitudes towards and expectation from the Decision Support Systems. In Proceedings of the Fourth Biennial Conference of the International Environmental Modelling and Software Society (iEMSs), Sànchez-Marrè M, Béjar J, Comas J, Rizzoli A, Guariso G (eds). iEMSs: Barcelona;925–931.

Mysiak J, Henriksen H-J, Sullivan C, Bromley J, Pahl-Wostl C. (eds). 2010. The Adaptive Water Resource Management Handbook. Earthscan: London.

Ospina J. E., 2009. Efectos del cambio climático en la generación hidroeléctrica con énfasis en proyecciones de generación-transmisión eléctrica en Colombia. Ph. D. Thesis on Atmospheric Physics. Programa de Posgrado en Ciencias de la Tierra. Universidad Nacional Autónoma de México. México D.F., 206 pp.

Ospina-Noreña J.E., C. Gay., C. Conde. and G. Sánchez, 2009a. "Analysis of the Water Supply-Demand Relationship in the Sinú-Caribe Basin, Colombia, Under Different Climate Change Scenarios". Atmósfera 22(4), 331-348.

Ospina-Noreña J.E., C. Gay., C. Conde. and G. Sánchez. 2011a. "A Proposal for a Vulnerability Index for Hydroelectricity Generation in the Face of Potential Climate Change in Colombia". Atmósfera 24(3), 329-346.

Ospina-Noreña J.E., C. Gay., C. Conde. and G. Sánchez. 2011b. "Water Availability as a Limiting Factor and Optimization of Hydropower Generation as an Adaptation

Strategy to Climate Change in the Sinú-Caribe River Basin". Atmósfera 24(2), 203-220.

Ospina-Noreña J.E., C. Gay., C. Conde., V. Magaña. and G. Sánchez, 2009b. "Vulnerability of Water Resources in the Face of Potential Climate Change: Generation of Hydroelectric Power in Colombia". Rev. Atmósfera. N° 22. Vol. 3. pp 229-252.

Ospina-Noreña, J.E., G. Sánchez Torres Esqueda y C. Conde Álvarez, 2010. Impactos del cambio climático en las regiones hidrológicas del Golfo de México, p. 73-88. En: E. Rivera-Arriaga, I. Azuz-Adeath, L. Alpuche Gual y G.J. Villalobos-Zapata (eds.). Cambio Climático en México un Enfoque Costero-Marino. Universidad Autónoma de Campeche Cetys-Universidad, Gobierno del Estado de Campeche. 944 p.

Pahl-Wostl C. 2002. Participative and Stakeholder-Based Policy Design, Evaluation and Modelling Processes. Integrated Assessment 3 (1):3-14.

Pahl-Wostl, C, Holtz, G., Kastens, B., Knieper, C., 2010. Analysing complex water governance regimes: the Management and Transition Framework. Environmental Science and Policy 13 (7), 571–581.

Pahl-Wostl, C., 2009. A conceptual framework for analyzing adaptive capacity and multi-level learning processes in resource governance regimes. Global Environmental Change 18, 354–365.

Pahl-Wostl. C. & Hare, M.P. 2004. Processes of social learning in integrated resources management. Journal of Community and Applied Social Psychology. 14: 193-206.

Peña del Valle, A.E., C. Gay, and M. Hare. (in submission). Using an adaptive approach for mainstreaming climate change adaptation into policy. Environmental Science and Policy.

Ridder D, Mostert A, Wolters H. 2005. Learning Together to Manage Together. Improving Participation in Water Management. HarmoniCOP Handbook on Social Learning. University of Osnabrück, Institute of Environmental Systems Research: Osnabrück.

Rotmans, J. 1998. Methods for IA: The challenges and opportunities ahead. Environmental Modelling and Assessment 3:155-180.

Sánchez-Torres., G., J.E. Ospina-Noreña, C. Gay. and C. Conde. 2011. "Vulnerability of water resources to climate change scenarios. Impacts on the irrigation districts in the Guayalejo-Tamesí river basin, Tamaulipas, México". Atmósfera 24 (1), 141-155.

Schlüter, M., Hirsch, D., Pahl-Wostl, C., 2010. Coping with change – responses of the Uzbek water management regime to socio-economic transition and global change. Environmental Science and Policy 13 (7), 620–636.

Urbano Terrón, P. (1995). "Tratado de fitotecnia general", 2ª edición, Ed. Mundi – Prensa, Bilbao, 885 p.

Videira N, Antunes P, Santos R, Lobo G. 2006. Public and stakeholder participation in European water policy: a critical review of project evaluation processes. Environmental Policy and Governance 16: 19–31.

Webler, T., Tuler, S., Dietz, T. 2011. Modellers' and Outreach Professionals' Views on the Role of Models in Watershed Management. Environmental Policy and Governance 21: 472-486.

Wigley, T.M.L. 1994. MAGICC (Model for Assessment of Greenhouse –gas Induced Climate Change): User's Guide and Scientific Research Manual. National Center for Atmospheric Research, Boulde.

Wigley, T.M.L., 2003. MAGICC/SCENGEN: User-friendly software for GCM inter-comparisons, climate scenario development and uncertainty assessment. Tom M.L. Wigley, National Center for Atmospheric Research, Boulder, CO 80307.

Chapter 7

Sustainable Utilisation of Groundwater Resources Under Climate Change: A Case Study of the Table Mountain Group Aquifer of South Africa

Anthony A. Duah and Yongxin Xu

Additional information is available at the end of the chapter

1. Introduction

In a global sense, the term "sustainability" may be applied to the environment (ecological sustainability), society (social sustainability), the economy (economic sustainability) or an organization or people (organizational or human sustainability respectively). To understand the concept one needs to identify the focus or what needs to be sustained then one can work out how to sustain the thing or the condition. The United Nations Brundtland Report (1987) defined "Sustainable Development" as development that meets the needs of the present without compromising the ability of future generations to meet their own needs. The United Nations Conference on Environment and Development (UNCED) held in June, 1992 in Rio de Janeiro, became the symbol for the common responsibility of all Governments in the world in achieving a sustainable development. The conference stated that, "The holistic management of freshwater as a finite and vulnerable resource, and the integration of sectorial water plans and programs within the framework of national economic and social policy, is of paramount importance for actions beyond the 1990s. Integrated water resources management (IWRM) is based on the perception of water as an integral part of the ecosystem, a natural resource and social and economic good". The Outcome Document of the 2005 United Nations World Summit in New York referred to the interdependent and mutually reinforcing pillars of sustainable development as economic development, social development, and environmental protection. Why sustainability? Sustainability ensures that resources especially natural resources are kept within nature's ability to replenish them.

A sustainable water service delivery is the thrust of the Millennium Development Goals on water for African governments. With 75 percent of African population using groundwater as its main source of drinking water (ECA *et al.*, 2000) and about 300 million people in sub-Saharan Africa still without access to safe water supplies – approximately 80 percent of them live in rural areas (Xu & Braune, 2010), not only will a sustainable management of the groundwater resources benefit countless populations but also provide hope for improved quality of life. Fractured rock aquifers account for about 40% of all aquifers and maintains livelihood for more than 200 million people in Africa (Xu and Braune, 2010). They are found extensively in sub-Saharan Africa and play a vital role in meeting MDGs on water in the southern, west and east Africa due to their availability and portability. The Table Mountain Group (TMG) aquifer is a major fractured rock aquifer system in South Africa which has already proven to be a bulk water supply for agriculture, industry and domestic use in the Eastern and Western Cape provinces.

2. Background

The Table Mountain Group (TMG) aquifer is a major fractured rock aquifer system in South Africa which has already proven to be a bulk water supply for agriculture, industry and domestic use in the Eastern and Western Cape provinces.

The reported decline of groundwater levels in the TMG aquifer system in the little Karoo (Klein Karoo Rural Water Supply Scheme, KKRWSS) area since 1984 as a result of pumping from production wells in the Vermaaks River Catchment (Jolly & Kotze, 2002; Wu, 2005) has called for sustainable measures in the management practices at the Well fields. The reduction of pumping rates in an attempt to halt the decline in water levels could not arrest the situation because it is believed that there has been an over-estimation of the recharge rates in the TMG and for that matter the catchment area. Jolly (2002) reported that some shallow TMG boreholes have been pumped at rates in excess of 30 *l*/s. Other reported abstraction schemes within the TMG included the Arabella Country Estate, near Botrivier where 4 production wells supply about 30 *l*/s of groundwater to the estate (Parsons, 2002); Botrivier water supply project with 6 wells supplying about 20 *l*/s (Weaver, 1999); Ceres Municipality wellfields where 7 boreholes from the TMG supply 48 l/s and 3 boreholes from the Bokkeveld supply 50 *l*/s (Rosewarne, 2002); the Hex River valley project and the CAGE project at Citrusdal are also documented. Some of these drilling projects have little or no management plans while others are well managed even though ecological and environmental concerns are hardly integrated in the management programmes. The impact of these groundwater projects on the surface water regimes i.e. the baseflow into rivers, wetlands and estuaries which also control the availability of water for plant use are hardly evaluated or monitored.

2.1. Literature review

Custodio (2005) refers sustainability to the use of natural resources without jeopardizing their use by future generations. According to him it goes along with the concept of human beings living in peace and harmony with the environment, both now and in the future. He strongly believes that in reality, scientific, technical, and social as well as space and time frameworks, under which sustainability is evaluated, are continuously changing and that what may be sustainable now may not be in the future, and what may appear unsustainable today may be sustainable in the future. Present-day decisions must therefore take the future into consideration but the influence given to the future, however, will depend on the credibility of scenarios considered and the weight given by society and politicians.

Devlin and Sophocleous (2005) however differentiate between sustainability and sustainable pumping rates as two different concepts that are often misunderstood and therefore used interchangeably. They argued that the latter term refers to a pumping rate that can be maintained indefinitely without mining an aquifer, whereas the former term is broader and concerns such issues as ecology and water quality, among others, in addition to sustainable pumping. Another important difference between the two concepts according to Devlin and Sophocleous is that recharge can be very important to consider when assessing sustainability, but it is not necessary to estimate sustainable pumping rates. Sophocleous (2000) had reported that, in the past, the volume of recharge to an aquifer was accepted as the quantity of water that could be removed from the aquifer on a sustainable basis, the so-called safe yield, but it is now understood that the sustainable yield of an aquifer must be considerably less than recharge, if adequate amounts of water are to be available to sustain both the quantity and quality of streams, springs, wetlands, and groundwater-dependent ecosystems. Sustainable resource management demands the managing of groundwater for both present and future generations, and providing adequate quantities of water for the environment and thus quantifying what these environmental provisions are is presently an urgent research need (*Sophocleous*, 2000). Sophocleous (2005) again affirms that sustainable use of groundwater must ensure not only that the future resource is not threatened by overuse and depletion, but also that the natural environment that depend on the resource are protected. One agrees with his opinion that there will always be trade-offs between groundwater use and potential environmental impacts, and therefore a balanced approach to water use between developmental and environmental requirements needs to be advocated. However, to properly manage groundwater resources, managers need accurate information about the inputs (recharge) and outputs (pumpage and natural discharge) within each groundwater basin, so that the long-term behavior of the aquifer and its sustainable yield can be estimated or reassessed. Thus, without a good estimate of recharge, the impacts of withdrawing groundwater from an aquifer cannot be properly assessed, and the long-term behavior of an aquifer under various management schemes cannot be reliably estimated (*Sophocleous, 2005*).

2.2. Objective of the study

The study addresses the measures that are essential and relevant in sustainable utilitzation of Groundwater Resources in line with the Africa Water Vision thus:

To evaluate the factors and variables determining the sustainable utilization of the TMG regional aquifer using the KKRWSS as a case study, i.e. to determine the balance between recharge to and discharges from the Wellfield that will ensure a reasonable sustainability with minimum or no adverse impacts on the environment. This is done in the context of climate change and climate variability.

2.3. Previous work

The majority of aquifer-tests conducted in the TMG aquifers are aimed at obtaining a first-order estimate of the sustainable yield of a production borehole, as well as the design of the pump equipment and abstraction schedule. Various evaluation methods are used by geohydrologists with varied degrees of success. The case study of the Little Karoo Rural Water Supply Scheme - borehole safe-yield versus the sustainable yield of the aquifer as reported by Woodford (2002) provides a test case for the sustainable development of the TMG aquifer. The gross overestimation of the long-term supply potential of the Vermaaks River wellfield when relying solely upon conventional methods of aquifer-test analysis serves to highlight a problem that is currently being experienced by many groundwater practitioners working in the TMG fractured-rock aquifers and has led some to question the value of such tests (Woodford, 2002). In concluding his report on the interpretation and applicability of pumping-tests in TMG aquifers, Woodford (2002) suggested further research in order to improve the aquifer-testing and analysis techniques and thereby the understanding of the flow dynamics of the TMG.

Jolly (2002) reported that the unscientific testing and evaluation of boreholes drilled in TMG aquifers has often created a false impression of the aquifer's long term sustainable potential. Often 'blow yields' measured at the end of drilling have been mistakenly taken as borehole yields. He affirmed that even the normal scientific assessment of the borehole's potential via step and 72 hour constant rate tests can grossly over-estimate sustainable yield, if assessments do not take into account issues like existing boundaries and matrix transmissivity and storativity. He believes that many schemes have failed because the abstraction has exceeded recharge, resulting in water levels declining to the depth of the pump. He however, advised that aquifer storage must be utilized before the next recharge event topped up the aquifer because not only is recharge low but it is also sporadic. According to Jolly (2002), the ultimate cause of borehole or wellfield failure in the TMG aquifers is the poor management of the resource and suggests that the decline in water levels must be carefully managed to make certain that water levels do not drop below the top water strikes in the hole. He believes the storativity in the TMG is lower than traditionally expected.

3. Methodology

3.1. Description of current research

In the current study the Vermaaks Well field Water Supply Scheme was used as a case study to determine the sustainable utilization of groundwater resources in the TMG.

The following analyses were done in reaching the desired objective:

- Determination of the recharge to the well field using different methods and comparing them with those obtained by previous studies;
- Analysis of the long-term climate trends in the study area on the backdrop of climate change and potential impact on groundwater resources of the TMG;
- Evaluate the abstraction rates of the production wells in the well field and the long-term effect on groundwater levels;
- Assess the effect of abstraction on hydrological and environmental resources (rivers, springs and groundwater dependent ecosystems).

3.2. The study area

The study area is located in the Little Karoo area of the TMG in the Western Cape Province (Figure 1). The project area between the towns of Calitzdorp and De Rust comprises a broad valley, with an elevation of approximately 500 m (amsl) and surrounded by mountain ranges, the Kammanassie Mountain range with elevation of up to 1950 m (amsl) on the east and the great Swartberg Mountain range in the north of up to 2150 m (amsl). The Rooiberg Mountains occur in the western part of the scheme and down south is the Outeniqua Mountains. The area is drained by two perennial rivers, the Olifants River to the north of the Kammanassie Mountain range and the Kammanassie River to the south. One minor but important river, the Vermaaks River drains the Vermaaks River wellfield which is the most important wellfield of the KKRWSS. The Marnewicks River drains the eastern part of the Vermaaks River wellfield. The two minor rivers are ephemeral in the steep upper reaches, with more sustained flow in the lower reaches, and drain northward into the Olifants River. Runoff from the mountains is captured in a number of dams and used for irrigation and water supply to Oudtshoorn. The area falls within seven quaternary catchments that controls the surface water drainage however, the groundwater flow regime is controlled by the boundaries of the geological formations. For example, the Cedarberg formation which is an aquitard and the contact between the Nardouw Subgroup and the Bokkeveld Group act as groundwater flow barriers. There is however an active interaction between the surface water and the groundwater regimes. Jia (2007) calculated the mean annual baseflow as approximately 22% of the mean annual river flow for this area. With the groundwater level decline over the years interaction will be restricted between rivers and the water in the shallow weathered zone.

The KKRWSS has two sections, the Western Section at Calitzdorp and the Eastern Section at the Kammanassie Mountain area near Dysselsdorp. The Eastern Section is the most productive section of the scheme and also where the highest declines in water levels have been recorded. The KKRWSS was designed to supply up to 4.7×10^6 m³/a of groundwater from the two sections. The eastern section initially had 13 boreholes of which 5 constituted the Vermaaks River wellfield. There are now 4 production boreholes in the Vermaaks River wellfield. The western section at Calitzdorp had 5 boreholes. Some 400 km of pipeline delivered the groundwater to two purification plants at Dysselsdorp and Calitzdorp before it is supplied to end-users (Jolly and Kotze, 2002).

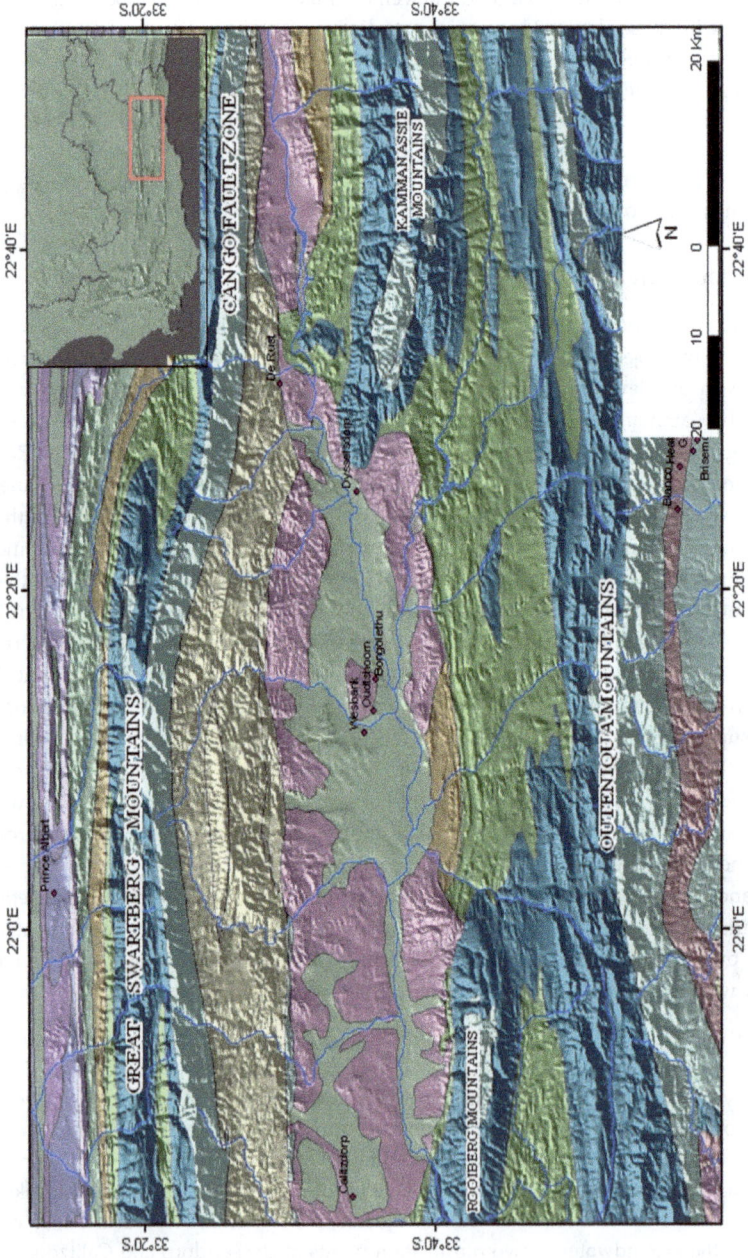

Figure 1. Location of KKRWSS area and Kammanassie Mountains

4. Results and discussions

4.1. Recharge to aquifers

Scanlon *et al.* (2002) and Xu and Beekman (2003) presented a comprehensive review on choosing the appropriate technique for quantifying groundwater recharge, including the applicable space and time scales, the range of recharge rates that have been estimated with each method, the reliability of the estimates, and important factors that promote or limit the use of the various methods have been outlined.

A comprehensive review and analysis of four methods namely, Chloride Mass Balance (CMB), Cumulative Rainfall Departure (CRD), spring flow and water balance, were evaluated and integrated resulting in a recommended recharge range of 1.63% - 4.75% of MAP for the Kammanassie area. The major factors limiting the accuracy of these recharge estimation methods were identified as the accurate measurements of the recharge area and aquifer storativity among other factors. These and other factors still present potential errors in recharge estimation especially for a fractured rock aquifer like the TMG. The results ranged from 0.2% to 12% of MAP. The upper range of values were derived from integrated approach such as water balance methods while the lower range of values were obtained from methods such as the CMB and CRD. For a given study area, the recharge rate is recharge-area dependent for the CMB mixing model and regression of Cumulative Flow (CF) methods, while the CRD method is dependent on the storativity of the aquifer. Errors would therefore depend on how accurate the recharge area and storativity values are obtained. Recharge rates are less than 5% of precipitation in contrast to the 15%-20% and over proposed by earlier researchers. The results indicate that the recharge rate varies from 0.24% to 7.56% of the MAP using a storativity of between 0.0001 and 0.001 with the CRD method. The average recharge rate was 5.38% of MAP or 48.67 mm, which equals to 3.88% of total precipitation of 1256 mm under storativity of 0.001 in considering impact of preceding rainfall. Most high recharge percentages are related to rainfall of 300 mm/a to 1100 mm/a. The recharge in terms of percentage of rainfall is lower if precipitation is greater than 1100 mm/a or less than 300 mm/a.

Recharge estimation methods have their uncertainties and inaccuracies arising from spatial and temporal variability in processes and parameter estimations, measurement errors and validity of assumptions. However, Xu *et al.* (2007) concluded that the following methods have been used in the study area with much certainty based on their reliability in space and time scales in the sub-region spanning a period of three decades: the Chloride Mass Balance (CMB), Cumulative Rainfall Departure (CRD), Rainfall Infiltration Breakthrough (RIB), Water Balance (WB) and Regression of Spring Flux.

Results from the several recharge studies conducted in the TMG area over the years by different individuals and groups of people using different methods indicated varying estimates in the range of 1% to 55% of MAP. Methods used included CMB, SVF, CRD, EARTH, Base Flow, Isotopes, Water Balance and GIS. Wide variations in estimates ranged between 165 and 2020 mm per annum from these methods however, it has been accepted that the scale of the recharge study often dictates the most appropriate methods to be used to determine aquifer recharge. Concerns were raised for review of recharge estimates in the Kammanassie area due to the continual decline of groundwater levels in the Vermaaks River Wellfield from 1994. The recharge estimate of 17% of MAP in the Wellfield was adjusted downwards several times to arrest the situation.

4.2. Climate trend in the study area

Temperature increases are worldwide however; precipitation decreases or increases are highly variable in time and space. The Mann-Kendall Trend test was applied on historical temperature and rainfall data records over several years. The Trend tests in the study area did show gradual increases in temperature over recent years while generally no significant trends have been observed with precipitation amounts in the TMG area generally. Using the Standard Precipitation Index (SPI) analysis variations in precipitation patterns were observed such as the intensity, duration and shifts in seasons. The SPI analysis clearly showed that there has been a persistent low rainfall in the KKRWSS area in particular since 2004. A particularly significant drought period was observed between 2004 and 2006 for three stations whiles in the fourth, Purification Works East, the drought is reflected between 2007 and 2010 period (figure 2). The low rainfall coupled with abstractions has greatly influenced the continual decline in water levels in the well fields in the KKRWSS area.

The results of the SPI analysis in this study have shown that there are wide variations in rainfall records in most of the stations which could be attributed to climate variability. It is still not very clear if the variations are increasing with time. With the results of the autocorrelation analysis the present trend cannot be used to predict the future trend of rainfall in the region. Higher temperatures would however, increase evapotranspiration which would in turn reduce direct recharge. In the catchment area there have been more drought effects in recent years resulting in reduced recharge which together with groundwater abstraction have resulted in massive groundwater level declines the area.

4.3. Effects of abstraction rate on groundwater levels

The major concern in the area has been the decline of water levels. It has been reported that groundwater levels of the production boreholes have been falling since 1984 in the Vermaaks catchment. Even though the Vermaaks River wellfield had a fairly good recharge, by 1999 the water level decline was approximately 20 m. By 2002 the decline had reached about 30 m and again approximately 40 m by 2006. Current records obtained from GEOSS (with permission from the Oudtshoorn Municipality) in 2010 showed that the water levels in boreholes VR6, VR7, VR8 and VR11 have declined by 70 m, 109 m, 134 m and 160 m respectively. Abstraction rates in the production boreholes have been reset a number of times since the scheme began its operations as the scheme managers battled the high demand for water and low recharge. In February 1993, after evaluating the step-drawdown and 72 hr constant discharge tests conducted upon the production boreholes in 1990-91, Mulder estimated the 24 hr production potential of the wellfield at 72 l/s – with a peak supply potential of 110 l/s.

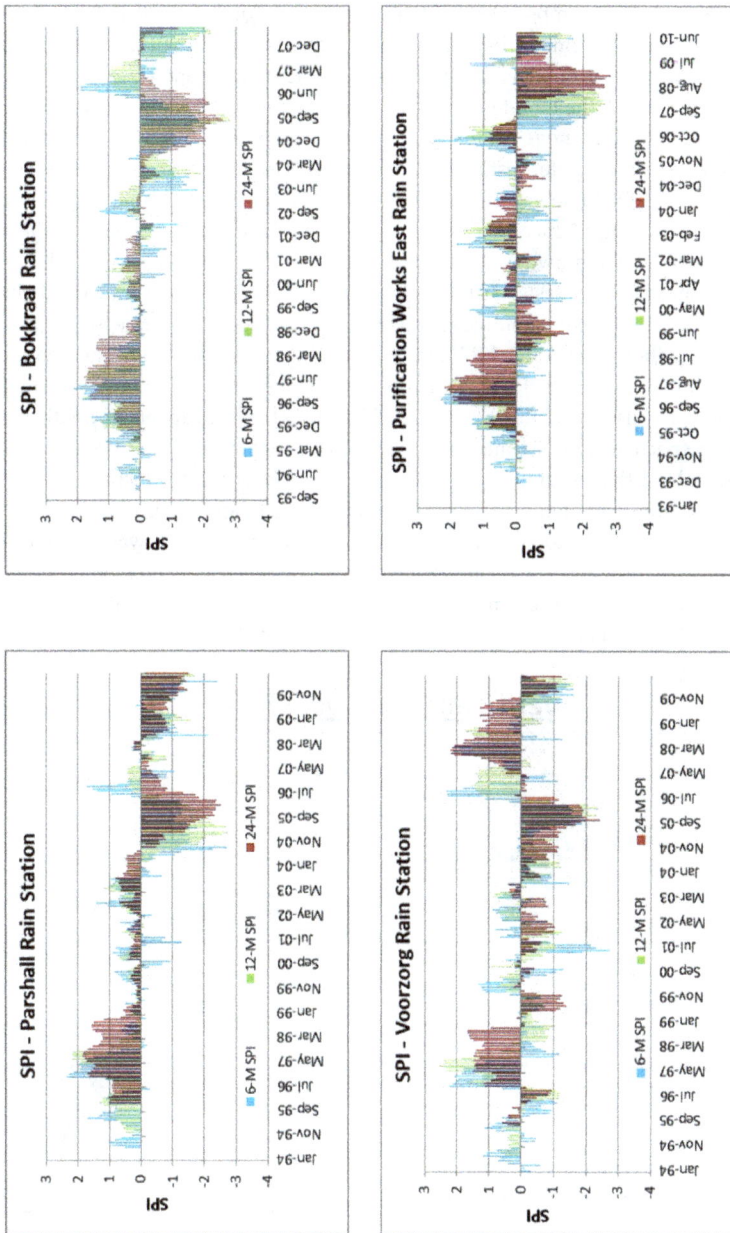

Figure 2. Comparative variations of 6-, 12- and 24-month SPIs for rainfall stations in the KKRWSS project area.

Costly, high-yielding pumps were installed in the production boreholes to meet this expected yield. In November 1993, after only eight months of production, Mulder re-evaluated the pump-test data in conjunction with the abstraction and water level monitoring data and down-scaled the long-term production potential of the wellfield to 40 l/s (peak 80 l/s). This indicated Mulder overestimated the production potential of the wellfield by at least 36% when using only aquifer-test information. In 1995, Kotze again re-adjusted the supply potential of the wellfield downwards to 20 l/s due to continual declines in water levels in the wellfield, representing a 72% downscaling of Mulder's original yield estimates. Jolly in 1998 conducted further step-drawdown tests on boreholes VR6, VR7, VR8 and VR11, as well as 72 hr constant-discharge tests on boreholes VR6 and VR7 and recommended that only boreholes VR7 and VR11 should be continuously pumped at a rate of 11 and 6 l/s, respectively, as boreholes VR6, VR7 and VR8 are interconnected with one another. Jolly added that this combined yield of 17 l/s is a conservative estimate upon the current water demand only. He also stated that boreholes VR7 and VR11 were capable of yielding up to 25 and 10 l/s respectively, on a continual basis, which could add an additional 18 l/s to the supply but added that accurate estimates of the volumes of rainfall recharge and a water-balance calculation were required in order to obtain the long-term sustainable yield of the Vermaaks River aquifer. Kotze in 2000 conducted such recharge and water-balance studies using 74 months of hydrological monitoring data, as well as a re-evaluation of the 1990 and 1997 aquifer-test data. Kotze estimated that the long-term supply potential of the wellfield is in the order of 8.5 l/s. Figure 3 is a plot of water levels in the Vermaaks wellfield.

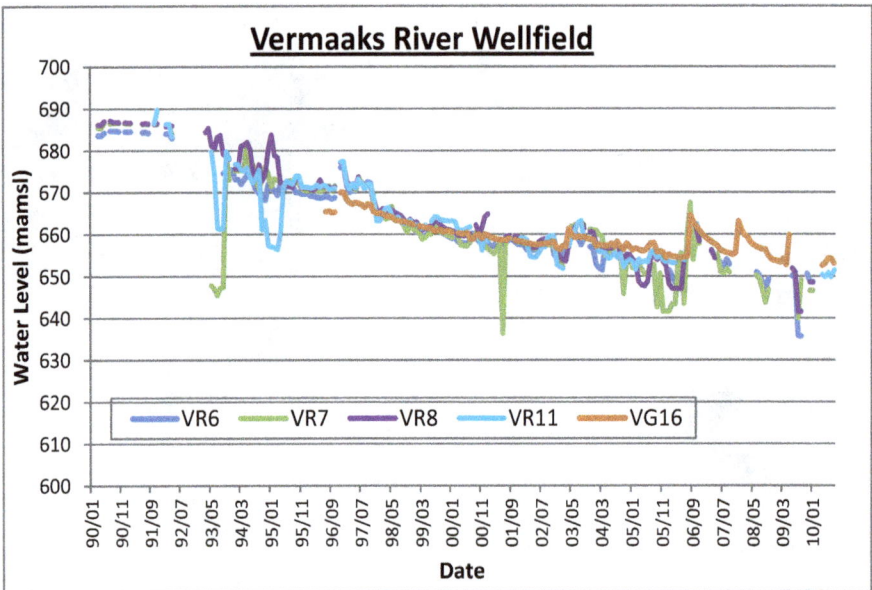

Figure 3. Decline of water levels in the production boreholes in the Vermaaks Wellfield

VG16 is the only observation borehole which reflects the decline of the production boreholes. With the continued adjustment in abstraction rates the decline of the production boreholes slowed down considerably from about 2001 except for the seasonal increases in abstraction rates at high demand periods (summer) as can be seen from figure 6.4. The wells are currently being pumped on the average at 5.5, 15, 4.5 and 5 l/s for VR6, VR7, VR8 and VR11 respectively on a 24 hour basis.

4.4. Effects of abstraction on hydrological and environmental resources

There are several springs in the KKRWSS catchment. Even though some of these springs are ephemeral their existences however have been affected by groundwater abstractions in the catchment. In a report to the Department of Water Affairs, Xu et al. (2002) recorded a number of springs that have dried up as a result of borehole construction in the vicinity of those springs. A spring G46083, 2 km downstream of borehole VG16 and located near the Cedarberg shale outcrop in the Vermaaks valley dried up after the construction of borehole G40175A in the vicinity in September 1999. Further downstream the spring G46084 was affected when a borehole G46077 was drilled through the Cedarberg shale into the Peninsula formation nearby in November 2001. The initial high pressure in the borehole was lost eventually. It has also been reported by Xu et al., (2007) that a hot spring which used to flow at a regional discharge area in Dysselsdorp has dried up. The hot spring was located at the intersection of two faults between the TMG and the Bokkeveld group. The drying up of the hot spring has been attributed to earthquakes (Tulbagh in 1969, magnitude 6.5 and Oudtshoorn in 2001, magnitude 3.6) and the large abstractions from the wellfield near the site (Jia, 2007). There were reported losses of many springs in 2001 and a few before, in 1999 and 2000 as a result of low rainfall and borehole constructions. Investigations by Xu et al. (2002) also established that flow in the Vermaaks River had abruptly dipped since the onset of the Water Supply Scheme in the catchment even though rainfall is the major contributory factor to the continual flow of the river.

The most comprehensive study on the impact of abstractions on spring flow in the Kammanassie area was done by Cleaver et al. (2003). The study grouped 53 springs in the area into 3 categories, 9 were considered most vulnerable to abstractions from the wellfield; 10 were considered as intermediate to vulnerability while the remaining 34 were considered least vulnerable. The study also confirmed that groundwater abstraction by the KKRWSS has impacted on the low-flow discharge in the Vermaaks River. On the impact of abstraction on vegetation, the study concluded that groundwater abstraction has a significant negative impact on plant water stress at the experimental sites in the Vermaaks River valley and recommended that changes in the water abstraction management could improve the situation. Spring losses caused localized impact on spring vegetation and ecosystems. Spring losses were also linked directly and indirectly to the death of four Cape Mountain Zebra on the Kammanassie Mountain between November 2000 and August 2001 as a result of inaccessibility to natural flowing water sources. Two artificial watering points were installed to protect the endangered species from extinction (Cleaver et al., 2003).

4.5. Sustainable management concept

The problems associated with the sustainable management of the KKRWSS are common to most abstraction schemes of such nature. There are a lot of positive impacts that such schemes bring to the beneficiaries and the general economy of the municipality and the nation as a whole. It is a laudable project that has brought a lot of improvements in the life of several communities, the right to access safe drinking water. Another positive indication from the project is the fact that the TMG has proven to be a good source of bulk water supply for many purposes. However, the challenges faced by the scheme need to be addressed not only to curb the negative impacts on the environment but also to ensure the long term survival of the scheme itself. In order to deal with the challenges facing the scheme there is a need to categorize them into what can be referred to as reversible and irreversible problems. The reversible problems are those that can be rectified or reversed because no permanent damage has been caused. On the other hand, an irreversible problem is that which causes permanent damage that cannot be reversed, life of species may be lost. Sometimes the full impact of a problem may not be immediately known until a thorough investigation has been done. There have been suggestions and remedies provided in the past to arrest the critical problems of the scheme but they have been on ad-hoc basis and it is important that a comprehensive approach is taken to maintain a sustainable project that will continue to improve the quality of life to humans and other forms of life. Some key goals related to groundwater sustainability in the United Kingdom are listed in figure 6.12. These goals must apply equally well in South Africa especially in the KKRWSS.

The most important attribute of the concept of groundwater sustainability is that it fosters a long-term perspective to management of groundwater resources. Several factors reinforce the need for a long-term perspective. First, groundwater is not a non-renewable resource, such as a mineral or petroleum deposit, nor is it completely renewable in the same manner and timeframe as solar energy. Recharge to groundwater from precipitation continually replenishes the groundwater resource but may do so at much smaller rates than the rates of groundwater withdrawals. Second, groundwater development may take place over many years; thus, the effects of both current and future development must be considered in any water management strategy. Third, the effects of groundwater pumping tend to manifest themselves slowly over time. For example, the full effects of pumping on surface water resources may not be evident for many years after pumping begins. Finally, losses from groundwater storage must be placed in the context of the period over which sustainability needs to be achieved. Groundwater withdrawals and replenishment by recharge usually are variable both seasonally and from year to year. Viewing the groundwater system through time, a long-term approach to sustainability may involve frequent temporary withdrawals from groundwater storage that are balanced by intervening additions to groundwater storage. The consequence of pumping should be assessed for each level of development; developments such as water-level declines, reduced streamflow, degradation of water quality and loss of dependent ecosystems through vegetation loss or spring loss.

The KKRWSS supplies water to meet demand from a population of 15,000 in addition to stock watering. The latter is estimated to account for about 20% of the water supplied from the scheme. A sustainable groundwater yield which will lead to a sustainable utilization should be expressed in the form of an extraction regime, not just an extraction volume. A regime in this context means a set of management practices that are defined within a specified time and place. Abstraction limits may be set in volumetric quantity terms or rates of extraction over a given period and/or impact, water level or quality trigger rules. In exceptional cases where draw beyond the rate of recharge may be acceptable, it may be only for a specified period, after which time the rate may be less than the rate of recharge to compensate for the loss. Also under specified circumstances (for example, high or low rainfall years) the amount of water that may be abstracted may be greater or less than the longer-term value. It has already been stated above that records from the KKRWSS site indicate less than average rainfall for the past few years and one would have expected that abstraction levels would be much less than normal. From figure 3, it appears that the Vermaaks wellfield is still experiencing water level decline. The four production boreholes in the Vermaaks wellfield are functioning well and appear to be producing the bulk of the water supply to the scheme. The following analysis is centered on the Vermaaks wellfield as the main area of concern. As has been stated elsewhere human activities such as groundwater withdrawals change the natural flow patterns and these changes must be accounted for in the calculation of the water budget. Because any water that is withdrawn must come from somewhere, human activities affect the amount and rate of movement of water in the system, entering the system, and leaving the system. For a sustainable utilization to occur, a sustainable pumping rate is defined by *Eqn. 1* as:

$$Ps \leq Ro + \Delta Ro - DR \tag{1}$$

and

$$DR = Do - \Delta Do \tag{2}$$

where Ro is natural recharge, ΔRo is induced recharge, DR is residual discharge, Do is natural discharge and ΔDo reduced discharge. Unless the borehole is drilled close to a reservoir, the induced recharge usually does not occur and reduced discharge dominates. Hence a sustainable pumping rate is given by:

$$Ps \leq Ro - DR \tag{3}$$

In the case of over-pumping the residual discharge could be reduced to nothing and pumping will be drawing on storage setting up a long-term decline in groundwater level. The yield of the groundwater system is at the expense of the groundwater discharge and storage components. This is the situation in the Vermaaks River wellfield where borehole levels have seen a long-term decline as shown above. The long-term decline of local groundwater level is an indication of unsustainable groundwater resource depletion. Table 1 is the results of groundwater budget analysis in the Vermaaks wellfield.between July 2009 and June 2010.

The average rainfall records for three stations (Parshall, Wildebeesvlakte and V-notch) from 2000 to 2010 were computed for the Vermaaks wellfield. The Peninsula formation window at the Vermaaks wellfield was used as the area of recharge given as 48 km². The recharge is taken as 3% of rainfall based on the most recent studies. The abstraction volume is the actual records obtained from the KKRWSS management for the period of July 2009 to June 2010. The water demand volume is obtained from a population of 15000 plus a 20% demand from stock watering. The legal provision of 25 litres per person per day is the basic human need requirement. The same quota was assumed for stock watering. The population figure was obtained from the scheme management. The historical recharge (Hist Rech) is based on 3% of long-term MAP for the area which is 50% more than the records for the period of analysis.

Date	Rainfall (mm)	Rech (3%)	Recharge (m³)	Abstraction (m³)	Popn & Stock Dmd (m³)	Historical Recharge (m³)	Storage Depletion (m³)
Jul-09	26.24	0.7872	37785.6	73275	11475	56678.4	-35489.4
Aug-09	29.13	0.8739	41947.2	71936	11475	62920.8	-29988.8
Sep-09	26.43	0.7929	38059.2	158025	11475	57088.8	-119965.8
Oct-09	32.08	0.9624	46195.2	12041	11475	69292.8	-34154.2
Nov-09	37.35	1.1205	53784	158630	11475	80676	-104846
Dec-09	21.11	0.6333	30398.4	81958	11475	45597.6	-51559.6
Jan-10	19.93	0.5979	28699.2	82494	11475	43048.8	-53794.8
Feb-10	24.93	0.7479	35899.2	47428	11475	53848.8	-11528.8
Mar-10	30.8	0.924	44352	39307	11475	66528	-5045
Apr-10	18.76	0.5628	27014.4	62774	11475	40521.6	-35759.6
May-10	28.75	0.8625	41400	35292	11475	62100	-6108
Jun-10	36.35	1.0905	52344	59501	11475	78516	-7157
Total	331.86	9.9558	477878.4	882661	137700	716817.6	-404782.6

Table 1. Results of Vermaaks groundwater budget

Results from the table are plotted in figure 4. The following conclusions may be drawn from the results:

- The water demand is more than the requirements for basic human needs
- The recharge in the last decade has been less than average due to low rainfall
- The abstractions from the wellfield exceed the recharge leading to decline in water levels
- There has been depletion of groundwater storage due to over-abstraction
- The over-abstraction has the potential to capture natural discharge to surface water bodies
- The over-abstraction leaves no provision for discharge to cater for environmental flows

In figure 4 the abstraction line is well over the recharge line. For the sustainable utilization of the groundwater resource, the abstraction level should be less than or at least equal to the recharge line as indicated in Eqn, (1). The graph shows a situation of unsustainable utilization of the resource.

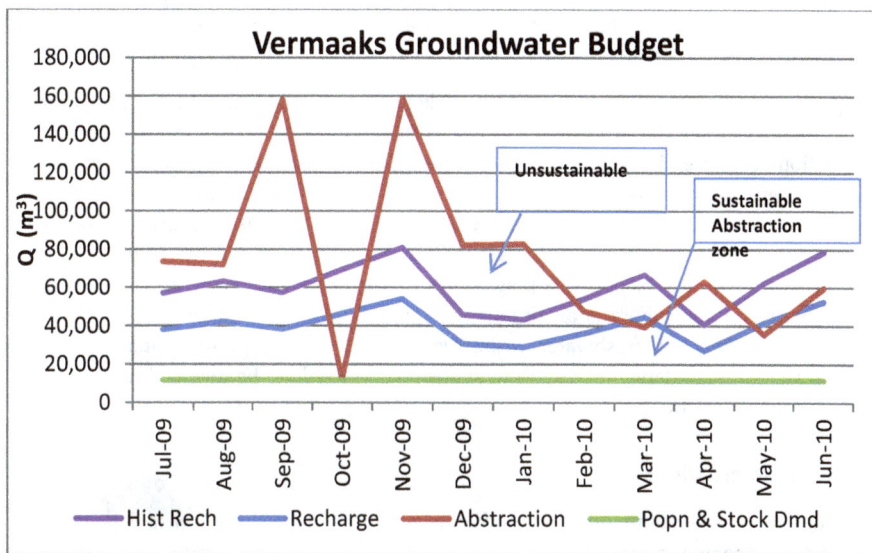

Figure 4. Vermaaks Wellfield groundwater budget

There are abstractions from the other wellfields notably Varkieskloof, Bokkraal and Calitzdorp to augment the total groundwater supply from the scheme. In addition surface water is drawn from the Olifants River to augment the scheme's supply. Between July 2009 and June 2010 a total of 1,518,425 m³ of water was supplied from the scheme made up of 1,275,920 m³ of groundwater and the rest from surface water sources.

5. Conclusion

From the above study it has been established that recharge to the TMG aquifers, and in particular to the Vermaaks Wellfield is much lower (about 3% of MAP) than has been estimated in the past. Climate trends generally have shown that temperature is gradually increasing everywhere while rainfall shows minor increases and decreases over different areas. In the KKRWSS region rainfall has generally declined in recent years. Rainfall variability also remains high with potential for causing droughts and floods. Recommendations on abstraction rates based on higher recharge estimates have resulted in serious decline in water levels in the wellfields posing serious threat to the water supply scheme itself. The water demand has risen far and above the basic human requirements and stock demand over the years and with abstraction rates based on demand requirements, the decline in water levels has persisted over a decade. The production wells in the Verimaaks wellfield do not have time to recover. Decline in water levels has resulted in decline in natural discharges from the aquifer to streams, springs and other wetlands with negative impacts on the hydro-ecological environment. The reduction in low-flows in the Vermaaks River, the loss of spring flows and the resultant loss of certain vegetation types as well as threat to local ecosystems have led to temporal and permanent damage to the hydro-ecological environment.

Temperature increases in the atmosphere is likely to cause droughts during the summer whiles in the winter the atmosphere can absorb more moisture and cause floods. The implications of these changes on freshwater resources either directly or indirectly are of much concern to water resources managers. Responses to climate variations by surface water resources are fast and intense however groundwater response is much delayed even though it is likely to be affected indirectly by the absence of fresh surface water resources. Semi-arid and arid regions of developing countries are particularly vulnerable to impacts of climate variations and there is an urgent call for mitigation and adaptation measures to sustainably utilize our freshwater resources particularly the groundwater. The high groundwater storage depletion due to over-abstraction needs to be reversed to restore the hydro-ecological environment.

5.1. Recommendations

The long-term sustainability of the groundwater resources in the study area is dependent on the time-management approach, a form of an adaptive management in which abstraction rates are adjusted according to the recharge patterns and not driven by demand as currently pertains in the area. Generally, abstraction rates have been higher than sustainable levels in almost all year round. If the abstraction rates are not reduced, unless there is a major increase in the rainfall pattern, the scheme is bound to fail in the near future.

The priorities for groundwater management should be:

- Sustainable long-term yield from aquifers;
- Effective use of the large volume of water stored in aquifers in critical periods,
- Preservation of groundwater quality;
- Preservation of the aquatic environment by prudent abstraction of groundwater and
- Integration of groundwater and surface water into a comprehensive water and environmental management system.

The trade-off between water for consumption and the effects of withdrawals on the environment often become the driving force in determining a good management scheme. The concept of adaptive management which treats management policies and actions as experiments, not fixed policies has been recommended. Management continually improves by learning from experiences. Changing technology and increasing knowledge and understanding change the perception of risk and priorities with regard to the acceptability of trade-offs, hence water resource management must be adaptive and flexible.

Acknowledgement

The authors are indebted to the DBBS/VLIR programme of UWC for sponsoring the research and CSIR WRI, Ghana for supporting this publication.

Author details

Anthony A. Duah
CSIR Water Research Institute, Accra, Ghana

Yongxin Xu
Department of Earth Science, University of the Western Cape, Bellville, South Africa

6. References

Cleaver, G., Brown, L.R., Bredenkamp, G.J., Smart, M.C. and Rautenbach, C.J. de W. (2003). Assessment of environmental impacts of groundwater abstraction from Table Mountain Group (TMG) aquifers on ecosystems in the Kammanassie Nature Reserve and environs. WRC Report No. 1115/1/03. Pretoria, South Africa.

Custodio, E. (2005). Intensive use of ground water and sustainability. (Guest Editorial), *Ground Water* 43 (3), 291.

Devlin, J.F. and Sophocleous, M. (2005). The persistence of the water budget myth and its relationship to sustainability. *Hydrogeology Journal* 13 (4), 549-554.

ECA, Organisation for African Unit and African Development Bank (2000). Safeguarding life and development in Africa. A vision for water resources management in the 21st Century. African caucus presentation – Second World Water Forum, The Hague, The Netherlands, 18 March 2000.

Jia, Haili. (2007). Groundwater resource evaluation in Table Mountain Group aquifer systems. Dessertation submitted to the University of the Western Cape in fulfillment for the degree of Doctor of Philosophy at the Dept. of Earth Sciences. November, 2007.

Jolly, J.L. (2002). Sustainable use of Table Mountain Group aquifers and problems related to scheme failure. In: Pietersen, K., Parsons, R. (Eds). *A synthesis of the hydrogeology of the table mountain group – formation of a research strategy*. WRC Report No. TT 158/01, 108-111.

Jolly, J.L. and Kotze, J.C. (2002). The Klein Karoo rural water supply scheme. In: Pietersen, K., Parsons, R. (Eds). *A synthesis of the hydrogeology of the table mountain group – formation of a research strategy*. WRC Report No. TT 158/01, 198-201.

Parsons, R. (2002). Recharge of Table Mountain Group Aquifer Systems. In: Pietersen, K., Parsons, R. (Editors), (2002). *A synthesis of the hydrogeology of the table mountain group – formation of a research strategy*. WRC report no. TT 158/01, 97-102.

Rosewarne, P. (2002). Case study: Port Elizabeth Municipal area. In: Pietersen, K., Parsons, R. (eds.) *A synthesis of the hydrogeology of the table mountain group – formation of a research strategy*. WRC report no. TT 158/01, 205-208.

Scanlon, B.R., Healy, R.W. and Cook, P.G. (2002). Choosing appropriate techniques for quantifying groundwater recharge. *Hydrogeology Journal* (2002) 10: 18-39.

Sophocleous, M. (2000). From safe yield to sustainable development of water resources: the Kansas experience. *Journal of Hydrology* 235, 27-43.

Sophocleous, M. (2005). Groundwater recharge and sustainability in the high plains aquifer in Kansas, USA. *Hydrogeology Journal* 13 (2), 351-365.

Weaver, J.M.C., Talma, A.S. and Cavé, L. (1999). Geochemistry and isotopes for resource evaluation in the fractured rock aquifers of the Table Mountain Group. WRC Report No. 481/1/99.

Woodford, A.C. (2002). Interpretative and applicability of pumping-tests in table mountain group aquifers in: Pietersen, K., Parsons, R. (Eds). A synthesis of the hydrogeology of the table mountain group – formation of a research strategy. WRC Report No. TT 158/01, 71-84.

Wu, Y. (2005). Groundwater recharge estimation in Table Mountain Group aquifer systems with a case study of Kammanassie area. PhD Thesis, Dept. of Earth Sciences, University of the Western Cape, August 2005.

Xu, Y. and Beekman, H.E. (Editors), (2003). Groundwater recharge estimation in Southern Africa. UNESCO IHP Series No. 64, UNESCO Paris. ISBN 92-9220-000-3.

Xu, Y. and Braune, E. (2010). Groundwater resources in Africa in: Y. Xu and E. Braune (Eds), 2010. Sustainable groundwater resources in Africa: Water supply and sanitation environment. UNESCO Paris and CRC Press/Balkema, Taylor & Francis Group.

Xu, Y., Wu, Y. and Duah, A. (2007). Groundwater recharge estimation of Table Mountain Group aquifer systems with case studies. WRC Report No. 1329/1/07.

Xu, Y., Wu, Y. and Titus, R. (2002). Influence of the Vermaaks Wellfield abstraction on groundwater levels and streams in the vicinity. Technical report prepared for the Department of Water Affairs and Forestry, Bellville, Cape Town.

Chapter 8

Effects on Summer Monsoon and Rainfall Change Over China Due to Eurasian Snow Cover and Ocean Thermal Conditions

Renhe Zhang, Bingyi Wu, Jinping Han and Zhiyan Zuo

Additional information is available at the end of the chapter

1. Introduction

Climate over China is mainly governed by the East Asian monsoon, which mainly arises from the seasonal variation of the land-sea thermal contrast. Climatologically, the East Asian continent is warmer than surrounding oceans in summer (June-August) and colder in winter (December-February), leading to the occurrence of the East Asian summer monsoon and winter monsoon, respectively. During summer monsoon period, warm and moist air over tropical oceans is transported to the East Asian region [1], forming plenty summer monsoon rainfall. Since the rainfall in China is mainly concentrated in summer, the change and variability of the summer monsoon rainfall have been the research focus because of their important effects on economy, society and human life.

In the past several decades, one predominant feature of global warming is that the increasing of air temperature over land is stronger than that over sea [2], implying strengthening of the East Asian summer monsoon. However, many researches have shown that the summer climate in eastern China is characterized by multi-timescale variations [3-5]. The long-term variation is featured by the distinguished inter-decadal variability. During the period of 1880-2002, over eastern China there were no long-term trends for both annual and seasonal mean rainfalls, exhibiting dry and wet cycles in a noticeable inter-decadal timescale [6]. There are significant differences in the rainfall change among the areas of northern China, the Yangtze-Huai River Valley and southern China. In eastern China the above-normal rainfall zone moves successively from northern to southern China in the past several decades. More summer rainfall appeared over northern China in 1960s and 1970s; the Yangtze-Huai River valleys experienced a wet period from the late 1970s to the late 1980s; from the late 1980s or early 1990s more rainfall zone shifted to southern China [7-9].

The above-normal rainfall over eastern China is associated with the warm and moist southerlies. In 1960s and 1970s the southerlies reached northern China and more rainfall appeared there [10]. The more rainfall over the Yangtze-Huai River valleys from the late 1970s to the early 1990s and that over southern China afterwards were also related with the southerlies appeared over these regions [8,9].

Under the background of the global climate warming, the observed summer rainfall change over the East Asian monsoon region in eastern China exhibits unique features. Many studies have revealed that the Eurasian snow cover and ocean thermal conditions have important impacts on the summer monsoon rainfall change over eastern China. In this Chapter, possible causes of the summer monsoon rainfall changes over eastern China will be discussed based on relevant studies. This chapter is organized as follows. The observed features of the summer monsoon rainfall change over eastern China in the past half century will be given in Section 2. Section 3 describes the impact of the spring (March-May) Eurasian snow cover on the summer monsoon rainfall change over China. The role played by the changes of the thermal condition in the Atlantic Ocean and the sea ice in the Arctic Ocean will be discussed in Section 4. Section 5 focuses on the influence of the changes of sea surface temperature (SST) in the tropical Pacific and Indian Oceans. A conclusion is given in Section 6.

2. Observed summer monsoon rainfall change over eastern China since 1950s

The geographic location of the East Asian monsoon region over East Asia is usually taken to be in the area to the east of 100°E [11-13]. In order to check the long-term variation of the summer monsoon rainfall over China, we average the observed summer (June-August) rainfall to the east of 100°E in China in each year from 1958 to 2011. The rainfall data is from the monthly rainfall dataset observed at 160 stations in China and the station locations can be found in [14].

2.1. Changes of summer rainfall in eastern China and East Asian summer monsoon

The variation of the summer monsoon rainfall averaged to the east of 100°E over eastern China with time is shown in Figure 1. From Figure 1 it can be seen that, besides the inter-annual variability, the inter-decadal variability is a predominant feature for the long-term variation. No clear trend can be found for the summer monsoon rainfall from 1958 to 2011. A weak decline was observed from 1950s to the late 1980s, when the rainfall began to increase until the middle 1990s. From then on the rainfall declined again to the early 2000s and kept stable afterwards.

Since the summer rainfall in eastern China is greatly influenced by the East Asian summer monsoon, to see the long-term variation of the East Asian summer monsoon, we calculated the western North Pacific-East Asian summer monsoon (WNP-EASM) index [15] in 1958-2011 by

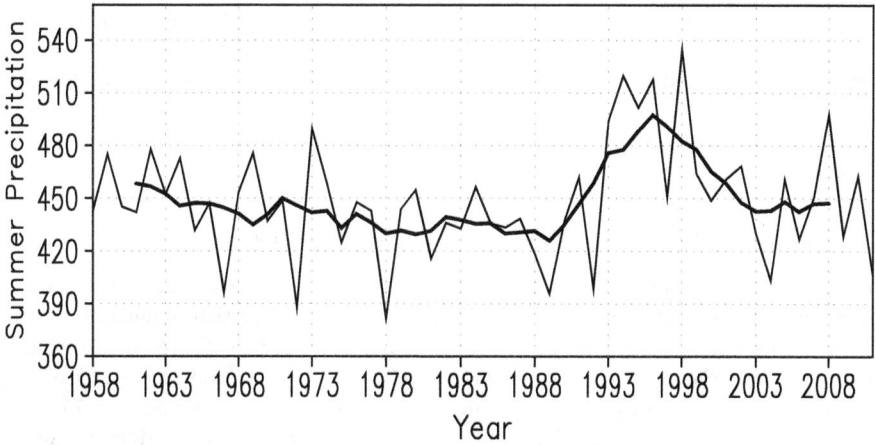

Figure 1. Observed summer (June-August) rainfall (thin line) and its 7-year running mean (Thick line) averaged in the area to the east of 100°E in China (Unit: mm).

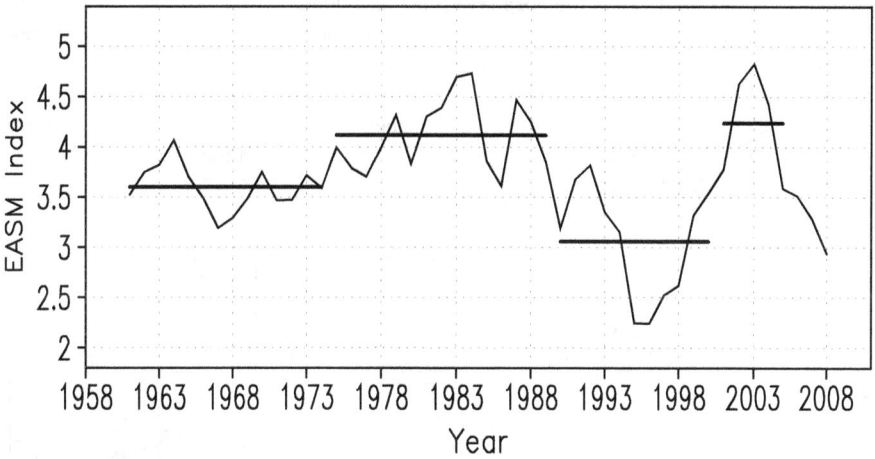

Figure 2. 7-year running mean of the western North Pacific-East Asian summer (June-August) monsoon (WNP-EASM) index (Unit: m/s). The thick lines indicate averaged WNP-EASM indexes in 1958-1974, 1975-1989, 1990-2000 and 2001-2005, respectively.

using the reanalysis data from National Center for Environmental Prediction (NCEP) and the National Center for Atmospheric Research (NCAR) [16]. Figure 2 shows the 7-year running mean of the WNP-EASM index. Same as the variation of the East Asian summer monsoon rainfall, there is also no clear trend for the WNP-EASM index, and strong inter-decadal variability appears. From Figure 2 we can see that climate shifts for the East Asian summer monsoon appeared at middle 1970s, late 1980s and early 2000s, respectively. In 1958-1974 the East Asian summer monsoon was weak. It turned to be strong in the period from 1975 to 1989, and became weak again in 1990-2000. In the period of 2001-2005 the East Asian summer monsoon strengthened and weakened afterwards. In addition to the inter-decadal variation, Figure 2 also shows that the larger amplitudes and shorter periods appeared since the late 1980s. Such feature can also be observed in the changes of the rainfall as shown in Figure 1. It implies a larger variability and more frequent variation of the East Asian summer monsoon have occurred since then.

2.2. Changes of abnormal rainfall pattern in eastern China

In order to check changes of abnormal pattern in summer monsoon rainfall over eastern China in association with the inter-decadal variability of the East Asian summer monsoon, in Figure 3 we show the differences of the averaged summer rainfall over eastern China between 1975-1989 and 1958-1974, between 1990-2000 and 1975-1989, and between 2001-2008 and 1990-2000, respectively. Corresponding to the climate shift in the middle 1970s (see Figure 2), the summer rainfall in 1975-1989 increased in the area over the middle and lower reaches of the Yangtze River valley and decreased over southern and northern China compared to the summer rainfall in 1958-1974 (Figure 3a). This shift of the summer rainfall anomalies caused droughts in the northern China and floods around Yangtze River valley [17,18]. After the weakening of the East Asian summer monsoon in the end of 1980s, more rainfall appeared in southern China in the period of 1990-2000 relative to 1975-1989 (Figure 3b), which was in accordance with the findings in [7]. Compared to the rainfall in 1990-2000, in 2001-2008 the East Asian monsoon strengthened and the rainfall decreased around Yangtze River valley, and increased to the south of about 25°N in the south of China and between about 31°N-36°N to the north of Yangtze River valley (Figure 3c).

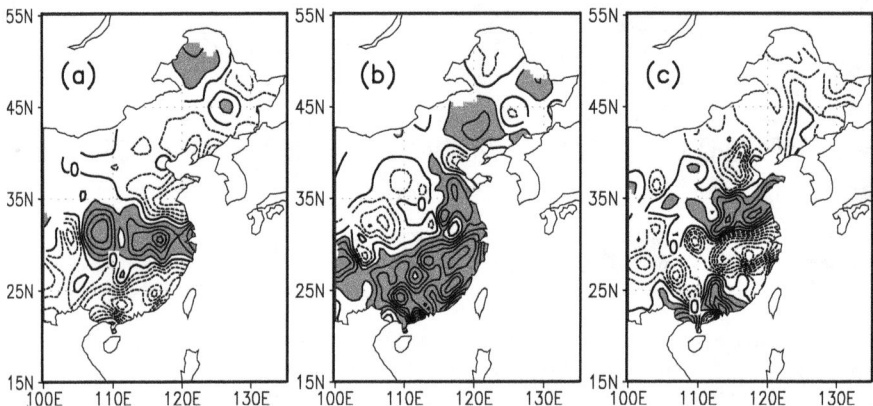

Figure 3. Averaged summer rainfall differences between (a) 1975-1989 and 1958-1974, (b) 1990-2000 and 1975-1989, and (c) 2001-2008 and 1990-2000. The dotted and real lines represent the negative and positive, respectively. The line interval is 30 mm. The thick line is the 0 isoline. The difference larger than 30 mm is shaded.

Here we can see that in association with the global climate warming in the recent half decade, although no clear trends for both the summer rainfall over eastern China and the East Asian summer monsoon, the abnormal rainfall pattern changes obviously. Compared to 1958-1974, more rainfall appeared around the middle and lower reaches of the Yangtze River valley in 1975-1989; the increased rainfall zone moved southward from 1975-1989 to 1990-2000 to the south of the Yangtze River valley in southern China; relative to 1990-2000, the rainfall increased in both the further southern area and around the area to the north of Yangtze River valley in 2001-2008. In following sections the possible physical reasons contributing to the changes of the abnormal rainfall pattern will be discussed.

3. Impacts of the spring (March-May) Eurasian snow cover

Snow cover is largely controlled by atmospheric circulation, while widespread snow cover affects local and large-scale atmospheric circulation and hydrological processes through changing water and energy flux. In addition to its high albedo and low thermal conductivity, snow cover acts as a heat sink through sublimation and melting processes. Snow cover cools the overlying atmosphere and warms the underlying ground [19].

3.1. Tibetan Plateau snow cover

The association between the Tibetan Plateau snow cover variability and Chinese rainfall has been extensively explored. The excessive snow cover in the Tibetan Plateau during winter/spring is associated with above-normal May-June rainfall in southern China [20,21]. Summer rainfall in central China along the mid- and low-reaches of the Yangtze River has a positive correlation, whereas that in northern and southern China has a negative correlation with the Tibetan Plateau snow depth in the preceding winter/spring [22-24].

Chinese summer rainfalls in the Yangtze-Huaihe River valleys and in the upper-lower reaches of the Yangtze River showed a remarkable transition from drought period to rainy period in the end of 1970's, in good correspondence with the decadal transition of the winter snow cover on the Tibetan Plateau [22]. It is further demonstrated that there is a close relationship among the inter-decadal increase of snow depth over the Tibetan Plateau during March–April, a wetter summer over the Yangtze River valley, and a dryer one in the southeast coast of China and the Indochina peninsula; the excessive snowmelt and increased surface moisture supply over the Tibetan Plateau lead to a wetter summer in the vicinity of the Yangtze River valley [24]. In fact, a moderate positive correlation between the Tibetan Plateau snow cover and spring rainfall in southern China contains El Niño-South Oscillation (ENSO) effects, and ENSO has larger impacts than the Tibetan Plateau snow cover on spring rainfall in southern China [25].

3.2. Spring Eurasian snow cover

Utilizing monthly mean 513-station rainfall data in China from the National Meteorological Information Centre of China spanning the period from 1968 to 2005 and monthly mean

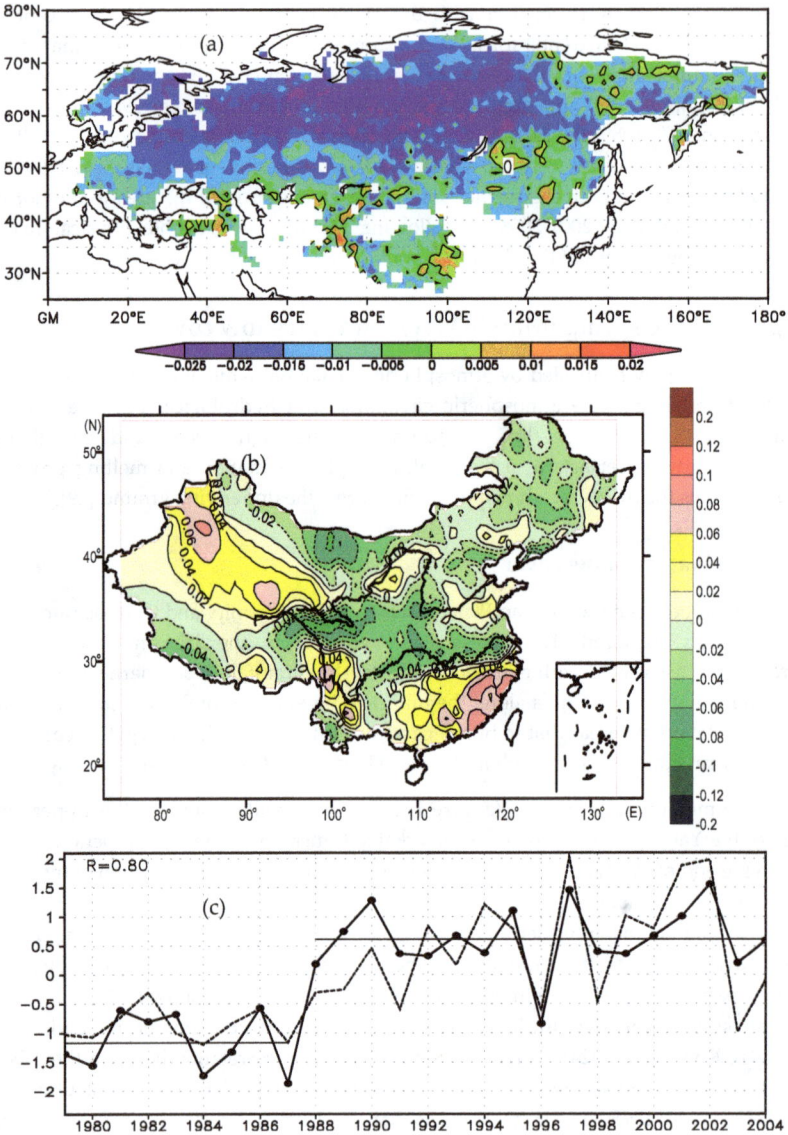

Figure 4. Spatial distributions of the SWE and rainfall fields of the leading SVD mode and their time series, derived from spring Eurasian SWE and the succeeding summer station rainfall in China during the period 1979-2004, (a) spring Eurasian SWE, only "0"-isoline is plotted, (b) summer station rainfall, and (c) normalized time series of spring SWE (solid line) and summer rainfall (dashed line), their correlation is 0.8. In (a) and (b), units are arbitrary. The straight lines indicate averaged time series in 1979-1987 and 1988-2004, respectively. (from [27])

snow water equivalent (SWE) data derived from National Snow and Ice Center during the period from 1979 to 2004 [26], the singular value decomposition (SVD) method is applied to calculate the coupled modes for the summer rainfall in China and spring (March-May) SWE during the period from 1979 to 2004 [27]. The leading SVD mode accounts for 36% of the co-variance, and its spatial distributions and corresponding time series are shown in Figure 4.

The leading SVD mode of the spring SWE variability shows a coherent negative anomaly in most of Eurasia (Figure 4a). It is seen that negative anomalies are dominant, particularly in the central Siberia, whereas positive anomalies are confined to some small areas including the most northeastern Russia, east and southeast to the Lake Baikal, the western and eastern Tibetan Plateau. Besides a strong inter-annual variability, the leading SVD mode of the spring SWE displays strong inter-decadal variation with persistent negative phases in 1979–1987 and frequent positive phases afterwards; an apparent inter-decadal shift occurred in the late 1980s with persistent negative phases during 1979-1987 (the mean value is -1.17) and predominant positive phases during 1988-2004 (the mean value is 0.62) (Fig. 4c). Figures 4a and 4c indicate that there was excessive spring SWE in Eurasia during 1979-1987, followed by abrupt decreases in the period from 1988 to 2004.

For the anomalous summer rainfall field corresponding to the leading SVD mode (Figure 4b), positive anomalies appear in southern China to the south of the Yellow River valley, and negative anomalies emerge in most parts of northern China to the east of 95°E. The relationship between spring rainfall variations in southern China and anomalous spring snow cover in western Siberia is in agreement to the previous results [25]. For the relation of the Eurasian winter snow cover with summer rainfall in China, the Eurasian winter snow cover is positively correlated with the following summer rainfall in northern and southern China and negatively in central, northeast, and western China [28]. Here we can see that the spring snow cover is mainly related with a reversed north-south summer rainfall anomalies in the eastern China, which differs from the rainfall anomalies associated with winter snow cover. The time evolution of the anomalous summer rainfall field shows that there is an upward trend, and coherent negative phases before 1989, followed by frequent positive phases afterwards, which are in good agreement with time series for spring Eurasian SWE (Figure 4c). The two fields are significantly correlated and their correlation coefficient is 0.8. The inter-decadal shift of the spring Eurasian SWE in the late 1980s corresponds well to that of the East Asian summer monsoon as shown by the WNP-EASM index in Figure 2, and the increasing summer rainfall in southern China shown in Figure 3b.

In order to check the summer rainfall difference between the periods before and after the late 1890s, Figure 5 shows the summer mean rainfall anomalies during 1979-1987 (Figure 5a) and 1988-2004 (Figure 5b), respectively, relative to the mean averaged for the period from 1979 to 2004. The pattern of rainfall anomaly in either Figure 5a or 5b resembles that in Figure 4b. In the inter-decadal time scale, accompanying spring excessive SWE in most of Eurasia during 1979-1987, summer rainfall decreased by about 40 mm in southern China, whereas it increased by about 20 mm along the Yangtze River valley (Figure 5a). In contrast, summer rainfall increased in southern China and decreased along the Yangtze River valley during 1988-2004 (Figure 5b). The results are same as those in that the East Asian summer monsoon experienced an inter-decadal shift in the late 1980s, and more rainfall moved from the Yangtze-Huai River valleys to southern China. The shift of the spring Eurasian SWE in the late 1980s may be one of reasons for the inter-decadal shift of the summer monsoon rainfall decrease around Yangtze River valley and increase in southern China.

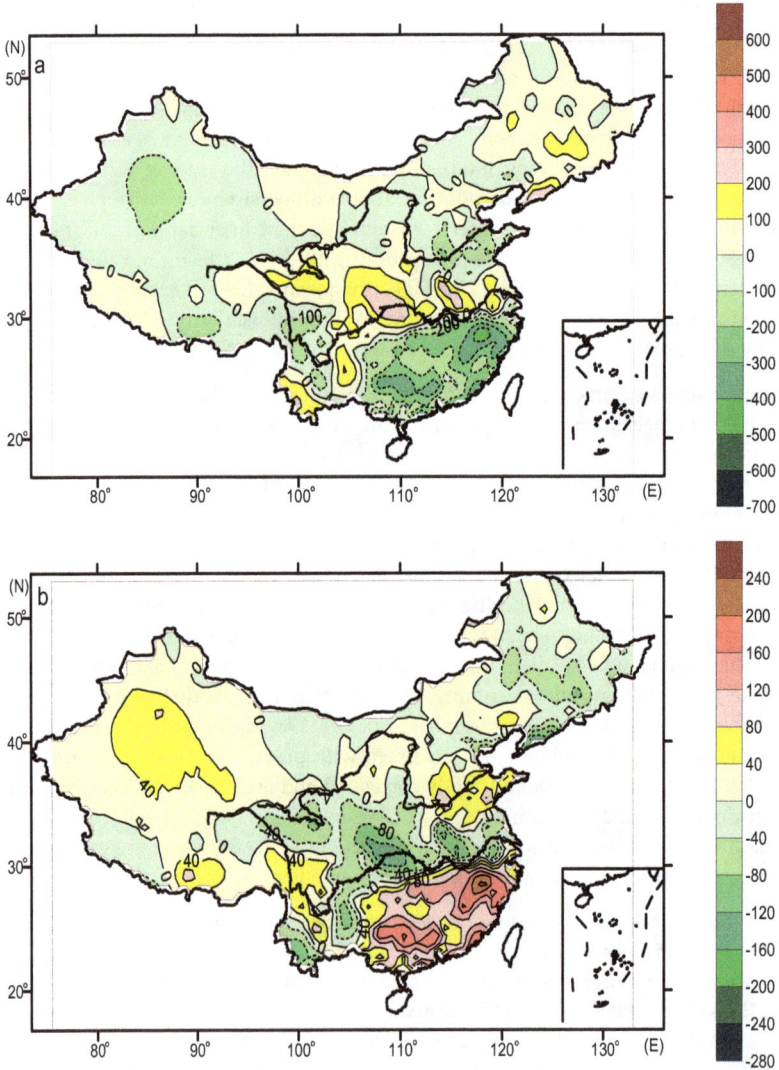

Figure 5. (a) Summer mean rainfall anomalies during 1979-1987 relative to the mean averaged for the period from 1979 to 2004, (b) same as (a) but for the period 1988-2004. In (a) and (b), the intervals are 100 and 40, respectively. Units are 0.1 mm.

3.3. Physical linkage between Spring Eurasian snow cover and summer rainfall

By using the NCEP/NCAR reanalysis data in 1960-2006, the physical process for the effects of spring Eurasian snow cover on the summer rainfall in China were diagnosed [27]. Figure 6 shows geopotential height anomalies at 500 hPa regressed by the leading SVD

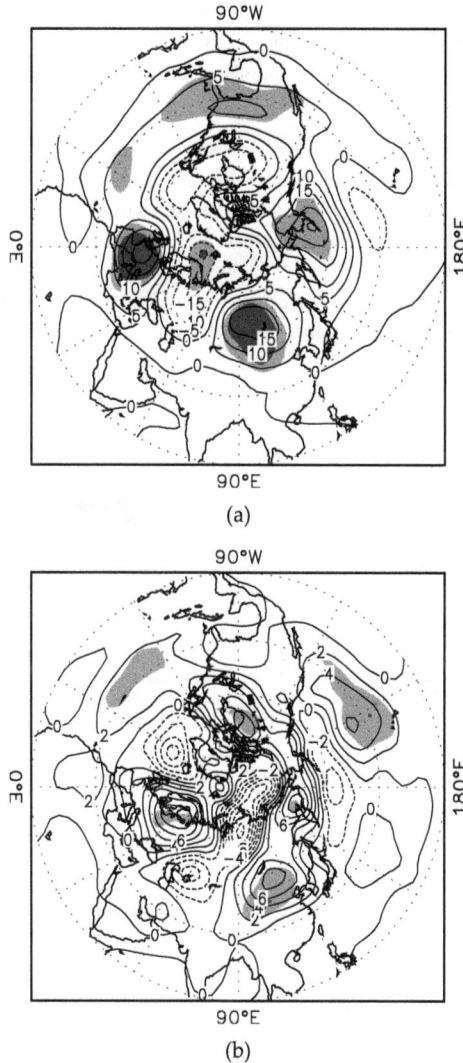

(a)

(b)

Figure 6. Geopotential height anomalies at 500 hPa in (a) spring and (b) summer regressed by the leading SVD mode of spring SWE (unit: gpm). The contour intervals are 5 and 2 gpm in (a) and (b), respectively. (from [27])

mode of spring SWE in spring and summer, respectively. In both spring (Figure 6a) and summer (Figure 6b), a wave train structure can be observed over Eurasian continent. There are two positive anomalous centers located over western Europe and northeastern Asia, respectively. The positive height anomalies over northeastern Asia between 30°N and 60°N are unfavorable for the rainfall to the north of the Yangtze River valley, and more rainfall appears to its south in southern China. The similarity between the regressed height anomalies in spring and summer implies that the atmospheric circulation anomalies associated with Eurasian snow cover persist from spring to summer, which exerts significant effect on the summer rainfall in China.

As shown in Figure 6, in spring and summer similar features of the atmospheric circulation anomalies appear over East Asia. In fact, in 1979-2004 the spring rainfall in southern China also exhibited a climate shift in the late 1980s with more rainfall in southeastern China and less in southwestern China before the late 1980s, and less rainfall in southeastern China and more in southwestern China after then [29,30]. Figure 7 shows the leading SVD mode between Chinese spring rainfall (Figure 7a) and spring Eurasian SWE (Figure 7b). It explains 31.5% square covariance fraction and correlation coefficient between the expansion

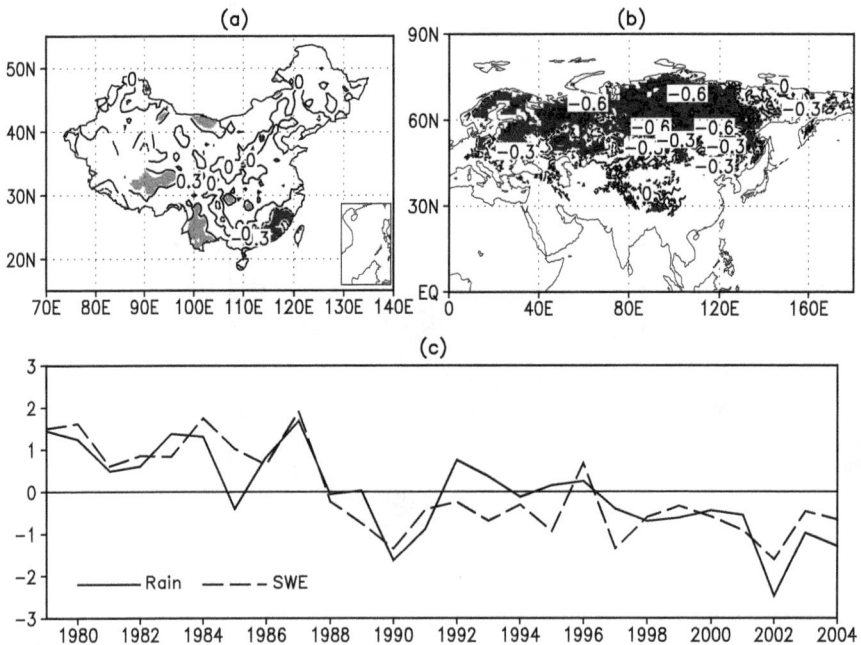

Figure 7. Spatial patterns of the leading coupled SVD mode for (a) spring rainfall and (b) Eurasian spring SWE and (c) corresponding time series (solid line for rainfall, dash line for SWE). Shaded areas in (a) and (b) denote the correlations exceeding the 0.05 significance level. (from [29])

coefficients of both variables is 0.81, exceeding the 0.01 significance level. The time series (Figure 7c) for both the Chinese spring rainfall and Eurasian SWE were mainly positive phase in 1979-1987 and frequently negative phase afterward. The decrease in Eurasian spring SWE is accompanied by the reduced rainfall over Southeast China and enhanced rainfall over Southwest China. The reduction in Eurasian SWE results in reduced upward and poleward wave flux activity, which alters the atmospheric circulation and thus affects the rainfall in southern China [30]. Actually, using the Monthly tabulated Scandinavia teleconnection index from NCEP (http://www.cpc.ncep.noaa.gov/data/teledoc/scand.shtml), the inter-decadal shift in the late 1980s can also be detected in the time series of the Scandinavian pattern [31] of atmospheric circulation variability. As shown in Figure 8, the positive phases of the Scandinavian pattern were dominant before 1988, followed by coherent negative phases. The abrupt shift of the atmospheric circulation possibly reflects the role of Eurasian spring SWE forcing.

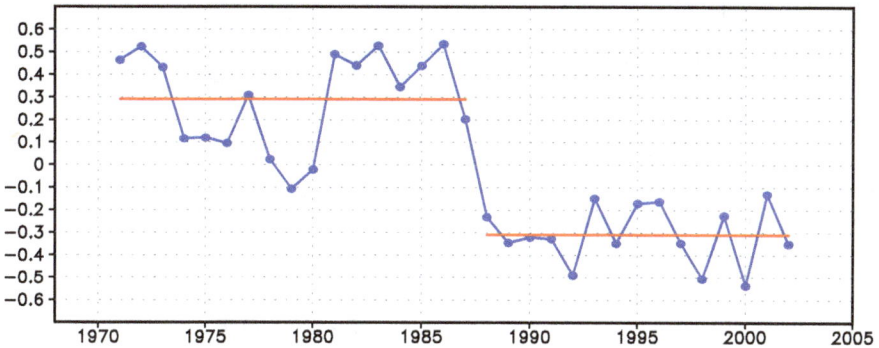

Figure 8. The 7-year running mean of the normalized time series of the Scandinavian pattern averaged over March to May (blue line). The red straight lines indicate averaged time series in 1971-1987 and 1988-2002, respectively.

4. Impacts of the Atlantic and Arctic Oceans

4.1. Atlantic Multidecadal Oscillation (AMO)

One of the predominant features in long-term SST variation in Atlantic is the Atlantic multidecadal oscillation (AMO). The AMO is the alternation of cool and warm SST anomalies (SSTAs) throughout the North Atlantic Ocean in the period of several decades [32,33]. A lot of studies have shown that the AMO has the influence in the global scale, and many regional multidecadal climate variability are related to the AMO [34].

To see the association of the summer precipitation in China with the AMO, the AMO positive phases of 1951-1965 and 1996-2000, and negative phase in 1966-1995 are selected [35]. The summer rainfall composites are made for the AMO positive phases and negative phase, respectively. The composite difference between positive and negative AMO phases is shown in Figure 9. It can be seen that except a small area in southwestern China, in the most areas in eastern China the summer rainfall is increased by 0.3-0.6 mm d^{-1}, implying that the warm AMO phase is favorable for more summer rainfall in eastern China [36].

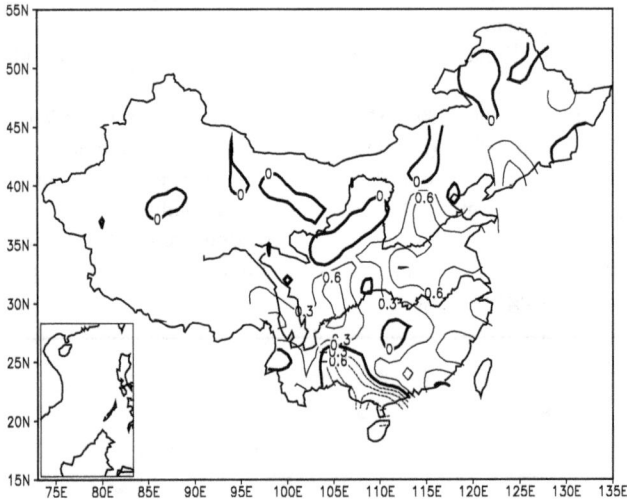

Figure 9. Difference of summer rainfall composites between positive and negative AMO phases (unit: mm d^{-1}). Thick lines are "0"-isolines. (from [36])

In order to verify the influence of the AMO on the summer rainfall in China, a coupled atmosphere-ocean general circulation model was utilized to perform numerical experiments [36], in which the SSTAs related with the positive or negative AMO phase in the Atlantic are specified, and the atmosphere-ocean interactions outside the Atlantic are allowed. The results of numerical experiments show in the positive AMO phase strong East Asian summer monsoon and more rainfall in eastern China, and reversed rainfall anomalies can also be found in the AMO negative phase, which resemble the observed summer monsoon rainfall in China in association with the AMO phases as shown in Figure 9. The model results suggested that, through coupled feedback, warm AMO phase result in the warmer SST in the eastern Indian Ocean and maritime continent. The warmer SST strengthens the convective heating over there and leads to an anticyclone anomaly over western North Pacific, which is responsible for the more rainfall over China.

4.2. Triple mode of North Atlantic SSTAs

In association with the seasonal march of the East Asian summer monsoon, climatologically the summer rainfall in eastern China moves northward in a stepwise way. The more rainfall zone exists mainly in southern China from mid-May to early June, which is referred to as the pre-Meiyu rainy season. In the period from mid-June to mid-July, the more rainfall zone shifts northward and stays around the middle and lower reaches of Yangtze River and Huai River valleys, which is named as the Meiyu. From late July to early August the more rainfall zone appears further northwards in North China.

By using long-term Meiyu dataset from 1885 to 2005 from the National Climate Center of China Meteorology Administration (CMA) and monthly SST data of HadISST from 1870 to date from the Hadley Center of the UK Meteorological Office [37], it is found that the decadal variability of the Meiyu precipitation amount and duration are closely correlated with a triple mode of North Atlantic SSTAs in the preceding winter [38]. The triple mode constitutes of three zonally elongated centers of SSTAs in the North Atlantic, with two positive SSTAs centered at about 50°N to the south of Greenland and at about 15°N in the subtropics, respectively, and one negative center to the east of the U. S. continent at about 30°N. An index I_{NA} of the triple mode is defined as the difference between normalized SSTAs averaged over the areas around two northern centers. Figure 10 shows the 9-16-yr band-pass filtered normalized Meiyu rainfall amount and duration as well as the I_{NA} index. In general, the Meiyu rainfall amount and duration coincide well with the I_{NA} index in the last century except short periods of 1930-1940 and 1985-1995. The triple mode of Atlantic SSTAs can persist from winter until late spring and excite a stationary wave-train propagating from west Eurasia to East Asia, exerting an impact on summer rainfall over eastern China.

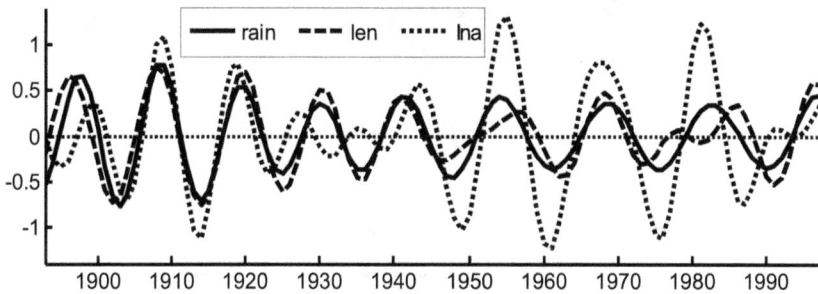

Figure 10. The 9-16-yr band-pass filtered normalized Meiyu rainfall amount (solid line) and duration (dashed line), and the I_{NA} index (dotted line). (from [38])

4.3. Spring Arctic sea ice

Utilizing the monthly mean Arctic sea ice concentration (SIC) dataset for the period of 1961–2007 obtained from the British Atmospheric Data Centre (BADC, http://badc.nerc.ac.uk/data/hadisst/), and a monthly 513-station rainfall dataset for China obtained from the National Meteorological Information Centre of China spanning the period from 1968 to 2005, a statistical relationship between spring Arctic SIC and Chinese summer rainfall is identified using singular value decomposition (SVD) [39]. The leading SVD mode accounts for 19% of their co-variance. As shown in Figure 11, both Arctic SIC and summer rainfall in the leading SVD mode display a coherent inter-annual variability (the correlation is 0.83, Figure 11a) and apparent inter-decadal variations with turning points occurring around 1978 and 1992, respectively. For spring SIC, negative phases were frequent during the periods 1968-1978 and 1993-2005, and separated by dominant positive phases. The

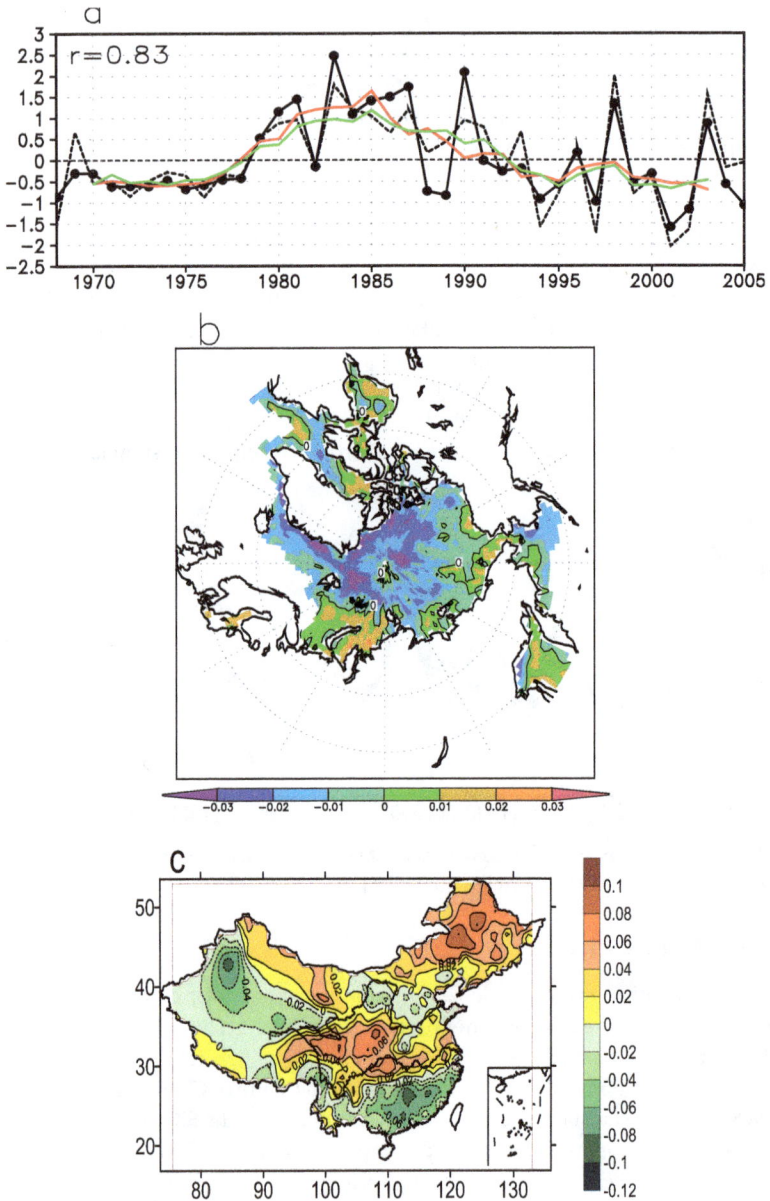

Figure 11. (a) Normalized time series of spring Arctic SIC (solid line) and summer rainfall (dashed line) variations in the leading SVD mode, the red and green lines denote their 5-year running means, respectively, (b) the spatial distribution of spring Arctic SIC in the leading SVD mode, (c) same as in (b) but for the summer rainfall. In (b) and (c), units are arbitrary. (from [39])

corresponding SIC anomalies were positive in most of Eurasian marginal seas, with negative SIC anomalies in the Arctic Basin, the Beaufort Sea, and the Greenland Sea during 1979-1992 (Figure 11b). Opposite SIC anomalies occurred during the periods 1968-1978 and 1993-2005. During the period 1979-1992, positive rainfall anomalies frequently appeared in northeast China and central China between the Yangtze River and the Yellow River (28°-36°N), with negative rainfall anomalies in south and southeast China (Figure 11c). Opposing spatial distribution of summer rainfall anomalies frequently occurred in the period 1968-1978 and 1993-2005.

The SVD analysis shows that in the decadal time scale the decreased (increased) spring SIC in the Arctic Ocean and the Greenland Sea corresponds to increased (decreased) summer rainfall in northeast China and central China between the Yangtze River and the Yellow River; and decreased (increased) rainfall in southern China. Figure 12 shows the regressed 500 hPa geopotential height in summer by the time series of the spring SIC leading SVD mode. The summer 500 hPa height anomalies associated with the spring SIC in the leading SVD mode show a Eurasian wave train structure, which originates in northern Europe and extends southeastwards to northeast China, and a south-north dipole structure over East Asia south to Lake Baikal. Compared to the regressed height anomalies in summer by the time series of the leading SVD mode for the spring Eurasian SWE (Figure 6b), the wave-train over the Eurasian continent is quite similar. In fact, the positive correlation exists between the time series of the leading SVD mode for the spring Eurasian SWE (Figure 4c) and that for the spring Arctic SIC (Figure 11a), and the correlation coefficient is 0.64 in the period of 1979-2004. Therefore, although a Eurasian wave train in association with the Arctic SIC can be found at 500 hPa lasting from spring to summer and extending southeastwards to northeast China, the spring Arctic SIC provides a complementary impact on the summer rainfall over China. The Eurasian wave train should result from combined effects of both spring Arctic SIC and Eurasian snow cover.

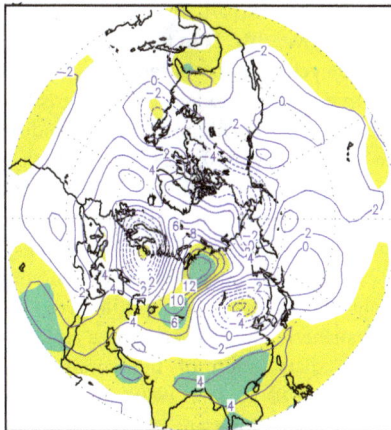

Figure 12. Regressed 500 hPa geopotential height in summer by the time series of the spring SIC leading SVD mode (unit: gpm). The yellow and blue shadings represent height anomalies exceed the 0.05 and 0.01 confidence levels, respectively. (from [39])

5. Impacts of the tropical Pacific and Indian Oceans

5.1. Central and eastern equatorial Pacific SSTAs

In the central and eastern equatorial Pacific, there exists large inter-annual variability of the SST. An El Niño is referred to as the period when the ocean temperature of upper layer in the central and eastern equatorial Pacific rises abnormally. In the atmosphere over the tropical Pacific, there is a widespread inter-annual oscillation in sea-level pressure between the area near northern Australia and that in the southeastern Pacific. This oscillation is called Southern Oscillation. Because of the intrinsic relationship between El Niño and Southern Oscillation, they are combined and referred to as ENSO (El Niño/Southern Oscillation). The El Niño is the most outstanding inter-annual variability in oceans. The occurrence of El Niño changes greatly the pattern of the thermal heating of the atmosphere, which leads to large circulation anomalies.

The El Niño has significant effects not only in the tropical region, but also in the extratropics. The influence of El Niño on the climate over East Asia has been intensively studied by many investigators [40]. It is found that the effect of El Niño on the summer rainfall in eastern China depends on the phase of the El Niño [41]. In the El Niño developing phase, more summer rainfall exists in the Yangtze and Huai River valleys and less over northern and southern China, whereas in the El Niño decaying phase opposite summer rainfall anomalies appear. The impact of El Niño on the East Asian climate is through an anomalous anticyclone appeared over the western North Pacific in the lower troposphere [12], which can significantly affect the rainfall in China by altering water vapor transport over East Asia [42]. This anomalous anticyclone acts as a bridge connecting the warm events in the eastern tropical Pacific and the East Asian monsoon [43].

The above-mentioned researches focus on the inter-annual time-scale. The El Niño also appears in the inter-decadal time-scale [44]. By examining the variation of the 5-year running mean of SSTAs averaged over 0-10°N in the central and eastern equatorial Pacific, a concept of 'inter-decadal ENSO cycle' was proposed [18]. An 'inter-decadal El Niño event' appeared in the period from middle 1950s to late 1960s when SSTAs in central and eastern equatorial Pacific was above normal, with highest 0.6°C in middle 1960s; an 'inter-decadal La Niña event' from early to late 1970s with -0.6°C below normal; and from late 1970s to late 1980s an 'inter-decadal El Niño event' again with 0.4°C above normal. Same as the impact of El Niño on the summer rainfall in eastern China in the inter-annual time scale, the impact of the 'inter-decadal ENSO cycle' on the inter-decadal change of the summer rainfall in eastern China also depends on the phase of the 'inter-decadal El Niño event' [18]. In the decaying phase of the 'inter-decadal El Niño event' from middle 1960s to the middle 1970s, more summer rainfall appeared in northern and southern China and less in the Yangtze and Huai River valleys, while opposite distribution of the summer rainfall anomalies in the developing phase from middle 1970s to late 1980s. Such inter-decadal changes of the summer monsoon rainfall in eastern China are in good agreement with those shown in Figure 3a. Therefore, the inter-decadal changes of SSTAs in the central and eastern equatorial Pacific is possibly an important factor for the inter-decadal changes of summer monsoon rainfall in eastern China.

5.2. Western North Pacific and Indian Ocean SSTAs

The empirical orthogonal function (EOF) analysis is applied to the summer SST in the area
of 0°-40°N and 100°-180°E in the western North Pacific during the period 1968-2002 [45]. The
leading EOF mode, which has a variance contribution of 30.5%, and its time series are
shown in Figure 13. The leading EOF mode shows a wholly consistent SST distribution
(Figure 13a). The time series indicates a sustained inter-decadal variability in the first
principle component (PC1) and an inter-decadal climate shift occurred in the late 1980s
(Figure 13b).

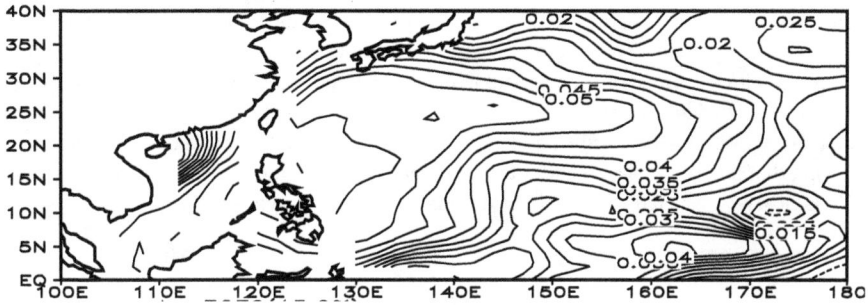

(a)

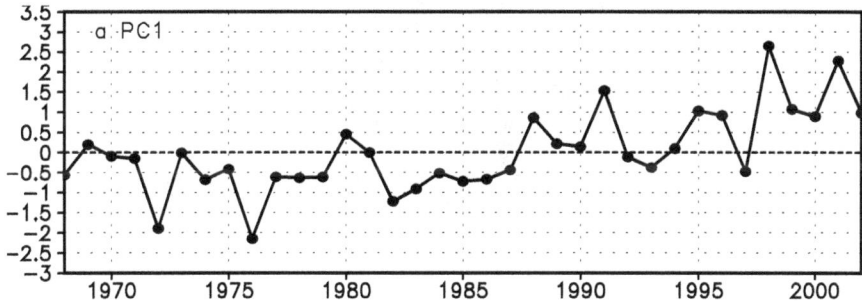

(b)

Figure 13. (a) Leading EOF mode of the sea surface temperature in western North Pacific and (b) its
time series. (from [45])

As seen from Figure 13b, the values of the PC1 were basically negative before 1988, after
then became almost positive. Referring to Figure 13a, we know that in the late 1980s, the
leading EOF mode of the uniformly consistent SST in western North Pacific had a significant
inter-decadal climate shift, namely the SSTAs were negative before the late 1980s but
positive afterwards. In North Pacific, the most striking inter-decadal variability is the Pacific
decadal oscillation (PDO), and a climate shift for PDO occurred in late 1970s [46]. It can be
seen from Figure 13b that the leading EOF mode of the western North Pacific SST did not

show a climate shift in the late 1970s. With correlation coefficient between the summer mean PDO index and the leading mode being -0.06, the leading mode is not related with the PDO. Therefore, the leading EOF mode of the summer SST in the western North Pacific is independent from the PDO.

Here we can see that the inter-decadal shift of the western North Pacific SSTAs in the late 1980s coincided well with the climate shift of the East Asian summer monsoon. To illustrate summer rainfall anomalies in China related to the inter-decadal variation of the western North Pacific SSTAs, a 5-year running mean is performed on both the summer rainfall and PC1 time series in 1968-2002, and their correlation coefficients [7] are shown in Figure 14. Positive correlation coefficients exist in southern China to the south of the Yangtze River valley, indicating that lower (higher) western North Pacific SST is associated with less (more) rainfall in southern China. Compared Figure 14 to Figure 5, the distribution of the correlation coefficients resembles the summer rainfall anomalies, especially in southern China. The western North Pacific had lower SST before the late 1980s and higher afterwards, implying that the inter-decadal shift of the summer rainfall to the southern China in the late 1980s is closely associated with the inter-decadal change of the western North Pacific SST.

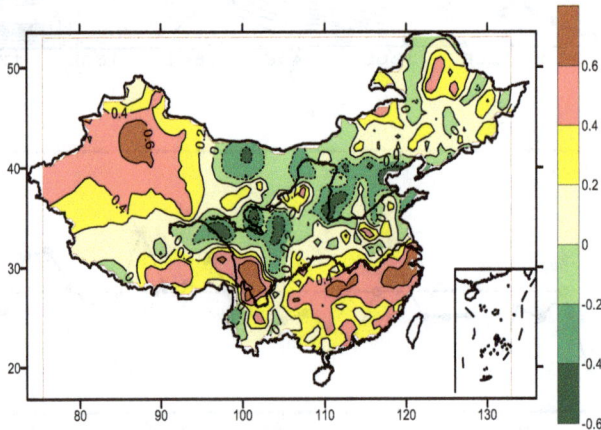

Figure 14. Correlation coefficients between 5-year running means of the summer rainfall in China and the PC1 time series of the western North Pacific SST. (from [7])

By checking the EOF modes of the summer rainfall in China in the period of 1958-2001, it is found the second EOF mode exhibits a north-south dipole distribution with opposite summer rainfall anomalies to the north and south of the Yangtze River valley [47]. An inter-decadal shift occurred in the late 1980s when less rainfall to the south of the Yangtze River valley and more to the north changed to be opposite distribution with more to the south of the Yangtze River valley and less to its north. The summer rainfall increasing after the late 1980s to the south of the Yangtze River valley is closely related with the warming of the summer SST in western North Pacific and northern Indian Ocean. The warmer summer SST in western North Pacific may affect the rainfall in eastern China through stimulating anomalous atmospheric circulations in the form of the EAP (East Asian-Pacific) teleconnection pattern over East Asia [48].

One of the most striking features of the inter-decadal shift of the East Asian summer monsoon is the westward extension of the western Pacific subtropical high (WPSH) [7,49,50], which affects the water vapor transport [1,51] and thus determines the location of the abnormal monsoon rainfall zone. Based on the ERA-40 Reanalysis data from the European Centre for Medium-Range Forecast (ECMWF) and taken the 5880 gpm isoline at 500 hPa in summer as the representative of the WPSH, Figure 15 shows the contour of the 5880 gpm at 500hPa in summer averaged in 1975-1989 and that averaged in 1990-2001, respectively, to illustrate the extent of WPSH [7]. It can be seen that, compared to that in the period of 1975-1989, the WPSH in the period of 1990-2001 became stronger, stretching farther westward with a larger south-north extent, which is favorable for the development of the southerlies over southern China. The strengthened southerlies are beneficial to the strengthening of the water vapor transport and therefore more precipitation in the south of the Yangtze River. Here we can see that, corresponding to the decadal climate shift of the East Asian summer monsoon in the late 1980s, both the summer circulation over East Asia and the summer rainfall in China changed significantly.

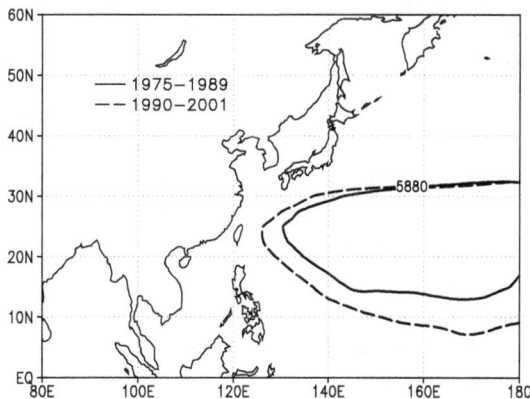

Figure 15. Summer (JJA) mean 5880 gpm contour at 500hPa averaged in 1990-2001 (dashed line) and in 1975-1989 (real line), respectively. (from [7])

Since the late 1970s, the SST in the tropical Indian Ocean-western Pacific (IWP) has increased over 0.4°C relative to that in the previous two decades [53]. By utilizing five atmospheric general circulation models (AGCMs), it is demonstrated that the forcing of the warming in IWP area is responsible for the westward extension of the WPSH [50]. All five AGCMs well reproduced the features that the WPSH extended further westward in warmer IWP period compared to that in the cooler period. Two ways for the IWP warming affecting the WPSH westward extension were proposed [50]. First, through Walker circulation the IWP heating leads to the weakening of convection in the central and eastern equatorial Pacific; the convective cooling anomalies force a response of the Gill-type anticyclone [54] in the North Pacific, which is favorable for the westward extension of WPSH. Second, the forced Kelvin wave response by the monsoon diabatic heating [55] reinforces the low-level equatorial flank of WPSH, while the poleward flow along the western flank of WPSH is intensified because of the Sverdrup vorticity balance.

6. Conclusion

Observations show that from 1958 to 2011 the predominant feature of the summer monsoon rainfall variation over eastern China is its inter-decadal variability besides the inter-annual variability, and no clear trend can be found for the summer monsoon rainfall. Same as the rainfall, strong inter-decadal variability also appears in the East Asian summer monsoon index (WNP-EASM index), and no trend can be found for the index. In addition to the inter-decadal variation, larger amplitudes and shorter periods for both East Asian summer monsoon index and monsoon rainfall appeared after the end of 1980s, implying a larger variability and more frequent variation of the East Asian summer monsoon have occurred since then.

In association with the inter-decadal variability of the East Asian summer monsoon, climate shifts for the East Asian summer monsoon appeared at middle 1970s, late 1980s and early 2000s, respectively. Corresponding to the climate shift in the middle 1970s, the summer rainfall increased in the area over the middle and lower reaches of the Yangtze River valley and decreased over southern and northern China. After the weakening of the East Asian summer monsoon in the late 1980s, more rainfall appeared in southern China. From the early 2000s, the East Asian summer monsoon strengthened and the rainfall decreased around Yangtze River valley, and increased to the south of about 25°N in South China and between about 31°N-36°N to the north of Yangtze River valley.

The inter-decadal variability of the Eurasian snow cover, Arctic sea ice, and SSTAs in Atlantic, tropical Pacific and Indian Ocean can be causes of the inter-decadal variability of the East Asian summer monsoon and associated monsoon rainfall over eastern China. By examine the leading SVD modes between springtime SWE over Eurasian Continent and summertime rainfall over China in the period 1979-2004, it shows an apparent inter-decadal shift occurred in the late 1980s. The leading SVD mode of the spring SWE displays strong inter-decadal variation with persistent negative phases in 1979–1987 and frequent positive phases afterwards. For the anomalous summer rainfall field corresponding to the leading SVD mode, positive anomalies appear in southern China to the south of the Yellow River valley, and negative anomalies emerge in most parts of northern China to the east of 95°E. In the inter-decadal time scale, the decreased (increased) spring SIC in the Arctic Ocean and the Greenland Sea corresponds to increased (decreased) summer rainfall in northeast China and central China between the Yangtze River and the Yellow River; and decreased (increased) rainfall in south China. The effects of spring Eurasian snow cover and Arctic sea ice on the summer rainfall in China are through a wave train of 500 hPa geopotential height anomalies, which persist from spring to summer and exert significant effect on the summer rainfall in China.

The composite difference of summer rainfall between positive and negative Atlantic AMO phases shows that in the most parts of eastern China the summer rainfall increases, implying that the warm AMO phase is favorable for more summer rainfall in eastern China. The results of numerical experiments by a coupled atmosphere-ocean general circulation mode suggest that, through coupled feedback, warm AMO phase result in the warmer SST in the eastern Indian Ocean and maritime continent. The warmer SST strengthens the convective heating over there and leads to an anticyclone anomaly over western North Pacific, which is responsible for the more rainfall over China.

The leading EOF mode of SSTAs in western North Pacific shows a significant inter-decadal climate shift in the late 1980s, namely the SSTAs were negative before the late 1980s but positive afterwards in the period 1968-2002. Lower (higher) western North Pacific SST is associated with less (more) rainfall in southern China. The inter-decadal shift of the summer rainfall over southern China in the late 1980s is closely related with the inter-decadal change of the western North Pacific SSTAs. The warmer summer SST in western North Pacific may affect the rainfall in eastern China through stimulating anomalous atmospheric circulations in the form of the EAP (East Asian-Pacific) teleconnection pattern over East Asia. The summer rainfall in eastern China also depends on the phase of the 'inter-decadal El Niño event'. In the decaying phase of the 'inter-decadal El Niño event' from middle 1960s to the middle 1970s, more summer rainfall appeared in northern and southern China and less in the Yangtze and Huai River valleys, while opposite distribution of the summer rainfall anomalies in the developing phase from middle 1970s to late 1980s. The location of the abnormal monsoon rainfall zone over eastern China is related with the westward extension of the western Pacific subtropical high, which is determined by the forcing of the warming in the tropical Indian Ocean-western Pacific since the late 1970s.

Author details

Renhe Zhang, Bingyi Wu, Jinping Han and Zhiyan Zuo
Chinese Academy of Meteorological Sciences, Beijing, China

Acknowledgement

The authors would like to thank the book editors for their constructive comments and suggestions. This work is supported by the National Natural Science Foundation of China (Grant No. 40921003) and the International S&T Cooperation Project of the Ministry of Science and Technology of China under Grant No. 2009DFA21430.

7. References

[1] Zhang R (2001) Relations of Water Vapor Transports from Indian Monsoon with Those over East Asia and the Summer Rainfall in China. Adv. Atmos. Sci. 18: 1005-1017.

[2]   IPCC (2007) Climate Change 2007: The Physical Science Basis. Contribution of Working Group I to the Fourth Assessment Report of the Intergovernmental Panel on Climate Change. S. Solomon et al., Eds., Cambridge: Cambridge Univ. Press. 996 p.

[3]   Wang B, Li T, Ding Y, Zhang R, Wang H (2005) Eastern Asian-Western North Pacific Monsoon: A Distinctive Component of the Asian-Australian Monsoon System. In: Chang C-P, Wang B, Lau N-C, Editors. The Global Monsoon System: Research and Forecast. WMO/TD No. 1266 (TMRP Report No. 70), pp. 72-94.

[4]   Ding Y (2007) The Variability of the Asian Summer Monsoon. J. Meteor. Soc. Japan 85B: 21-54.

[5]   Huang R, Chen J, Huang G (2007) Characteristics and Variations of the East Asian Monsoon System and its Impacts on Climate Disasters in China. Adv. Atmos. Sci. 24, 993-1023.

[6]   Wang S, Gong D, Ye J, Chen Z (2000) Seasonal Precipitation Series of Eastern China since 1880 and the Variability. Acta Geographica Sinica 55: 281-293. (in Chinese)

[7]   Zhang R, Wu B, Zhao P, Han J (2008) The Decadal Shift of the Summer Climate in the Late 1980s over Eastern China and Its Possible Causes. Acta Meteor. Sinica 22: 435-445.

[8]   Ding Y, Wang Z, Sun Y (2008) Inter-decadal Variation of the Summer Precipitation in East China and its Association with Decreasing Asian Summer Monsoon. Part I: Observed Evidences. Inter. J. Climatol. 28: 1139–1161.

[9]   Huang R, Chen J, Liu Y (2011) Interdecadal Variation of the Leading Modes of Summertime Precipitation Anomalies over Eastern China and its Association with Water Vapor Transport over East Asia. Chin. J. Atmos. Sci. 35: 589-606. (in Chinese)

[10] Huang G (2006) Global Climate Change Phenomenon Associated with the Droughts in North China. Climatic and Environ. Res. 11: 270-279. (in Chinese)

[11] Tao S, Chen L (1987) Review of Recent Research on the East Asian Summer Monsoon in China. In: Chang C-P, Krishnamurti T N, Editors. Monsoon Meteorology. Oxford: Oxford University Press. pp. 60–92.

[12] Zhang R, Sumi A, Kimoto M (1996) Impact of El Niño on the East Asian Monsoon: A Diagnostic Study of the '86/87 and '91/92 Events. J. Meteor. Soc. Japan 74: 49–62.

[13] Wang B, LinHo, Zhang Y, Lu M-M (2004) Definition of South China Sea Monsoon Onset and Commencement of the East Asia Summer Monsoon. J. Climate 17: 699-710.

[14] Zhang R, Sumi A, kimoto M (1999) A Diagnostic Study of the Impact of El Niño on the Precipitation in China. Adv. Atmos. Sci. 16: 229-241.

[15] Wang B, Wu R, Lau K M (2001) Interannual Variability of the Asian Summer Monsoon: Contrasts between the Indian and the Western North Pacific-East Asian Monsoon. J. Climate 14: 4073-4090.

[16] Kalnay E, Coauthors (1996) The NCEP/NCAR 40-year Reanalysis Project. Bull. Amer. Meteor. Soc. 77: 437-471.

[17] Chen L (1999) Regional Features of Interannual and Interdecadal Variations in Summer Precipitation Anomalies over North China. Plateau Meteor. 18: 477-485. (in Chinese)

[18] Huang R, Xu Y, Zhou L (1999) The Interdecadal Variation of Summer Precipitations in China and the Drought Trend in North China. Plateau Meteor. 18: 465-476. (in Chinese)

[19] Ellis A W, Leathers D J (1998) The Effects of a Discontinuous Snow Cover on Lower Atmospheric Temperature and Energy Flux Patterns. Geophys. Res. Lett. 25: 2161-2164.

[20] Chen L, Yan Z (1979) Impact of Himalayan Winter-Spring Snow Cover on Atmospheric Circulation and Southern China Rainfall in the Rainy Season. In: Yangtze River Regulating Office, Editor. Collected papers on Medium- and Long-term Hydrological and Meteorological Forecasts (Vol. 1). Beijing: Water Conservancy and Power Press. pp. 185-194. (in Chinese)

[21] Chen L, Yan Z (1981) A Statistical Study of the Impact of Himalayan Winter-Spring Snow Cover Anomalies on the Early Summer Monsoon. In: Yangtze River Regulating Office, Editor. Collected Papers on Medium- and Long-term Hydrological and Meteorological Forecasts (Vol. 2). Beijing: Water Conservancy and Power Press. pp. 133-141. (in Chinese)

[22] Chen L, Wu R (2000) Interannual and Decadal Variations of Snow Cover over Qinghai-Xizang Plateau and Their Relationships to Summer Monsoon Rainfall in China. Adv. Atmos. Sci. 17: 18-30.

[23] Wu T, Qian Z (2003) The Relation between the Tibetan Winter Snow and the Asian Summer Monsoon and Rainfall: An Observational Investigation. J. Climate 16: 2038-2051.

[24] Zhang Y, Li T, Wang B (2004) Decadal Change of the Spring Snow Depth over the Tibetan Plateau: The Associated Circulation and Influence on the East Asian Summer Monsoon. J. Climate 17: 2780-2793.

[25] Wu R, Kirtman B P (2007) Observed Relationship of Spring and Summer East Asian Rainfall with Winter and Spring Eurasian Snow. J. Climate 20: 1285-1304.

[26] Armstrong R L, Brodzik M J, Knowles K, Savoie M (2005) Global Monthly EASE-Grid Snow Water Equivalent Climatology. National Snow and Ice Data Center, Boulder, CO. Available: http://nsidc.org/data/nsidc-0271.html.

[27] Wu B, Yang K, Zhang R, (2009) Eurasian Snow Cover Variability and its Association with Summer Rainfall in China. Adv. Atmos. Sci. 26: 31-44.

[28] Yang S, Xu L (1994) Linkage between Eurasian Winter Snow Cover and Regional Chinese Summer Rainfall. Int. J. Climatol. 14: 739-750.

[29] Zuo Z, Zhang R, Wu B (2012) Inter-decadal Variations of Springtime Rainfall over Southern China Mainland for 1979-2004 and Its Relationship with Eurasian Snow. Sci. China Earth Sci. 55: 271-278.

[30] Zuo Z, Zhang R, Wu B, Rong X (2012) Decadal Variability in Springtime Snow over Eurasia: Relation with Circulation and Possible Influence on Springtime Rainfall over China. Int. J. Climatol. 32: 1336-1345.

[31] Barnston A, Livezey R E (1987) Classification, Seasonality, and Persistence of Low-frequency Circulation Patterns. Mon. Wea. Rev. 115: 1083-1126.

[32] Delworth T L, Mann M E (2000) Observed and Simulated Multidecadal Variability in the Northern Hemisphere. Clim. Dyn. 16: 661-676.

[33] Kerr R A (2000) A North Atlantic Climate Pacemaker for the Centuries. Science 288: 1984-1985.

[34] Knight J R, Folland C K, Scaife A A (2006) Climate Impacts of the Atlantic Multidecadal Oscillation, Geophys. Res. Lett. 33: L17706.

[35] Dong B, Sutton R T, Scaife A A (2006) Multidecadal Modulation of El Niño– Southern Oscillation (ENSO) Variance by Atlantic Ocean Sea Surface Temperatures. Geophys. Res. Lett. 33: L08705.

[36] Lu R, Dong B, Ding H (2006) Impact of the Atlantic Multidecadal Oscillation on the Asian Summer Monsoon, Geophys. Res. Lett. 33: L24701.

[37] Rayner N A, Parker D E, Horton E B, Folland C K, Alexander L V, Rowell D P, Kent E C, Kaplan A (2003) Global Analyses of Sea Surface Temperature, Sea Ice, and Night Marine Air Temperature since the Late Nineteenth Century. J. Geophys. Res. 108: 4407.

[38] Gu W, Li C, Wang X, Zhou W, Li W (2009) Linkage between Mei-Yu Precipitation and North Atlantic SST on the Decadal Timescale. Adv. Atmos. Sci. 26: 101–108.

[39] Wu B, Zhang R, Wang B, D'Arrigo R (2009) On the Association between Spring Arctic Sea Ice Concentration and Chinese Summer Rainfall. Geophys. Res. Lett. 36: L09501.

[40] Huang R, Chen W, Yan B, Zhang R (2004) Recent Advances in Studies of the Interaction between the East Asian Winter and Summer Monsoons and ENSO cycle. Adv. Atmos. Sci. 21: 407–424.

[41] Huang R, Wu Y (1989) The Influence of ENSO on the Summer Climate Change in China and its Mechanisms. Adv. Atmos. Sci. 6: 21-32.

[42] Zhang R, Sumi A (2002) Moisture Circulation over East Asia during El Niño Episode in Northern Winter, Spring and Autumn. J. Meteor. Soc. Japan 80: 213-227.

[43] Wang B, Wu R, Fu X (2000) Pacific-East Asian Teleconnection: How does ENSO Affect East Asian Climate? J. Climate 13: 1517-1536.

[44] Zhang Y, Wallace J M, Battisti D S (1997) ENSO-like Interdecadal Variability: 1900–93. J. Climate 10: 1004–1020.

[45] Wu B, Zhang R (2007) Interdecadal Shift in the Western North Pacific Summer SST Anomaly in the Late 1980s. Chin. Sci. Bull. 52: 2559-2564.

[46] Trenberth K E, Hurrell J W (1994) Decadal Atmosphere-Ocean Variations in the Pacific. Climate Dyn. 9: 303-319.

[47] Han J, Zhang R (2009) The Dipole Mode of the Summer Rainfall over East China during 1958–2001. Adv. Atmos. Sci. 26: 727–735.

[48] Huang R, Sun F (1992) Impacts of the Tropical Western Pacific on the East Asian Summer Monsoon. J. Meteor. Soc. Japan 70: 243–256.

[49] Gong D, Ho C H (2002) Shift in the Summer Rainfall over the Yangtze River Valley in the Late 1970s. Geophys. Res. Lett. 29: 1436.

[50] Zhou T, Yu R, Zhang J, Drange H, Cassou C, Deser C, Hodson D L R, Sanchez-Gomez E, Li J, Keenlyside N, Xin X, and Okumura Y (2009) Why the Western Pacific Subtropical High Has Extended Westward since the Late 1970s? J. Climate 22: 2199-2215.

[51] Zhou T, Yu R (2005) Atmospheric Water Vapor Transport Associated with Typical Anomalous Summer Rainfall Patterns in China. J. Geophys. Res. 110: D08104.

[52] Simmons A J, Gibson J K (2000) The ERA-40 Project Plan. ERA-40 Project Report Series 1. ECMWF, Reading, United Kingdom, 62p.

[53] Webster P J, Moore A, Loschnigg J P, Leben R R (1999) Coupled Ocean–Atmosphere Dynamics in the Indian Ocean during 1997–98. Nature 401: 356–360.

[54] Gill A E (1980) Some Simple Solutions for Heat Induced Tropical Circulation. Quart. J. Roy. Meteor. Soc. 106; 447–462.

[55] Rodwell M J, Hoskins B J (1996) Monsoons and the Dynamics of Deserts. Quart. J. Roy. Meteor. Soc. 122: 1385–1404.

Chapter 9

Effect of Water Resources in the Queretaro River: Climate Analysis and Other Changes

E. González-Sosa, N.M. Ramos-Salinas, C.A. Mastachi-Loza and R. Becerril-Piña

Additional information is available at the end of the chapter

1. Introduction

Due to the imminent change of the continental surface and ocean temperature caused by the increase of greenhouse gases it is estimated that in the near future there may be a temperature increase of 3-4°C. The average temperature increase could have repercussions in vast regions of the planet affecting both socio-economic and living conditions, infrastructure, development and ecosystems. Some scientists already predicted changes in the water cycle since the 70's, including an increase in temperature of 1.2 to 2.0°C [1]. The work conducted by the IPCC [2], reports alarming results; however, there is uncertainty about whether the effects are part of the climate variables or product of the effects of global climate changes also accentuated in certain regions of the world. In semi-arid regions as well as in other regions, the hydrological cycle is disturbing the regime even if the temperature changes were minor. An increase in the annual measured temperature of 1-2°C could reduce surface runoff between 40-70% and a 10% of the precipitation [3-4]. The consequences of such changes should also be considered as key and the link to other problems such as increased and drought severity, heat waves, floods, the water quality and the subsidence in urban centers, changes in land use and the availability of water in peri-urban watersheds. Furthermore, to investigate the changes of peri-urban watersheds and subsidence in large urban centers is of great relevance to the so-called emerging countries. The urbanization process transforms the land into different uses, and the spread urbanization, in many cases, cause forest and agricultural lands to disappear. Urbanization also affects the natural behavior of hydrological processes. Scientists recognize, however, that such impact is greater, for example, the global expansion of croplands since 1850 has transformed some 6 million km² of wood forestlands and 4.7 million km² of savannas/grasslands/steppes. Within these categories, 1.5 and 0.6 million km² of croplands have already been abandoned [5]. By means of identifying the changes, land use can be the answer to identifying the main components of the hydrological processes that have been altered by human activities [6].

Throughout history, the management of water resources has played an important role in the development and civilization settlements. Within the last years, urban and peri-urban areas in Mexico have grown exponentially. Widespread urbanization generates land heterogeneity and fragmentation, diminishing natural resources and water supply. In fact, the settlements around of great cities produce changes in peri-urban catchments; therefore, the comprehension of their evolution is crucial and necessary for successful water resources management and territorial land use planning of high urbanization regions. The new settlements and land use changes are the articulation elements with the evolution of water resources [7]. The main effect produced by concentration in urban areas may be the decreasing in infiltration and increasing in surface runoff [8]. Fitzharris [9] found that the changes of land uses can alter the evapotranspiration, and hence runoff. The increase of runoff is produced by an increase of urbanization. These may be the study elements in order to solve water security concerns and water scarcity. Forefront, the increasing demands and decreasing supplies are the evaluation of disturbed level of the catchments to long term scenarios that are likely to be faced in the future [10]. In Brazil, it has been shown that forested areas have ten times the infiltration capacity of pastures deforestation, thus leading to a decrease in infiltration and increase in runoff. In Tanzania, the annual runoff from cultivated catchments is 30%-60% higher than that from similar catchments with original evergreen forest coverage [11]. The subsidence is another collateral problem in the world as well as in other large cities in Mexico caused by aquifers overexploitation. This vision seems consistent from the geological point of view; however, the subsidence problem has a greater importance in large urbanized areas due to its interaction with the environment and the collateral effects of climate change which generate great concentrated runoff pollutant loads and heavy metals, from cleaning products of streets and avenues, both negative as well as health matters. That is, the fractures caused by subsidies are common aquifer contamination ways, such are preferential flow paths to increase a point-source contamination in very short time, even hours, depending on the network connectivity among aquifer free surface. In other words, it is important to point out that recharge wells are part of efforts to mitigate soil cracks formation caused by subsidence; nevertheless, such become adverse because they also are sources of aquifers contamination by injecting contaminants throughout the internal flows. Those are generated through connections between soil cracks and because the deep recharge takes place in very short-terms compared to natural recharge times. In some cases, recharges take over hundreds of years, depending on the depth of the unsaturated zone and hydrodynamic properties of the soil. Mainly, it is very important if precipitation patterns are modifying by the effects of climate change.

2. Water resources in Queretaro river basin

Since the settlement of Queretaro City (Figure 1) in 1532 by the Otomi Kho-nin, also known as Fernando de Tapia, the population increase had a lien agricultural and industrial development. In 1743, the population was 5,849 habitants and after remained constant over 30,000 habitants during 1910 through 1940. Progress grew rapidly after the end of the Revolution. In 1950 the population reached 50,000 & 130,000; by 1970: 641,386 in 1980, and by 2010 the city had overflowed other counties surpassing one million in habitants. During the same period, the territorial expansion produced an increase in the population density and modified substantially the land uses. The availability of water resources has also changed. Currently, 83% of the population is urban and 95% is regarded as rural–peri-urban. Figure 2 shows a trend of population, the spread urban surface and water supply evolution. Visibly, the trend is exponential from 1950, a similar condition for urban surface and water supply. The water supply flow in the beginning of the colonial period and the industrial development of Queretaro in 1700 was of 30 l s⁻¹, from pumping springs located 100 km North of the city, and increased to 2360 l s⁻¹ which was transferring from 200 km of neighbor basin, such representing a 147.5 times increase (Figure 2). The most important hydraulic infrastructure of water supply was carried out between 1726 and 1735, the 1280 m long aqueduct. The population estimated at the time was about 46 472 habitants.

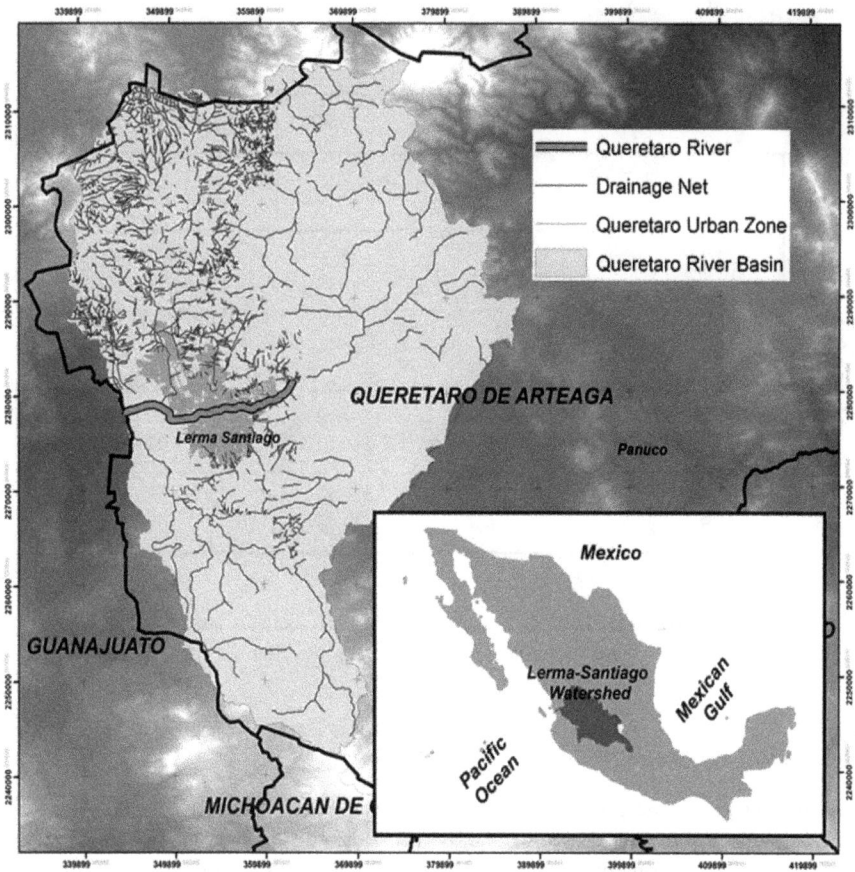

Figure 1. Location of Queretaro River Basin.

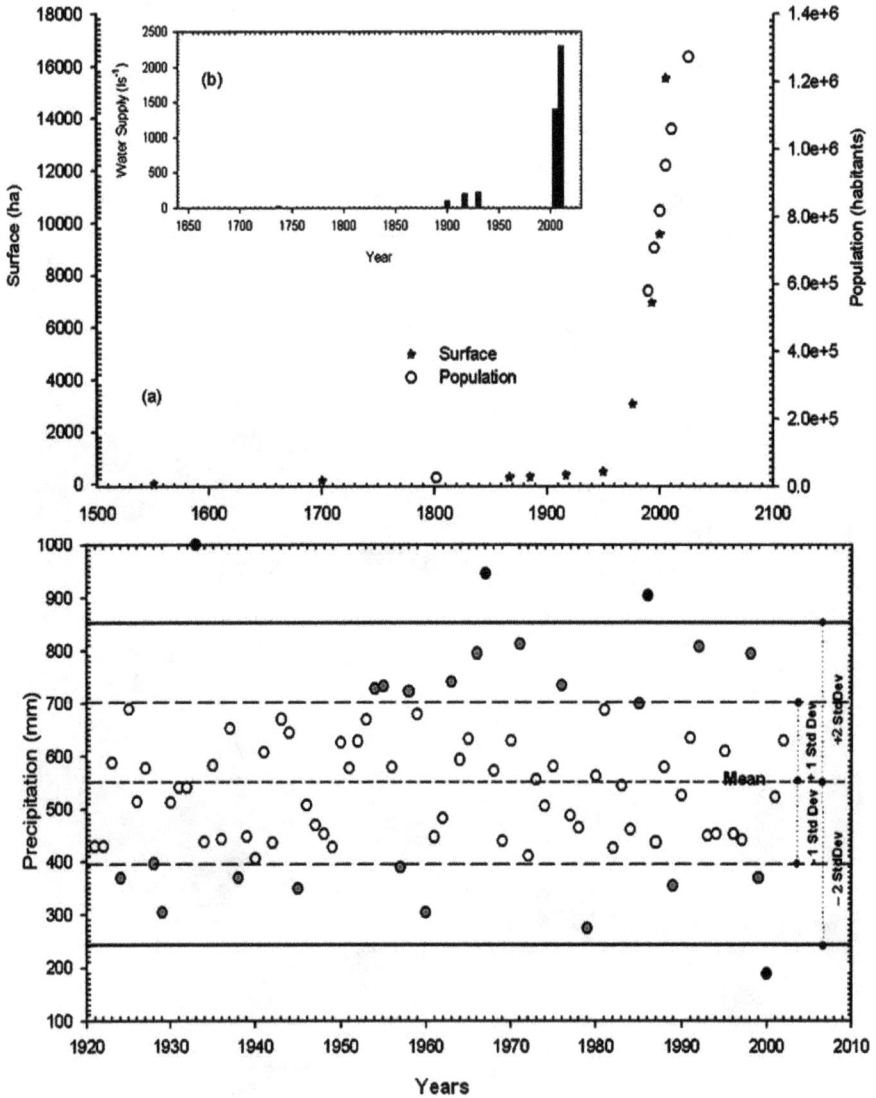

Figure 2. Population, urban surface and water supply evolution (above). Average annual precipitation in the "Observatorio Station", 1921-2007 (low).

Also, for this period there are existing reports that the water quality of the Queretaro River was polluted and people became ill after drinking such water as indicated by Von Humboldt in 1803 during his visit to Queretaro [12]. Official reports indicate that the water comes from the extraction of Queretaro's aquifers of Buena vista, Amazcala-Chichimequillas and Huimilpan located under the basin all being overexploited, as shown on Table 1. Agriculture generally consumes 54%, except in the Queretaro aquifer where most extractions are for urban public consumption. The water demand within the basin is estimated at 106 million m^{-3} Table 1 shows the five aquifers located in the Queretaro River's Basin where a deficit is observed due to high pumping for the supply of agriculture and urban water consumption. The Queretaro Water Commission (CEAQ) reported an annual average water reduction of 3 m y^{-1} and in some sites twice as much; in the 40's the water table was at the surface. Furthermore, it was reported that for each 103 m^3 pumping, only 70% is recovered through deep recharge.

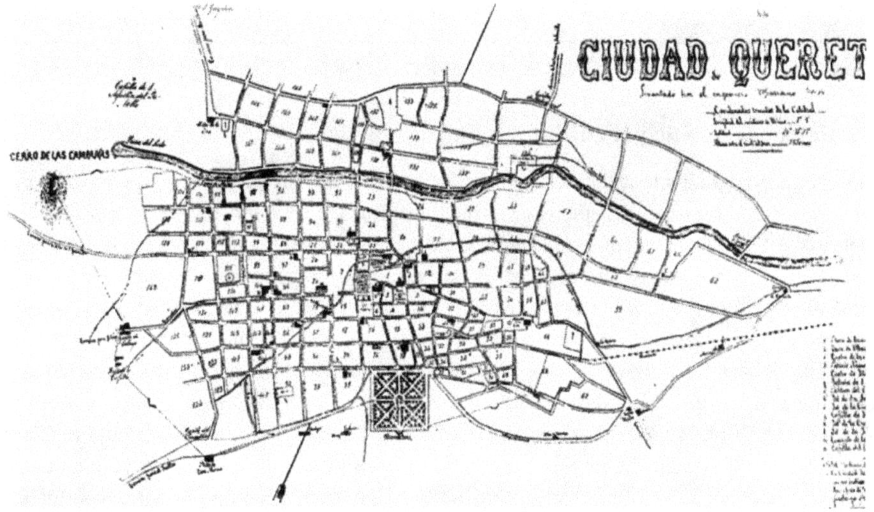

Figure 3. Plot outline of the main roads that strangle the water paths Mexico-Queretaro, railway Guadalajara-México (Pan American Highway embankment) railway. In the ditch, grey blue black road and railway tracks 1897.

The water pathway has been transformed through the evolution of the urban architecture from 1778 to the present time. The network of streets, acequias (irrigation ditch) and drains has defined the urban scheme of the city, and the urban architecture was instituted during the settlement of Querétaro in 1531 and grew at the same rate as Mexico City. "The water of the River coming from the Cañada village trough of channels network to irrigated a large number of orchards fields that made Querétaro famous for their jolly gardens and great variety of fruits"[13]. At the beginning of the War of Independence, as one may appreciate on Figure 3, the city still kept its original stroke, heart-shaped quasi-symmetric (West) and developed along the Queretaro River. However, the city began to extend towards the Hercules village (Villa Cayetano Rubio), Northeast zone of the Hacienda of Pate and South in limits of the posts along San Miguel de Allende and El Pueblito, result of industrial development. Such growth marked the limits of the city's outline, from an extended conformation.

Component water balance (mm)	Mexico	Queretaro River basin	Volume 106 m3	Percentage (%)
Average Precipitation	772	557	1528	100 Mex. 100 Qro.
Evaporation (soil+vegetation)	558	464	1106	72.3 Mex. 83.3 Qro.
Runoff	200	52	397	25.9 Mex 9.3 Qro.
Groundwater Recharge	14	41	27	1.8 Mex. 7.3 Qro.

Table 1. Comparison of water balance between Mexico and Queretaro River Basin.

2.1. Regional climate variability

Jáuregui [14] explains that the behavior of the tropical waves, are more frequent from the East between August and September, which is influenced by the topography across the Volcanic Transversal System, generated the back of the wave, the clouds grow up to 7 or 8 km producing abundant rainfall. For example, during September 2003, the rains were generated by frontline of (setting in which from the center of a low pressure the isobars deform to move away from him). Aguilar *et al.* [15] already marked in the 1970s the potential risks of flooding in the Queretaro Valley of Such identified the areas susceptible to flooding and the possible solution of the problem. Future storages with greater feasibility: Atongo, Los Vega, Pueblito, La Cañada, Ixtla, Olvera, Bolaños and Menchaca, could regulate of runoff by 18% under the time of the study, with a 9 million U.S. Dollars investment of 9 million of dollars. The studies distinguished part of the basin flooding problems: confluence of the Querétaro and Amazcala rivers. In the lower part of the Valley at the confluence of the rivers Querétaro, el Pueblito and the creek of the Arenal occur floods of great magnitude affecting much of the city of Querétaro, area Industrial Benito Juárez, Jurica, also the Obrera, San Pablo, Carrillo Puerto Santa María Magdalena, San Antonio de la Punta, San Juanico, San Pedro Martir and Jurica subdivisions (colonies). Moreover, the "global climate change" should be mentioned, regardless of the relevance today, the impact of the damage by weather disasters, water shortages or water excess address the political country's transformation of the Mexican plateau. The impacts of climate change on the regional components of the hydrological cycle of Queretaro River Basin serves as evidence of the behavior and temperature and precipitation trends, such detailed below, based on the analysis of the historical record of the Querétaro station between 1921 and 2007. Finally it should be mentioned, although the relevance today "global climate change" the impact of the damage by weather disasters, lack or excess water, is close by the social wealth and available technologies to address the political transformations of country and the Mexican Plateau. Analysis of annual precipitation in the urban Queretaro's area with a continuous 86 years record (1921-2007) comprising two periods of observation in one single: 60 years of records in central zone (Cerro de Las Campanas) and 30 years in the southern area, current location of the Observatory station (at the foot of the Hill of the Cimatario) was allowed to identify the oscillations of the annual water cycle. It is pertinent to point out however that the urban growth of the Queretaro Valley did not take into account the local effects where the Observatory station is located; therefore, it was considered as a regional influence to assess temporary dynamic precipitation, the annual cycle of the water and the temperature increase. Annual rainfalls were grouped into three bands of occurrence assuming a Gauss distribution, 257 mm < P (95%) < 857 mm, 407 mm < P (68%) < 707 mm, 500 mm < P (5%)< 600 mm. Figure 3 (low) shows three bands with reference to the average annual precipitation (557 mm) and normalized anomalies. The presence of extreme drought is limited to the year 2000 with 188 mm and 4 years (1933, 1967, 1986, 2002) considerably wet with 857 mm < P, the last three with approximately every 20 years. A significant presence of wet years in recent decades with 707 mm < P < 857 mm among the bands of occurrence 95% - 68%. A greater persistence of dry years with 257 mm < P < 407 mm with occurrence similar to the wet years; 17%, the events show a 5% occurrence environment than the historical average of 557 mm minimum and maximum precipitation are affected by a factor of 2 on the historic average, respectively. The rainfall is concentrated between June and August, with 73% of the total precipitation. In relation to Mexico's seasonal behavior a 4 month period was reported between June and September concentrating 67% precipitation [16]. Furthermore a pronounced relation to the temporal oscillation was reported in 2001 and 2002, whereas in the same period the concentration was of 79% and 59% respectively, allowing to see that there is a seasonal annual water cycle in the Queretaro Valley. It should be pointed out that the analysis was carried out by one single station because the records were not continuous at other stations located within the region, as it may be observed on Figure 4. Table 2 also shows the annual precipitation and evaporation average inside the Queretaro River Basin, as well as daily average temperatures for the 2010.

3. Climate and climate change

The weather conditions are typically described in terms of local temperature atmospheric pressure, humidity, wind speed are often described on the basis of the average, a period of 30 years average atmospheric conditions, to scales levels of study. Instead, climate variability refers to variations in the average climate scale condition spatial-time outside the individual weather conditions, i.e. out of context drought, prolonged floods and the effects generated by the El Niño and La Niña Phenomenon. This is why we conceive as climate changes the state average climate and its variability that persists for an extended period. The analysis also relies based on processes at various levels, as it is listed below:

1. Micro-scale Meteorology in agriculture < 100 m
2. Topo-scale or scale local 100 m – 3 Km, pollution and tornadoes
3. Meso-scale 3-100 km., storms, sea breezes of mountain
4. Global-scale 100-3000 Km, fronts, cyclones, cluster of clouds

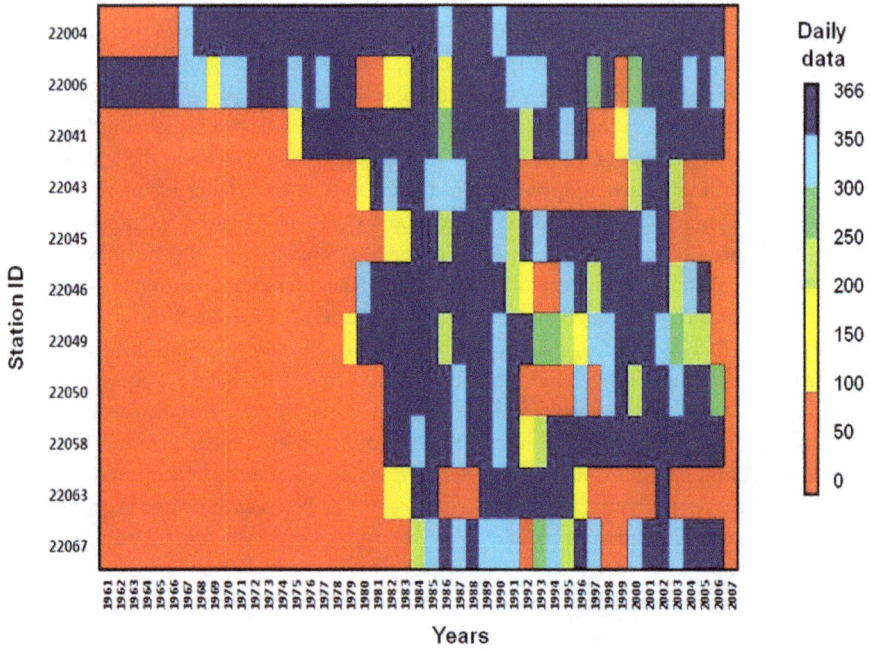

Figure 4. Data matrix of records observed by year and by each one of the weather stations under study.

To establish the vulnerability levels of climate change exposure, susceptibility and capacity for adaptation to the effects of climate change, and differences between climate variability and climate change on a scale space-time scenarios , and its seasonal variation by the effects of temperature increase in the case of the stage critical A2 defined by IPCC. The effects of the spatial variation of natural climate variability on large scale and mesoscale can be observed, and apparently, such are not so evidence that climate variability is greater than the A2 scenario. In the case of the central plateau where Queretaro State and Queretaro City are located, the temperature change in the southern sector is less susceptible than the North; however, the Central region presents a significant variability (Table 2). This could accentuate when compared to mesoscale level and could be magnify to a micro-scale. In order to scale the seasons, apparently the rainy season, summer-autumn, will have little significant change, however it is accentuated during the spring-summer period. In this regard, increased flood risk, by the increase in precipitation is contradictory. It is pertinent to note that this analysis does not provide the rain behavior patterns: duration, intensity and foil. Likewise, the thermal fluctuations at the daily level certainly modify the distribution components of the water balance. The synthesis should identify the relationship of causal effects of climate variability, both at the top-scale levels and micro-scale, in order to specify actions and specific adaptations for each sector and each region.

Climato-logical station	Altitude msnm	Temperature (°C)			Precipitation (mm)			Evaporation (mm)	
		Average	Máximum	Mínimum	Average	Bias	Rainny days	Average	Bias
Tres Lagunas	1610	17.8±8	26±8	9.7±9	691	±172	80	1446.7	±105
La Lagunita	1087	20.9±8	29.1±12	12.7±7	711	±139	86.6	1452	±87
Arroyo Seco	996	21.8±10	29.1±11	14.6±9.7	540.6	±140	64.4	1430	±104
Ayutla	791	23.6±9	30.9±9	16.2±10	713	±176	61.8	-	-
Jalpan	760	23.8±10	31.7±11	15.9±9.6	916.5	±184	88.3	-	-
Central Zone									
C. Campanas	2568	15±4.7	23.0±5	7±3.5	611.4	±115	57.3	1782	±75
Nogales	2053	16.5 ±6.4	23.4±7	9.5±6.4	362	±91	39.9	1714	±104
Cadereyta	2044	15.9 ±6.8	24.2±7	7.6±6.3	489	±88	50.4	1680	±91
El Salitre	1981	18.4±2	28.7±2.5	8.1±6.4	270	±67	28.2	1451	±57
El Comedero	1749	17.4±8	24.3±8	10.5±5.5	432	±88	52.3	1886.6	±70
Toliman	1720	23.6 ±8.5	28.5±9	11.2±9	377	±61	36.1	-	-
Higerrillas	1597	18.9±5.4	27.1±5	10.8±6	302	±50.4	38.4	2170	±76
Gillen	1370	22.4±8	31.1±8	13.7±9	399	±139	60.3	2061	±122
South Zone									
Amealco	2629	15±5.7	22.7±5	7.5 ± 4.7	820	±115	77.9	-	-
Huimilpan	2271	15.5 ±5.7	23.0±5.4	8.1±6.4	761	±188	72.2	1762	±98
Santa Teresa	2092	17.1 ±6.0	24.60±7	7±5	420	±91	35	1810	115
Juriquilla	1885	17.7±7	26.0±6	9.3±9.5	526	±100	52.4	1696	±103
Carrillo	1806	18.8±8	27.4±7	10.3±7	551	±105	53.5	2262	±120

Table 2. Study data matrix of the observed temperatures, daily, for a one year period, and for each one of the climatologic locations

4. Analysis of the precipitation

From the 24 year period (1950-1973) of the "Cerro de las Campanas" station produced a 40 % coefficient runoff (modified 32.23%), for an average 619 mm rainfall, and a coefficient variation of 40.4%. Therefore, the average annual runoff for the entire basin (1486.8 km²) was equivalent to 255 x 106 to 110. 42 mm sheet of rain. Yet, on the contrary, the historical precipitation can be reconstructed through the years without the same records that could be associated with the historical reconstruction of the Queretaro City's floods, based upon the statistical analysis of historical precipitation series, and its comparison with series of sites in the central region of the Mexican high plateau integrated by the States of Queretaro and Guanajuato. Double mass analysis is considered for the uniformity assessment of the series, taking as reference the Queretaro station [17]. Figure 5 shows a comparison of the precipitation stations Celaya, San Miguel of the State of Guanajuato and the Observatory of the city of Querétaro station. In addition, to observe that Queretaro has an average rainfall in comparison to the other two sites, similar and well defined precipitation persistence at stations of Celaya and Queretaro, may point out terms of the oscillations that can be inferred in the rainfall registered in Celaya, despite of being a period shorter.

The simple rainfall analysis curve mass from the three sites located in Mexico's central region support the zoning hypothesis precipitation of the Mexican high plateau, and the persistence and seasonal behavior of such, as Giddings *et al.* [18] also noted with the SPI study (Standardized precipitation Index) with 3313 (21 of the Queretaro) stations for the 1940. Such index is widely considered as an indicator to assess the droughts severity or excessive rainfall.

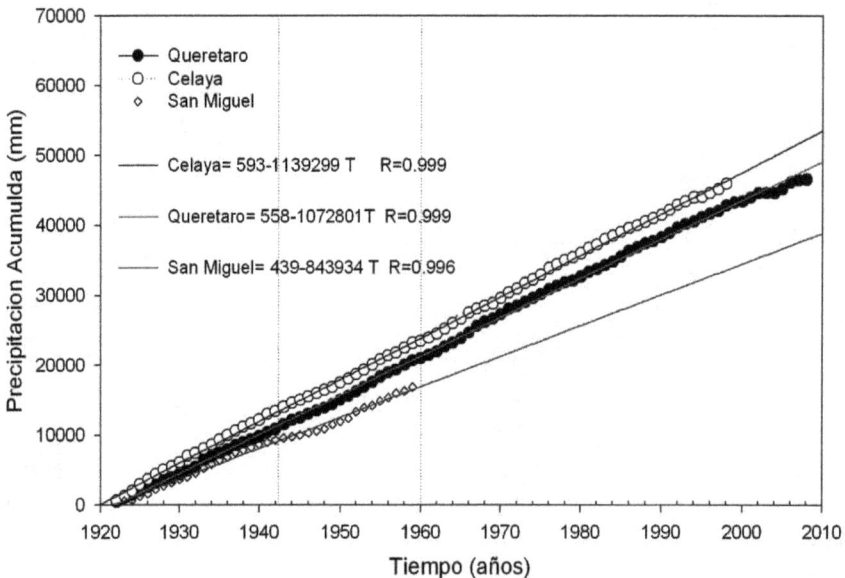

Figure 5. Curve mass of stations of Queretaro, Celaya and San Miguel.

4.1. Precipitation in Queretaro Valley (QV)

Reconstruction of the precipitation is based through mainstreaming the fluctuations of the annual water cycle represented by the observation of precipitation it corresponds to a stationary stochastic process by swings in rainfall registered in the last 90 years (1921-2009), the same can be reconstituted under the hypothesis of persistence of the annual cycle governed by this process.

4.1.1. The Hurst Exponent

Wallis and Mandelbrot [19] showed in his auto synthetic work similarity in hydrology, depicting this technical comparison of Markov analysis, in the best way processes hydrological as droughts and floods through. The Hurst Exponent [20]: If the process has a variance and a finite memory, a good process measure can be established:

$$R(n) / \sigma \propto n^{H} \tag{1}$$

where R (n) the range of the sample of size n and the standard deviation of the sample if the process is independent, produces (whitenoise) Gaussian not correlated, by analogy with the conduct of the spectrum of white light, its random behavior (Brownian noise) produces H=0.5. However, the hydrological series has a H>0.5, usually. Before the evidence, the Hurst exponent is a tool for characterization of nonlinear systems, and the fact that H is different from 0.5 envisions an underlying consequence of a non-linear dynamics. A study with the precipitation historical data of the period between 1901 and 1995, in Ghana and Venezuela reported Hurst values of 0.638 and 0.586 exponent, respectively [21]. In the particular case of the historical series of three sites considered within the study, threw values between 0.516<H< 0.982, values with the range reported by Van de Giesen and Mata [21] and the Mandelbrot and Wallis [22]. Based upon the Gaussian noise, such can be constructed with the moving average of a white noise, which remains the story of events with lasting effects. On the other hand, it was derived from a model of annual base of precipitation with sinusoidal behavior of four parameters, with the moving average of the historical precipitation of the Observatory in Queretaro stations, and the series of stations at Celaya and San Miguel de Allende, Guanajuato. On Figure 5, one can see that the behavior of the various models is led by the time space or lag (s). A stationary stochastic permanence for the three sites is clear, no matter the lag. However, the behavior of the historical model series of San Miguel de Allende is 30 years, Queretaro is the dual 60 years between extreme periods of wet and dry, passing through the 1332-1543 Climatic period, Therrel *et al.* [23]: such study proved the presence of 13 drought events that coincide with the chronological climate rebuilding, with the use of the so-called Dendrology technique. The said study is based on the analysis of the tree's trunk rings in central Mexico, area referred as the States of Puebla, Hidalgo, Tlaxcala, Morelos Federal District and State of Mexico being widely scrutinized between 1450 and 1900 [24]. The historical series of the historic catalogue of natural disasters [25] shows 388 reports about 70 droughts for the central zone without mentioning the State of Queretaro and the city of the same name. The drought analysis reports 3-5 years frequencies, 15, 24 and 55 years, it even mentions a 60 year period associated with solar activity. Between 1970 and 2003, 1744 events were recorded in the country: 232 (13.2%), 231 (13.1%), 183 (10.5%), 86 (5%), 10 (0.6%) in the States of Veracruz, Mexico, Federal District, Chiapas and Queretaro, respectively. Several studies attempted to explain the origin and the rain patterns of Central Mexico ([26], [14], [27], [24]) who analyzed six centuries of historical series (1400-1990) found a drought frequency in the central region of Mexico: 3, 5, 15, 24 and

50 years, compatible with solar activity, possible to reconstitute the annual rainfall of the last 200 years in order to associate it with floods or droughts. The attention that the persistence in San Miguel de Allende is almost half of Querétaro, 31 and 65.5 years, respectively. While the precipitation extent is of such magnitude, environment to the 50 mm, apparently a local effect causes a greater frequency of catastrophic events in the region North of Guanajuato. However, for the Queretaro Valley region, the greatest amplitude of oscillations indicates the presence of a static stochastic process.

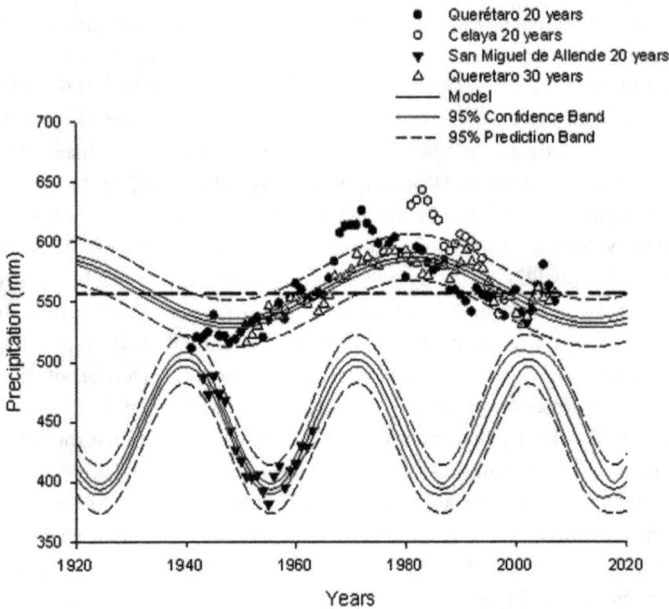

Figure 6. Possible Changes: annual variability or climate changes.

4.1.2. Possible changes: Annual variability or climate changes

The rapid urbanization growth directly affects the temperature of the QV and persist the urban growth and deforestation for land use change such condition will cause the average annual temperature further raise and increase the deficit of water supply to the urban and bulk of the QV, without taking into account the changes in the components of the water balance the evaporation increase and heat flow sensitive by the local albedo change. Hunt [28], Hunt and Elliot [29] simulated the climate of a 10 000 year period in order to investigate the existence and genesis of mega droughts in the Mexican region suggesting that Mexico went through a 19 year mega drought episode in 1550 DC, causing diseases and the disappearance of 80% of the indigenous population. Also Hunt and Elliot [29] identified episodes of major droughts that showed a period of return of 1000 years and a reduction in annual rainfall between 20% and 40%. The said study detected 13 droughts lasting 10-years, with some wet years, expressing 5 lasting drought events. They conclude that droughts can form independently by the phenomenon of Niño (ENSO) or stochastic processes. The mechanisms associated with El Niño can be identified by the Southeast oscillation pressure abnormal conditions generated by changes in the Walker circulation during El Niño. Both events and stochastic processes produce low surface pressure on the Mexican region, reducing moisture entering the territory and resulting in droughts. In context with the drought episodes previously outlined in Figure 6, shows the historical retrospective of the last 86 years concerning the presence of wet and dry years supported in the drought index, Palmer [30], [31] and normalized anomalies of precipitation [32] 1929, 1960 and 1979 were severely dry years based upon the historical annual average, with rates: 0.53, 0.52 and 0.47, respectively. In contrast 1933, 1967 and 1986, the drought index surpassed the 1.93, 1.83 and 1.75 units corresponding to extremely wet years. It is clear that aspects mixed with global warming generated by urban growth anthropogenic impact on the region's annual water cycle. This situation is necessary to permanently document

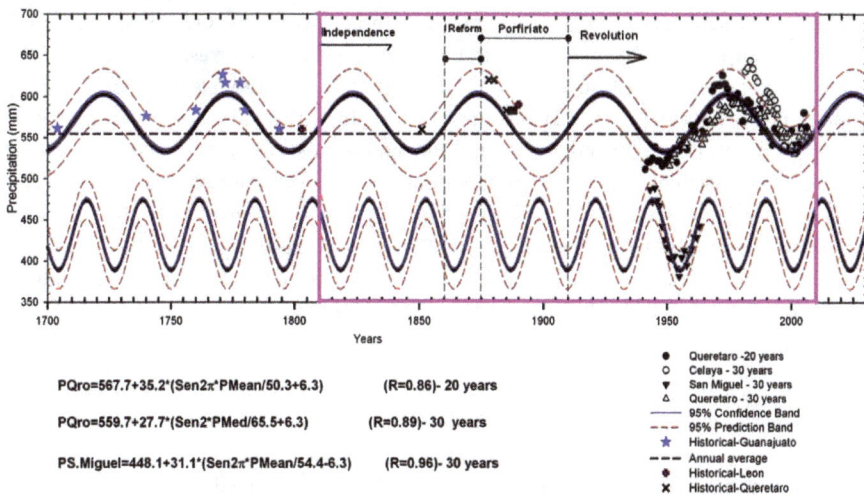

PQro=567.7+35.2*(Sen2π*PMean/50.3+6.3)   (R=0.86)- 20 years

PQro=559.7+27.7*(Sen2*PMed/65.5+6.3)   (R=0.89)- 30 years

PS.Miguel=448.1+31.1*(Sen2π*PMean/54.4-6.3)   (R=0.96)- 30 years

Figure 7. Variability of annual water cycle to 300 years last. Queretaro and San Miguel Station from sinusoidal model.

climatic variables and components of the hydrological cycle and he fluctuation of the annual water cycle that distinguishes the exchange mechanisms between the biosphere and the atmosphere In addition such conditions develop numerical models adapted to semi-arid environments or ecosystems anthropogenic that assimilate the critical scenarios and establish contingency measures for our own physical environment and water resources management.

4.1.3. Impacts on water availability

Within support of the sinusoidal model type shown on Figure 7 and proceeding in reverse form, it was possible to reconstitute the annual rainfall of the last 300 years in order to associate it with floods or droughts. The attention that the persistence in San Miguel de Allende is almost half of Querétaro, 31 and 65.5 years, respectively. While the precipitation extent is of such magnitude, environment to the 50 mm, apparently a local effect causes a greater frequency of catastrophic events in the region North of Guanajuato. However, for the Queretaro Valley region, the greatest amplitude of oscillations indicates the presence of a static stochastic process.

5. Conclusion

The Querétaro Valley is a region of large agro-industrial development which has allowed the increase of population density and with it the change in the land use and the availability of water resources. Despite the five aquifers are there in the valley of Querétaro there is a reduction in groundwater levels of 3 m per year. In the case of the Querétaro aquifer the urban consumption is the major demand of groundwater resources. Furthermore, the Querétaro city due to water infrastructure and topography is susceptible to flooding, mainly in the confluence area of Amazcala and Querétaro rivers, with an average annual rainfall of 557 mm, concentrated in the months of June, July and August (73%). The historical analyses of the climate variability in the region as well as the documentation of urban growth and water demand are required for decision-making under scenarios of the physical environment of a semi-arid zone as it is the Querétaro Valley region. In this sense, the historical analysis of the Querétaro station displays behavior and tendency of temperature, precipitation, and the occurrence of droughts and floods in the region

The oscillation evidences of the oscillations within the annual water cycle and the temperature and drought periods increase, including severe ones as Hunt (2001, 2001) points out allows us to see, that it is fundamental to generate efficient mechanisms for taking advantage of the rational water resources. At the same time, these characteristics in the pattern demonstrate if there are connections not yet identified between the natural climatic variables and the general alterations due to the global climatic change. In order to identify which are the alterations in the distribution of the hydraulic balance components and the hydrological processes in the river's basin caused by the accelerated dynamic growth within the last 50 years.

Among the mitigation schemes there is the identification of the driving mitigation zones and the aquifer recharge, taking into account the "Environmental Protection Agency" (EPA, 1993) indications for paved surfaces between 35 and 50% of the total area where the recharge represents 15% of the precipitation. In this sense the urban parks could represent an adequate option for the recharge and aquifers recovery of the highly urbanized zones as possible alternate to the contamination and subsidence problems. The great question is to know what would be the effect of the annual water cycle, above all, investigate the precipitation seasonal-space distribution suffering a radical change due to the global climatic change, for there would be dramatic changes to the physical ambience, transformation of the agricultural system components and the water reservoirs, including the water quality. For such reason, the adaptation of climatic change implies the historical, current and future knowledge of the seasonal-space precipitation distribution in order to identify the effects of possible extreme events, heat waves and droughts or great intense storm precipitations. One of the possible solutions for taking advantage of the current hydraulic resources is the direct rain recovery capitation, as being done by different countries and international non-governmental institutions promoted by policies and new technical-legal schemes for the exploitation of rain water. Even though in the immediate future of Queretaro Valley the Aqueduct II project shall mitigate the Queretaro Valley water problem, it is necessary to create policies in order to support the sustainable development and generate an efficient water resource management. At the same time, the water required to satisfy the deficit and its impact within society's development established the concept of "virtual water" which is economically invisible and politically silent, without really making the climatic changes effects visible.

Author details

E. González-Sosa* and N.M. Ramos-Salinas
Hydraulics Laboratory, Engineering Faculty,
Universidad Autónoma de Querétaro. Ciudad Universitaria,
Cerro de las Campanas. Querétaro State, México

C.A. Mastachi-Loza and R. Becerril-Piña
Centro Interamericano de Recursos del Agua (CIRA), Engineering Faculty,
Universidad Autónoma del Estado de México. Atlacomulco-Toluca Street,
México State, México

Acknowledgement

This work was conducted with the support of: CONACYT (FondosSectoriales CONAGUA-CONACYT, 2010-2, NO 148159). Also the student's participation is acknowledged.

---

* Corresponding Author

6. References

[1] Budyko, M.I. The influence of man on climate. Hydrometeoizdat. Leningrad; 1972.

[2] IPCC. Climate Change: The IPCC Scientific Assessment of Climate Change. Eds. J.T. Houghton, G.S. Jenkins and S.S. Ephrains.Cambridge University Express. Cambridge for WMO/UNEP; 1990

[3] Nemec J. and Shaake J. Sensivity of water resources systems to climate change in climatic imputs. Nordic. Hydrology 1982; 23 257-72.

[4] Stockton C.W. and Bogges, W.R. Geohydrological Change Implications Climate Change on Water Resources Development. US Army Coastal Engineering Research Center, Fort Belvoir, VA; 1979.

[5] Lambin, E., B. Turnerb, H. Geista, S. Agbolac, A. Angelsend, J. Oliver T. Coomesf, R. Dirzo, G. Fischerh, C. Folkei, P.S. George, K. Homewood, J. Imbernon, RikLeemansm, X. Lin, E. Morano, M. Mortimore, P.S. Ramakrishnan, J. Richards, H. Skaness, W. Steffent, G. Stoneu, U. Svedinv, T. Veldkampw, C. Vogelx and J. Xuy. 2001. The causes of land-use and land-cover change: moving beyond the myths. Global Environmental Change 2001; 11 261–269

[6] Pickett, Steward T., Burch Jr, Wiliam R., Dalton, Shawne E., Foresman, Timothy W., Grove J. Morgan., Rowntree Rowan. A conceptual framework for the study of human ecosystem in urban areas. Urban Ecosystems 1997 185-199.

[7] Ducrot, R. Water and land management in the periurban catchment of Sao Paulo: a first conceptual framework. Facilitating Negotiations Over land and Water Conflicts in Latin-American Peri-Urban Upsteram Catchment: Combining-Agent-Based Modeling with Role Playing Game Project Negowat. CIRAD-TERA 2005.

[8] Paul, M. and Meyer J. Streams in the Urban Land Scape. Annu. Rev. EcolSyste 2001; 32 333-365.

[9] Fitzharris, B. Land-use change and water balance an example of an evapotranspiration simulation model. Journal of Hydrology (N,Z.) 1974; 13(2) 98-114.

[10] Walton, B., Nawarathna, B., George, B., Malano M. 2009. Future water supply and demand assessment in peri-urban catchments using dynamics approach. 18th World IMACS/MODSIM Congress, Cairns, Australia 13-17 July. 3872-3878.

[11] Steffen, W., Saanderson, A., Tyson, P.D., Jager, J., Matson, P.A., More III B., Richardson K., Shellnhuber, H.J., Turner II, B.L., Wasson, R. J. Reverberations of change: The Reponses of the Earth System to Human Activities. Global Change and the Earth System. Springer-Verlang. Berlin 2005.

[12] Cordera, R., Tello C. La desigualdad en México. Siglo XXI. México 1984.

[13] Septién M., Septién. La historía de Querétaro: desde los tiempos prehistóricos hasta nuestros días. Gobierno del Estado de Querétaro 1999.

[14] Jáuregui, E. Las ondas del este y los ciclones tropicales en México, Ingeniería Hidráulica en México 1967; 21(3) 197-208.

[15] Aguilar U. G., Venegas M. J. G., Ramírez T. R. Estudios técnicos sobre cimentaciones e inundaciones en el Valle de Querétaro. Tesis para obtener el titulo de Ingeniero Civil. Escuela de Ingeniería. Universidad Autónoma de Querétaro. México; 1974.

[16] Comisión Nacional del Agua (CONAGUA). Estadísticas del agua en México. http://www.conagua.gob.mx/CONAGUA07/Publicaciones/Publicaciones/EAM_2008.pd f (accessed 20 April 2012)

[17] Kohler, M.A. Double-mass analysis for testing the consistency of records and for making adjustments. Bulletin of the American Meteorological Society 1949; 30 188–189.

[18] Giddings, L., Soto, M. Rutherford, B. M. Maarouf, A. Standardized Precipitation Index for México. Atmósfera 2005; 33-56.

[19] Wallis, J.R., Mandelbrot, B. B. Self-similar synthetic hydrology. The use of analog and digital computers in hydrology: proceedings of the Tucson Symposium, UNESCO; 1969.

[20] Hurst, H.E. Long term storage capacity of reservoirs. Trans. Am. Soc. Civ. Eng. 1951; 116 770-779

[21] van de Giesen N., Mata J. L. Comparison of the Hurst exponents of historical and GCM rainfall time series. Hydrology days; 2002.

[22] Mandelbrot, B.B Wallis J. R. Some long-run properties of geophysical records. Water Resour. Res. 1969; 5 321-340.

[23] Therrel M. D., Stahle D. W., Acuña S. R. Aztec Drought and the "Curse of One Rabit".American Meteorology Society 2004; 1263-1274.

[24] Mendoza B., Jáuregui E., Velasco V., García-Acosta V. Possible solar signals in historical droughts in central and southeastern México. 29th International Cosmic Ray Conference Pune 2005; 2 369-372.

[25] García Acosta V. La perspectiva histórica en la antropología del riesgo y del desastre. Acercamientos metodológicos. El Colegio de Michoacán. México. Relaciones 2004; 97

[26] Wallen C. C. Some characteristics of precipitation in México. Geografiska Annales 1995; 37 51-85.

[27] García E. Nuevas técnicas de análisis en la climatología. 1997. Memorias del VII Congreso Nacional de Meteorología. Universidad de Chapingo, Estado de México, pp. 39-73.

[28] Hunt B. G. A description of persistent climatic anomalies in a 1000-year climatic model simulation. Climate Dyn. 2001; 17 717–733

[29] Hunt B. G. Elliott T. I. Mexican megadrought. Climate Dyn 2002; 20 1–12

[30] Palmer W.C. Meteorological drought. Research Paper No. 45.U.S. Weather Bureau. (NOAA Library and Information Services Division, Washington, D.C. 20852) 1965.

[31] Willeke G., Hosking J. R. Wallis J. R., Guttman N. B. The national drought atlas. Institute for water resources report 94-NDS-4, U.S. Army Corps of Engineers 1994.

[32] Ogallo L. J. The spatial and temporal patterns of East African rainfall derived from principal component analysis. Int. J. Climatol. 1989; 9 145–167.

Chapter 10

Effect of Climate Change on Spatio-Temporal Variability and Trends of Evapotranspiration, and Its Impact on Water Resources Management in The Kingdom of Saudi Arabia

Mohammad Elnesr and Abdurrahman Alazba

Additional information is available at the end of the chapter

1. Introduction

Recently, climate change is receiving much attention. Changes in the world's climate have significant effect on water resources which affect the livelihood of people especially in hyper arid regions such as the Kingdom of Saudi Arabia (KSA). The KSA suffers an enduring water shortage problem, despite the fact that the agricultural activities consume up to 90% of the water amount in the Kingdom. Reference Evapotranspiration (ETo) is an agro-climatic property that involves temperature, humidity, solar radiation, and wind speed. Identifying changes in ETo can also help in future planning of agriculture-water projects and identify lower and higher ETo zones for proper planning and management of agricultural projects in arid regions.

1.1. Water resources and climate change

Water shortage is a swelling problem in the arid and semi-arid regions. Affected by its geographic location and its climate, the Kingdom of Saudi Arabia (KSA) suffers a severe water deficit. Even rain, which is the only renewable water source, comes in flash short duration storms of high intensity and most of it vanishes to evaporation. Thus, almost all agriculture of the kingdom is irrigated. Irrigation water, though, consumes 80 to 88 % of the total water consumption (Abu-Ghobar, 2000; Abderrahman, 2001). In addition to these water scarcity conditions, but it seems getting scary by the effects of climate change on the hydrological cycle and water supply. The quantity of irrigation water is determined initially by identifying the reference evapotranspiration (ETo). Several researches was conducted to detect climate changes, trends and variability in various parts of the world using some climate parameters such as air temperature, rainfall depth, ETo, and pan evapotranspiration ETp (Shwartz and Randall, 2003; Garbrecht, et al., 2004; Hegerl, et al., 2007; Fu, et al., 2009; Hakan, et al., 2010; Elnesr and Alazba, 2010; Elnesr et al., 2010a; and Elnesr et al. 2010b). The ETo parameter has a special importance because it combines changes in many other climate parameters including temperature, radiation, humidity, and wind speed. It has, however, direct influence on hydrologic water balance, irrigation and drainage canal design, reservoir operation, potentials for rain-fed agricultural production, and crop water requirements (Dinpashoh, 2006).

1.2. Climate change effect on evapotranspiration worldwide

Several studies conducted in North America have shown that some climate parameters are on the rise including ETo (Fehrman, 2007; Garbrecht et al., 2004; Szilagyi, 2001). Fehrman, 2007 found an increasing trend in ETo over the Mississippi area and that most of ETo increase can be attributed to the increase in July. He also found that the rate of ETo increase was 0.29 mm/years when his study period extended from 1940 to 1999 compared to 0.88 mm/year when the study period was limited to 1950 to 1999 records. The accelerated ET over North America is presumed to be due to a rise in temperature over the past century (Myeni et al. 1997, Milly and Dunne 2001). In the contrary ETo and pan evaporation has shown to decrease in China (Thomas, 2000, Liu et al., 2004) and at a rate of 1.19 mm/year (Song et al., 2010) despite the rise in maximum daily temperature. In the Tibetan Plateau ETo decreases as well at a rate of 1.31 mm/year or 2.0% of the annual total evapotranspiration (Shenbin et al., 2006). The decrease in ETo has been attributed to the decrease in wind speed and net radiation. In another study Gao et al., (2007) found that the actual evapotranspiration had a decreasing trend in most of the eastern part of china and there was an increasing trend in the western and the northern parts of northeast China and that the change in precipitation played a key role for the change of estimated actual evapotranspiration. Similar negative trends in pan evaporation were found in 24 out of 27 observation stations in a 19-year study in Thailand (Tebakari et al., 2005). In India, a significant decreasing trend was found in ETo all over the Indian plateau during the past 40 years, which was mainly caused by a significant increase in the relative humidity and a consistent significant decrease in the wind speed throughout the country (Bandyopadhyay et al., 2009). In Australia, Roderick and Farquhar (2004) found a decreasing trend in pan evaporation and conclude that Australia is becoming less arid. However, there is enough evidence now that a decrease in pan evaporation is an indicator to an increase in actual evaporation. This is what known now as the evaporation paradox (Hobbins et al., 2004).

Some researchers developed a hypothetical scenario to study the effect of possible increase on temperature over the ETo and subsequently on water supply. A study conducted by Abderrahman et al. (1991) concluded that in the KSA, a 1°C increase in temperature would increase ETo from 1 to 4.5%. In another study, that includes selected cities in KSA, United Arab Emirates and Kuwait, Abderahman and Al-Harazin (2003) concluded that an increase in temperature by 1°C would increase ETo over these area by a maximum of 20%. In general, studies involving ETo calculation seemed to be more limited worldwide compared to other climate parameters. In the other hand, regarding other climatic parameters, Hakan et al. (2010) reported an increasing trend in temperature and ETo in most of stations they analyzed in Turkey using Mann-Kendall analysis. Cohen and Stanhill, (1996) studied rainfall changes in the Jordan Valley/Jordan and found a tangible but insignificant decrease at a rate of -0.47 and -0.16 mm/year for two different stations. Similar conclusions were observed by Al-Ansari et al (1999) who observed a general decrease in rainfall intensity. Smadi (2006), and Smadi and Zghoul (2006) found a prompt shift in rainfall and temperature in Jordan. ElNesr et al (2010b) concluded that the Saudi Arabia and the Arabian Peninsula are suffering from a considerable warming trend form year 1980 to 2008. Still, Elnesr et al. (2010a) concluded that the percentage land area with annual ET_o>4000 mm increased from about 20% to 40% in the period they studied. On the other hand, lower ET_o values, less than 3600 mm, contracted from about 30% to 12%.

1.3. Objective of the study

This study aims to trace the ETo values over time throughout all the area of the Saudi Arabia, then to quantify the future of water demand according to the ETo trends

2. Material and methods

2.1. Geography of the Saudi Arabia

Saudi Arabia is the largest country of the Arabian Peninsula; it occupies about 80% of its area (Wynbrandt, 2004). The country lies between latitudes 16°21'58"N, and 32°9'57"N, and longitudes 34°33'48"E and 55°41'29"E, as illustrated in Fig. 1. Saudi Arabia has a desert dry climate with high temperatures in most of the country. However, the country falls in the tropical and subtropical desert region. Winds reaching the country are generally dry, and almost all the area is arid. Because of the aridity and the relatively cloudless skies, there are great extremes in temperature, but there are also wide variations between the seasons and regions (AQUASTAT, 2008).

2.2. Evapotranspiration calculation

Evapotranspiration was calculated using Food and Agricultural Organization (FAO) Penman- Monteith (PM) procedure, FAO 56 method, presented by Allen et al. (1998). In this method, ETo is expressed as follows:

$$ET_o = \frac{0.408\Delta\left(R_n - G\right) + \frac{900}{\left(T_a + 273\right)}\gamma U_2\left(e_s - e_a\right)}{\Delta + \gamma\left(1 + 0.34U_2\right)} \quad (1)$$

where ET_o is the daily reference evapotranspiration [mm day-1], R_n is the net radiation at the crop surface [MJ m-2 day-1], G is the soil heat flux density [MJ m-2 day-1], T_a is the mean daily air temperature at 2 m height [°C], U_2 is the wind speed at 2 m height [m s-1], e_s: saturation vapor pressure [kPa], e_a: actual vapor pressure [kPa], Δ is the slope of vapor pressure curve [kPa °C-1], and γ is the psychometric constant [kPa °C-1].

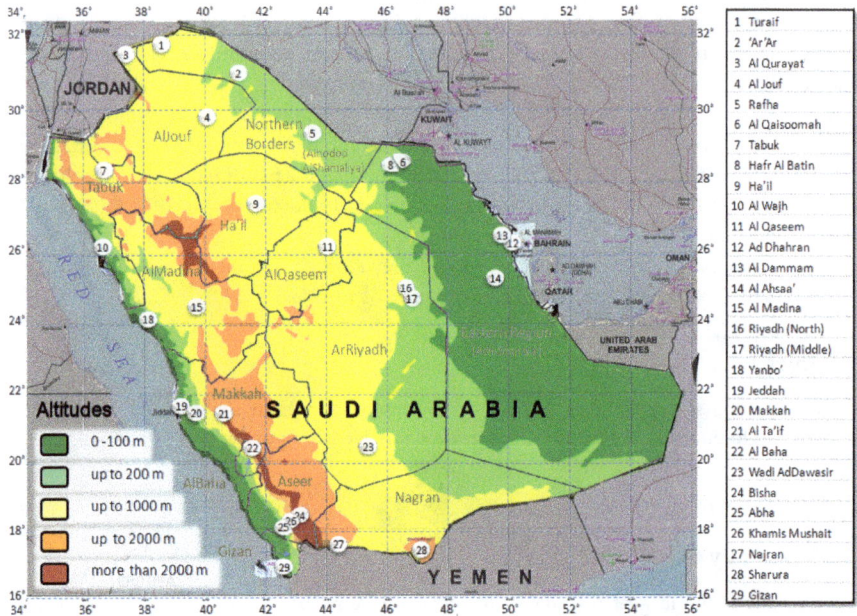

Figure 1. Geographic map of Saudi Arabia, showing 13 districts and 29 meteorological stations. *Base map Src: NIMA (2003). Districts Src: MOMRA (2007), Topography Src: Albakry et.al. (2010)*

The measured meteorological data available were T_a, Relative humidity (RH) and U_2 whereas soil heat flux (G) was taken equal to zero, (Allen et al, 2005). The slope of the saturation vapor pressure curve (Δ) is computed by the following equation as in Murray (1967):

$$\Delta = \frac{4098 \times e_o \left[T_a \right]}{\left(T_a + 237.3 \right)^2} \tag{2}$$

where $e_o \left[T_a \right]$ is calculated according to (Tetens, 1930):

$$e_o [T] = 0.611 \exp \left(\frac{17.27\, T}{T + 237.3} \right) \tag{3}$$

The net radiation R_n was estimated as the difference between the net short wave incoming radiation R_{ns} and the net long wave outgoing radiation Rnl. The calculation of R_{ns}, and R_{nl}, followed the procedures outlined in Allen et al. (1998) and Doorenbos and Pruitt (1977). All radiation was computed in daily energy flux units (MJ m^{-2} day^{-1}). Allen et al (1998) reported a validated formula to calculate the incoming solar radiation R_s from air temperature difference:

$$R_s = c R_a \sqrt{T_x - T_n} \tag{4}$$

where R_a: extraterrestrial radiation [MJ m-2 d-1], c: an adjustment coefficient =0.19 for coastal stations and 0.16 for inland stations; T_n: minimum dry bulb air temperature [°C], T_x: maximum dry bulb air temperature [°C]. The psychometric constant γ is evaluated as:

$$\gamma = 0.00163 \frac{P}{\lambda} \tag{5}$$

where P : atmospheric pressure [kPa], λ: latent heat flux [MJ kg-1]. The atmospheric pressure is expressed as in Burman et al. (1987)

$$P = 101.3 \left(\frac{293 - 0.0065z}{293} \right)^{5.26} \tag{6}$$

where z: altitude [m]. The latent heat λ depends on the average temperature, Eqn(7) , while it can be taken as an approximate value of 2.45 as reported by Harrison (1963) for T_a=20 °C. In the current study, we chose to calculate the latent heat using Eqn(7) .

$$\lambda = 2.5 - 0.00236 \, T_a \tag{7}$$

The saturation vapour pressure, e_s, and actual vapour pressure, e_a, are calculated according to Allen et al (2005) as:

$$e_s = 0.5 \left(e_o \left[T_n \right] + e_o \left[T_x \right] \right) \tag{8}$$

$$e_a = 0.005 \left(RH_x e_o \left[T_n \right] + RH_n e_o \left[T_x \right] \right) \tag{9}$$

where RH_x, RH_n: maximum and minimum relative humidity [%] respectively.

The average daily ET_o in a specific month was calculated by taking the arithmetic average of the daily values in that month. The summation of all ET_o daily values in a year for a station will give the total annual ET_o for that station.

2.3. Climatic data source and description

Basic climatic data were taken from the Presidency of Meteorology and Environment in KSA, the official climate agency in the country. The data set is the most accurate one in KSA and used by all other governmental and academic agencies for climate research and prediction. Weather stations are equipped with up-to-date monitoring devices and subjected to regular inspection and replacement for defected devices (personal communication with the Presidency of Meteorology and Environment). Data represents 29 meteorological stations as shown in Fig. 1. These stations represent all the 13 districts of the KSA. The data covers 29 years of daily meteorological records for 20 stations, 24 years for 6 stations, and 3 stations with less than 20 years as shown in Table 1. All of the data ends in 2008 and started at 1980 and 1985 for the 29 and 24 years logging.

| District | Station | | Station coordinates | | | Recroded Years** |
	ID	Name	Latitude Deg. N.	Longitude Deg. East	Altitude m	
Northern Borders	1	Turaif	31.41	38.4	818	29
	2	Arar	31.00	41.00	600	29
	5	Rafha	29.38	43.29	447	29
AlJouf	3	Guraiat*	31.50	37.50	560	4
	4	Al Jouf	29.47	40.06	671	29
Tabuk	7	Tabuk	28.22	36.38	776	29
	10	Wejh	26.12	36.28	21	29
Ha'il	9	Hail	27.26	41.41	1013	29
AlQaseem	11	Gassim	26.18	43.46	650	29
Eastern Region	6	Qaisumah	28.32	46.13	358	29
	8	Hafr Al-Batin	28.20	46.07	360	19
	12	Dhahran	26.16	50.10	17	29
	13	Dammam*	26.42	50.12	1	9
	14	Ahsa	25.30	49.48	179	24
Riyadh	16	Riyadh North	24.42	46.44	611	24
	17	Riyadh Middle	24.63	46.77	624	29
	23	W-Dawasir	20.50	45.16	652	24
Madina	15	Madina	24.33	39.42	636	29
	18	Yenbo	24.09	38.04	6	29
Makkah	19	Jeddah	21.30	39.12	17	29
	20	Makkah	21.40	39.85	213	24
	21	Taif	21.29	40.33	1454	29
Baha	22	Baha	20.30	41.63	1652	24
Aseer	24	Bisha	19.59	42.37	1163	29
	25	Abha	18.14	42.39	2093	29
	26	Khamis Mushait	18.18	42.48	2057	29
Nagran	27	Nejran	17.37	44.26	1210	29
	28	Sharurrah	17.47	47.11	725	24
Gizan	29	Gizan	16.54	42.35	3	29

*: Stations having less than 10 years of data.

**: average error ratio in data recording is less than 0.7% including missing records if any.

Table 1. Geographical information of the meteorological stations included in this study.

2.4. Data grouping and contouring

After correction the data sets, daily ETo values were calculated for each station, then aggregated to annual and monthly values. Annual ETo value (mm/year) for each station was calculated by summation of the daily ETo for the entire year. On the other hand, the monthly average ETo value was calculated by taking the average of the daily ETo values during each month.

Evapotranspiration data were graphically represented by contour maps irrespective of stations altitude. Analysis of ET variations with stations' altitude for each of the 30 years under study revealed no trends. Other researchers have also found no correlation between ET and station altitude in China (Thomas, 2000). Contour maps present clearly zones of common ET values as well as clarify vividly ET differences between zones and viability a long months or years. This approach has also been adopted by other researchers to study ET variability in China (Thomas, 2000; Shenbin et al., 2006).

Data was arranged in three columns format namely, longitude, latitude, and ETo. Each set of data was gridded separately using the ordinary point-Kriging method which estimates the values of the points at the grid nodes (Abramowitz and Stegun, 1972, and Isaaks and Srivastava, 1989). This procedure is used by SURFER™ Software which has been used in our calculations. The resulted grid was blanked outside the political borders of the KSA. The political borders' information of the KSA was grabbed from electronic map of NIMA (2003). The electronic map was digitized and converted to DMS geographic coordinate system. The blanked grid was plotted as a contour map using Surfer™ 8.0 software (Surfer, 2002). Sample plots for the average daily ET_o during a month, June in this case and the ET_o, in a year, 1991 in this case, is shown in Figure (#2a, b), respectively, where darker areas represent smaller magnitudes of ET_o.

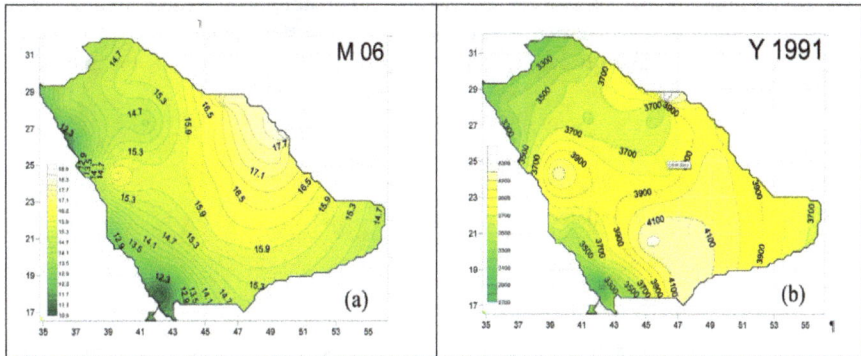

Figure 2. Sample contour map of daily and annual evapotranspiration in the KSA.
(a): average daily evapotranspiration for the month of June over 30 years period (mm/d). (b): annual evapotranspiration of the year 1991 (mm/y)

All of the data are daily values, the obtained climatic data records were carefully inspected for missing and erroneous reading. Very few errors were found, (median value of 00.45%). Errors were classified into four categories: Errors because of mistaken extreme values such as a relative humidity exceeds 100% or below 0%. Illogical errors such as the recorded maximum daily temperature (T_x) was less than the minimum daily temperature (T_n) in the same day, or if $T_x = T_n$. Missing values; i.e. T_x is present but T_n is missing. Recording an error-indication number (like 999 or 777) if the sensor is not functioning. On analyzing the data record, any value contains one or more errors was considered missing record unless the missing record could be predicted with minimal error, i.e. if the average temperature (Ta) is missing while T_x and T_n are logged with no errors; in this case Ta = ($T_x + T_n$)/2. However, the amount of missing data in the recorded period could be considered negligible in most of the stations, where the average amount of missing data is 0.78%.

2.5. Non-parametric trend analysis methods

2.5.1. Mann-kendall test

The Mann-Kendall test is a non-parametric test used for identifying trends in time series data. The test compares the relative magnitudes of sample data rather than the data values themselves Both Kendall tau coefficient (τ) and Mann-Kendall coefficient (s) are nonparametric statistics used to find rank correlation. Kendall (τ) is a ratio between the actual rating score of correlation, to the maximum possible score. To obtain the rating score for a time series, the dataset is sorted in ascending order according to time, and then the following formula is applied:

$$s = \sum_{j=1}^{j=n-1} \sum_{i=j+1}^{i=n} \text{Sign}\left(x_i - x_j\right) \tag{10}$$

where s: the rating score (also called the Mann-Kendall sum); x: the data value; i and j: counters; n: number of data values in the series; Sign is a function having values of +1, 0, or -1 if (x_i-x_j) is positive, zero, or negative, respectively. According this formula, the maximum value of s is:

$$s_{max} = \frac{1}{2}n(n-1) \tag{11}$$

Hence, the Kendall (τ) is calculated as:

$$\tau = \frac{s}{s_{max}} \tag{12}$$

A positive value of s or τ is an indicator of an increasing trend, and a negative value indicates a decreasing trend. However, it is necessary to compute the probability associated with s or τ and the sample size, n, to quantify the significance of the trend statistically. Kendall and Gibbons (1990) introduced a normal-approximation test that could be applied on datasets of more than ten values with s variance (σ^2):

$$\sigma^2 = \frac{1}{18} n(n-1)(2n+5) - CF_R \tag{13}$$

$$CF_R = \frac{1}{18} \sum_{k=1}^{g} m_k (m_k - 1)(2m_k + 5) \tag{14}$$

where CFR: repetition correction factor, to fix the effect of tied groups of data (when some of the data values appear more than one time in the dataset, this group of values are called a tied group); g: number of tied groups; k: a counter; m: number of data values in each tied group. Then normal distribution parameter (called the Mann-Kendall statistic, Z) is calculated as follows:

$$Z = \begin{cases} \frac{1}{\sigma}(s-1) & \rightarrow \quad s > 0 \\ 0 & \rightarrow \quad s = 0 \\ \frac{1}{\sigma}(s+1) & \rightarrow \quad s < 0 \end{cases} \tag{15}$$

The last step is to find the minimum probability level at which the parameter Z is significant, this could be found using two-tailed t statistical Tables or as mentioned by Abramowitz and Stegun *(1972)*:

$$\alpha_{min} = \left(b_0 \, e^{-0.5Z^2} \right) \sum_{q=1}^{q=5} b_q \cdot \left(1 + b_6 \, \text{ABS}(Z) \right)^{-q} \tag{16}$$

where αmin: Minimum level of significance; q: counter; bx: constants: b0= 0.3989, b1= 0.3194, b2= -0.3566, b3= 1.7814, b4= -1.8213, b5= 1.3303, b6= 0.2316, ABS(Z): the absolute value of Z. Kendall tau is considered significant when alpha min is less than a specified alpha value, i.e 0.05.

2.5.2. Sen-slope estimator test

Sen's statistic is the median slope of each point-pair slope in a dataset (Sen, 1968). To perform the complete Sen's test, several rules and conditions should be satisfied; the time series should be equally spaced, i.e. the interval between data points should be equal. However, Sen's method considers missing data. The data should be sorted ascending according to time, and then apply the following formula to calculate Sen's slope estimator (Q) as the median of Sen's matrix members.

$$Q = \text{Median} \left\{ \left[\left[\frac{x_i - x_j}{i - j} \right]_{j=1}^{j=n-1} \right]_{i=j+1}^{i=n} \right\} \tag{17}$$

Its sign reflect the trend's direction, while its value reflects how steep the trend is. To determine whether the median slope is statistically different than zero, the variance is calculated using Eqn. (13), to obtain the confidence interval of Q at a specific probability level, e.g 95%. The area (Z) under two-tailed normal distribution curve is calculated at the level (1-α/2), where α=1-confidence level. For example, for a confidence level of 95%, Z should be evaluated at 0.975, hence Z= 1.96. Next, the parameter C_α is calculated as follows:

$$C_\alpha = Z_{1-\alpha/2}\sqrt{\sigma^2} \qquad (18)$$

The upper and lower confidence boundaries for Q are then calculated as follows:

$$M_u = \text{int}\left(0.5\left(n_q - C_\alpha\right)\right)$$
$$M_l = \text{int}\left(0.5\left(n_q + C_\alpha\right)\right) + 1 \qquad (19)$$

where int() represents the integer value; M_u and M_l are the upper and lower boundaries for Q at 1-α probability level; nq is the number of Sen's matrix members calculated from Equation (17), equal to n_q=n(n-1). The median slope is then defined as statistically different from zero for the selected confidence interval if the zero does not lie between the upper and lower confidence limits.

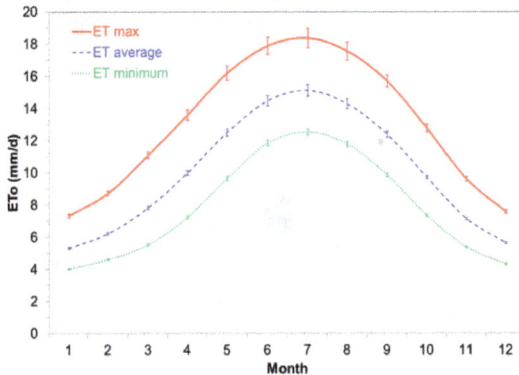

Figure 3. The monthly average ET_o variations during the year in KSA, vertical bars indicate standard deviation values.

3. Results and discussion

The daily ET_o data for each of the studied stations were calculated for the study period. Then we summarize the data on monthly basis to find the average, maximum and minimum values per month for the whole country, Figure 3. The lowest values of ET_o occurred in the winter season, December and January whereas the highest values occurred during summer months; June, July and August. The results showed high variation in ET_o from about 5 mm/day to 15 mm/day in July whereas the absolute minimum and maximum ET_o showed even higher variation and ranged from 3.9 mm/day in January to as high as 18.5 mm/day in July. However, Figure 2 indicates the high variability of climate conditions over Saudi

Arabia, which highlights the major challenge for agricultural development and water resources planning in the country.

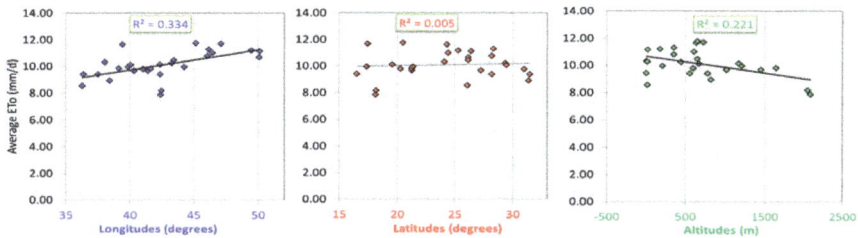

Figure 4. Relationship between the average annual ET_0 (mm/d) as related to station altitudes, latitudes, and longitudes.

The variation of ET_0 over the area of Saudi Arabia is large, as it should be due to its large surface area and large differences in altitudes. Therefore, the average ET_0 values in the study period for the studied stations were plotted with their altitude, latitude and longitudes and the results are shown in Figure 4. As expected ET_0 decreased with station altitude as weather stations varied from sea level, Yenbo' and Jeddah stations to as high as 2300 m above sea level in Nejran. Similar relationships were found with temperature and stations altitudes in the study of ElNesr et al., (2010b), which may be a significant reason for rising ET_0 values. However, large variabilities were also observed among stations at the sea level suggesting that other factors may have compound effect on ET_0 in Saudi Arabia such as the geographic location (latitude and longitude). While latitudes seemed to have no effect on the variability of ET_0, longitudes have tangible effect on it,. ET_0 increased steadily with stations longitudes as we travel toward the east. The concentration of oil industries and refineries in the eastern parts of KSA may have affected air temperature, while the nearness to the Arabian Gulf increased the relative humidity, thus leads to raising ET_0 in the eastern parts of KSA relative to the other parts.

The changes in ET_0 with time was examined by calculating the average daily evapotranspiration (mm/d) over the whole study period of each station, and plotting time series ET_0 for the study period, Figure 5. The Figure shows clearly a positive trend of ET_0 with time during the study period. The ET_0 has increased from about 9.6 in 1980 to about 10.4 mm/day in 2008 at a rate of 0.02 mm/day. Regression analysis between ET_0 and time has confirmed the positive trend and showed that the slope of the line was 0.020 with $R^2 = 0.50$. However, this relationship was not statistically significant at 95% probability level. Longer period of data analysis is needed to confirm this result. Nevertheless, the present analysis indicates clearly that climate variability is indeed affecting the country and the evapotranspiration demand is increasing with time.

Due to some restrictions in Man-Kendall and Sen's methods, two stations out of the 29 stations were omitted from calculations due to the small number of years they had (less than 10 years); those were stations #3 (Guraiat) and # 13 (Dammam). Mann-Kendall and Sen

Slope statistics were performed on the rest 27 stations on monthly basis to confirm trends direction and test its significance. Two parameters were calculated namely Kendall τ and Sen Slope Q and their confidence limits at 95% and 99% probability level as described in Materials and Methods. A group of selected results is shown in Figure 6 where the parameters of Mann-Kendall and Sen Slope and their significant tests are presented. The Figure represents ET trends in January for four stations, Tabuk, Sharurrah Yenbo, and Hail, showing possible combinations of Mann-Kendall (MK) and Sen Statistic, in addition to their significance under increasing or decreasing ET_o conditions. That is, Figure 6a showing a downtrend with MK and Sen significant at 95% and 99%. Figure 6b showing a downtrend with only MK is significant at 95%. Figure 6c showing an uptrend with all statistics was significant. Figure 6d showing an uptrend with only MK is significant at 95%.

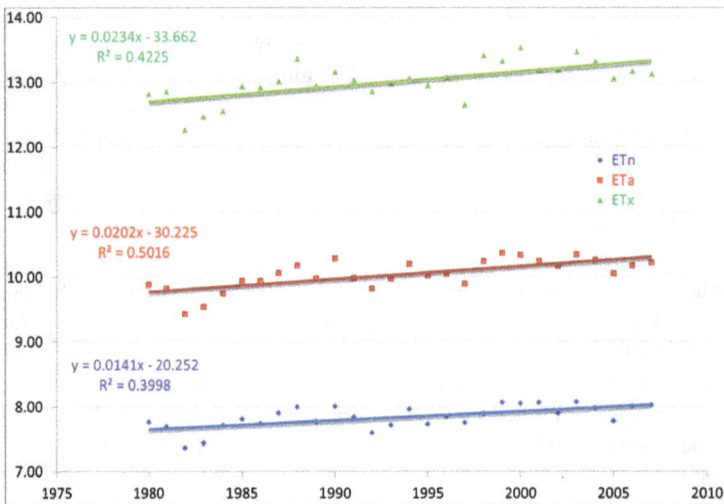

Figure 5. Temporal change of the average, maximum, and minimum ET_o through the study period.

The two tests gave similar results in all of these cases but Sen Slope test were found to be more conservative. A positive sign in τ or Q indicates an increasing trend, Figure 6 C and D while a negative value indicates a decreasing trend, Figure 6 A and B. The significance of τ was tested by comparing the calculated α_{min} with $\alpha = 0.05$ or 0.01 for 95% and 99% confidence level, respectively; $\alpha_{min} < 0.05$ or 0.01. The corresponding significant test for Q was carried out by calculating its confidence intervals at 95% and 99% indicated by ($Q_{min}^{95\%}$, $Q_{max}^{95\%}$) and ($Q_{min}^{99\%}$, $Q_{max}^{99\%}$), respectively. If the two limits have similar sign, then the calculated Sen Slope Q value is confirmed not to be zero and therefore the slope is significantly different from zero, indicating a positive or negative trend for $+Q$ or $-Q$, respectively.

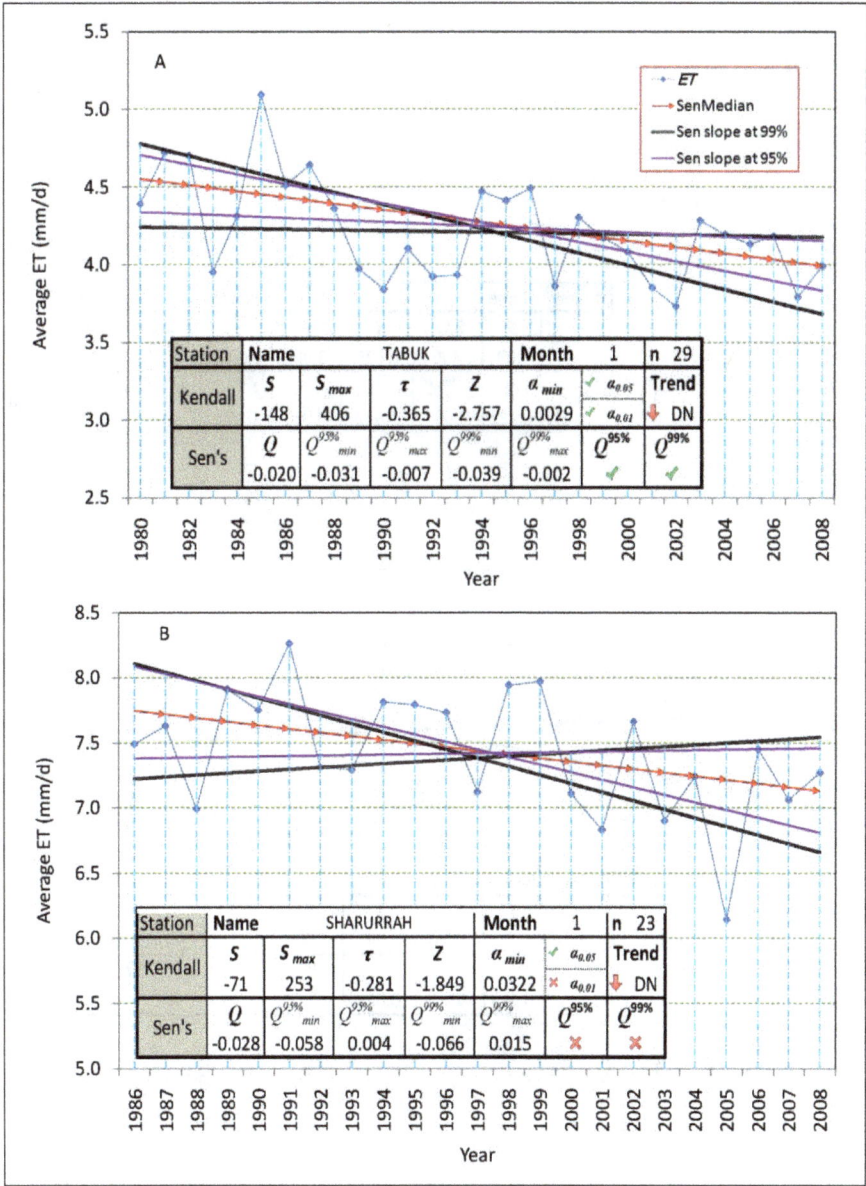

Figure A — Station TABUK

Station	Name	TABUK		Month	1	n	29

	S	S_{max}	τ	Z	α_{min}	✔ $a_{0.05}$	Trend
Kendall	-148	406	-0.365	-2.757	0.0029	✔ $a_{0.01}$	⬇ DN

	Q	$Q^{95\%}_{min}$	$Q^{95\%}_{max}$	$Q^{99\%}_{min}$	$Q^{99\%}_{max}$	$Q^{95\%}$	$Q^{99\%}$
Sen's	-0.020	-0.031	-0.007	-0.039	-0.002	✔	✔

Figure B — Station SHARURRAH

Station	Name	SHARURRAH		Month	1	n	23

	S	S_{max}	τ	Z	α_{min}	✔ $a_{0.05}$	Trend
Kendall	-71	253	-0.281	-1.849	0.0322	✖ $a_{0.01}$	⬇ DN

	Q	$Q^{95\%}_{min}$	$Q^{95\%}_{max}$	$Q^{99\%}_{min}$	$Q^{99\%}_{max}$	$Q^{95\%}$	$Q^{99\%}$
Sen's	-0.028	-0.058	0.004	-0.066	0.015	✖	✖

to be continued in the next page.

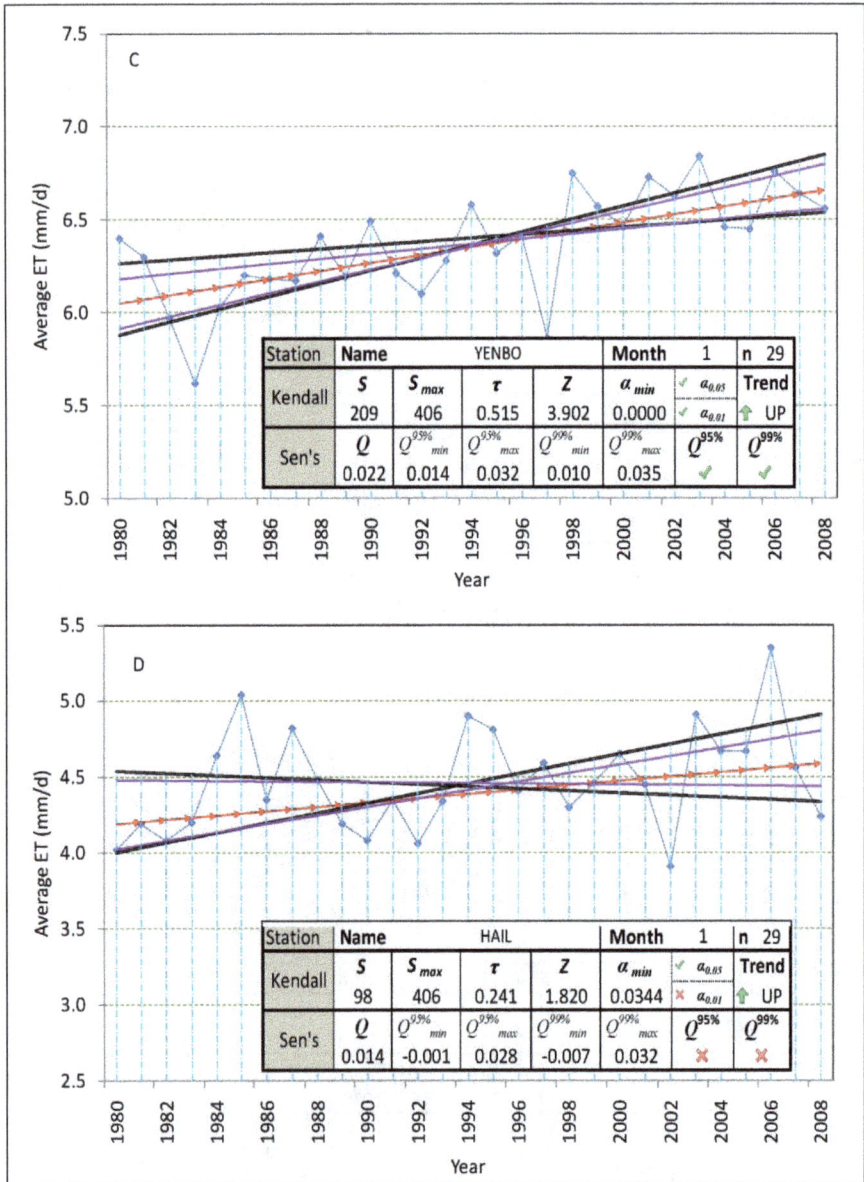

Figure 6. Analysis of monthly average ET_o trends using Mann-Kendall and Sen slope estimator and their significant tests parameters: A: Tabuk area, B: Sharurrah area, C: Yenbo area, and D: Hail area.

Figure 6A-D represents four possible cases of Q and τ and their significance. In Tabuk and Sharurrah, both Q (-0.020, -0.028) and τ (-0.365, -0.281), respectively, were negative indicating a decreasing trend for ET_o. However, this decrease is significant for Tabuk and not significant for Sharurrah at 99% probability level as confirmed by both statistical methods. In the case of Tabuk , the value of calculated α_{min} = 0.0029 is less than 0.01 indicating a significant trend according to Mann-Kendall test; for Sen slope test the 99% minimum and maximum values of the confidence interval are both negative indicating that the slope Q is not zero and therefore a negative trend is confirmed. However, for Sharurrah stations α_{min} = 0.03 is less than 0.05 but larger than 0.01 indicating that the positive trend is significant at 95% but not at 99% level according to Mann-kendall test. A slightly different result was found in Sharurrah with Sen slope test. The upper limits for the Q confidence interval at 95% and 99% were both larger than zero, $Q_{max}^{95\%} = 0.004$, $Q_{max}^{99\%} = 0.015$ indicating that the median slope of the ET_o series, Q, can actually be zero and therefore the negative trend is not significant at both levels. This results show that Sen Slope test can be more conservative than Mann-Kendall test.

Both Yenbo and Hail have increasing ET_o trends, as shown in Figure 6C and D, because τ and Q values were both positive (Q= 0.022, τ= 0.515 for Yenbo; while Q= 0.01 and τ = 0.24 for Hail). Both tests showed that ET_o in Yenbo is increasing significantly at 95% as well as at 99% level since τ and Q were positive and α_{min} = 0.000 < 0.05 and $Q_{min}^{95\%}$ > 0.0. However, in Hail this uptrend was significant according to Mann-Kendall (α_{min} = 0.000 < 0.05) at 95% level but not significant according to Sen Slope since $Q_{min}^{95\%} = -0.001$ indicating the possibility of the slope being zero. These results indicated the validity of these two statistical methods to detect trends in a time series data.

The previous analyses shown in Figure 6 were carried out for the 27 stations out of 29 under study. Man-Kendall, and Sen's methods' can deal with data series with 10 or more data points. However, Gurrayat and Dammam have less than 10 years of data and they were excluded from trend analysis. Average ET_o time series were analyzed for each month and the resultant Sen slope Q and Kendall τ are shown in Table 2 and Table 3, respectively followed by up or down arrows to indicate their significance. Up-arrows in light or dark black indicate significance up trend while similar light and dark black down arrows indicate a significant down trends at 95% and 99% probability level, respectively. Numbers without arrows are not statistically significant. The total number of stations with a decreasing or increasing trend in each month were calculated and shown at the bottom of the Table. whereas the number of months at which stations showed a decreasing or increasing trends were shown for each station at the right side of the Table. Numbers between brackets indicates the number of months or stations with the corresponding significant trend.

The tests were carried out for maximum, minimum and average monthly ET_o but only the average ET_o is shown in the Tables, because extreme ETo showed similar behavior to that of average ETo.

No	Station	\multicolumn{12}{c}{Month Number}												Number of months	
		1	2	3	4	5	6	7	8	9	10	11	12	Increasing	Decreasing
1	TURAIF	0.009	0.017	0.038	0.025	0.059	0.064	0.055	0.079	0.034	0.036	0.015	0.012	12 (7)	0 (0)
2	ARAR	0.010	0.010	0.046	0.044	0.063	0.099	0.089	0.085	0.058	0.062	0.024	0.015	12 (8)	0 (0)
4	ALJOUF	0.009	0.016	0.025	0.032	0.040	0.044	0.042	0.060	0.021	0.033	0.008	0.008	12 (6)	0 (0)
5	RAFHA	-0.018	-0.018	0.012	0.006	-0.007	0.025	0.019	-0.002	-0.018	0.009	-0.002	-0.016	5 (0)	7 (0)
6	QAISUMAH	-0.013	0.008	0.006	0.007	-0.012	-0.059	-0.090	-0.065	-0.046	-0.006	-0.009	-0.021	3 (0)	9 (4)
7	TABUK	-0.020	-0.007	-0.012	-0.020	-0.005	-0.018	-0.011	0.001	-0.025	-0.002	-0.010	-0.021	1 (0)	11 (3)
8	HAFR AL-BATIN	0.015	0.065	0.132	0.139	0.119	0.094	-0.005	0.068	0.023	0.039	0.014	-0.038	10 (4)	2 (0)
9	HAIL	0.014	0.032	0.039	0.049	0.049	0.072	0.060	0.074	0.047	0.035	0.020	0.017	12 (9)	0 (0)
10	WEJH	-0.016	-0.010	-0.017	-0.018	-0.016	-0.027	0.005	-0.013	-0.006	0.012	0.004	-0.001	2 (0)	10 (1)
11	GASSIM	-0.008	0.025	0.037	0.062	0.038	0.048	0.025	0.036	0.039	0.012	0.015	0.010	11 (4)	1 (0)
12	DHAHRAN	0.011	0.030	0.063	0.045	0.062	0.032	0.047	-0.026	0.030	0.020	0.026	0.009	11 (7)	1 (0)
14	AHSA	-0.018	-0.012	0.010	0.014	-0.077	-0.070	-0.101	-0.110	-0.053	-0.058	-0.033	-0.014	2 (0)	10 (7)
15	MADINA	-0.011	0.016	0.005	0.024	-0.010	0.013	-0.006	0.020	0.025	-0.003	0.012	0.006	8 (0)	4 (0)
16	RIYADH (New)	-0.008	0.005	-0.002	0.050	0.020	0.021	-0.023	0.016	0.030	0.026	0.002	0.002	9 (0)	3 (0)
17	RIYADH (Old)	0.006	0.025	0.033	0.047	0.051	0.021	-0.003	0.006	0.026	0.013	0.005	0.010	11 (2)	1 (0)
18	YENBO	0.022	0.040	0.049	0.037	0.058	0.070	0.053	0.072	0.071	0.050	0.036	0.032	12 (12)	0 (0)
19	JEDDAH	-0.001	-0.002	0.002	0.001	0.034	0.023	0.023	0.016	0.008	0.013	0.009	-0.006	9 (3)	3 (0)
20	MAKKAH	-0.002	0.005	0.000	-0.005	-0.002	-0.006	-0.019	-0.024	-0.023	-0.024	-0.018	-0.007	1 (0)	10 (1)
21	TAIF	0.010	0.040	0.028	0.026	0.046	0.063	0.055	0.033	0.030	0.022	0.014	0.008	12 (9)	0 (0)
22	BAHA	-0.027	-0.012	-0.015	0.017	0.013	0.020	0.000	0.008	-0.015	-0.013	-0.031	-0.021	5 (0)	7 (1)
23	W-DAWASIR	0.000	0.033	0.052	0.009	0.002	0.006	-0.026	0.033	0.014	0.013	0.029	0.012	10 (3)	1 (0)
24	BISHA	-0.001	0.018	0.017	0.024	0.023	0.037	0.020	0.037	0.036	0.025	0.015	0.007	11 (3)	1 (0)
25	ABHA	0.000	0.032	0.036	0.037	0.024	0.005	0.000	0.028	0.021	0.025	0.008	0.006	10 (5)	0 (0)
26	KHAMIS MUSHAIT	0.009	0.028	0.027	0.031	0.039	0.034	0.013	0.028	0.035	0.030	0.015	0.012	12 (8)	0 (0)
27	NEJRAN	-0.011	-0.004	-0.009	-0.002	-0.024	-0.044	-0.042	-0.032	-0.017	-0.016	-0.017	-0.007	0 (0)	12 (2)
28	SHARURRAH	-0.028	-0.044	-0.052	0.008	-0.062	-0.058	-0.091	-0.078	-0.072	-0.081	-0.026	-0.023	1 (0)	11 (4)
29	GIZAN	0.000	0.000	0.001	0.000	-0.002	0.003	0.007	-0.006	-0.003	-0.012	-0.002	0.000	5 (0)	5 (1)
	No. of stations Increasing	10 (1)	19 (9)	20 (11)	23 (9)	17 (11)	20 (10)	15 (8)	18 (9)	17 (10)	17 (8)	18 (3)	15 (1)		
	No of stations Decreasing	14 (3)	8 (0)	6 (1)	4 (0)	10 (1)	7 (3)	11 (4)	9 (2)	10 (4)	10 (4)	9 (1)	11 (1)		

Symbols Key: ↑ Increasing 90%   ↑ Increasing 95%   ↑ Increasing 99%   No trend   → Decreasing 95%   → Decreasing 95%   ↓ Decreasing 99%

Table 2. Monthly trends in the average ETo as estimated from Sen's slope statistics (Q) for various meteorological Station. Numbers in parenthesis indicate number of significant values.

Table 3. Monthly trends in the Average ETo as estimated from Mann-kendall statistics (τ) for various meteorological Station. Numbers in parenthesis indicate number of significant values.

No	Station	1	2	3	4	5	6	7	8	9	10	11	12	Increasing	Decreasing
1	ARAR	0.16	0.23	0.39	0.24	0.38	0.46	0.37	0.41	0.28	0.29	0.13	0.14	12 (9)	0 (0)
2	ALJOUF	0.19	0.14	0.36	0.32	0.44	0.54	0.33	0.40	0.40	0.43	0.24	0.22	12 (9)	0 (0)
4	RAFHA	0.13	0.20	0.33	0.22	0.37	0.50	0.33	0.48	0.21	0.30	0.26	0.12	12 (8)	0 (0)
5	QAISUMAH	-0.24	-0.17	0.09	0.04	-0.07	0.17	0.05	-0.01	-0.14	0.08	-0.01	-0.15	5 (0)	7 (1)
6	TABUK	-0.16	0.08	0.03	0.04	-0.10	-0.39	-0.46	-0.47	-0.34	-0.06	-0.06	-0.17	3 (0)	9 (4)
7	HAFR AL-BATIN	-0.36	-0.08	-0.14	-0.20	-0.06	-0.20	-0.14	0.01	-0.28	-0.02	-0.16	-0.33	1 (0)	11 (3)
8	HAIL	0.09	0.43	0.57	0.49	0.43	0.32	-0.02	0.30	0.12	0.20	0.07	-0.13	10 (6)	2 (0)
9	WEJH	0.24	0.43	0.39	0.40	0.47	0.50	0.44	0.57	0.33	0.28	0.15	0.22	12 (10)	0 (0)
10	GASSIM	-0.31	-0.16	-0.23	-0.20	-0.18	-0.16	0.12	-0.10	-0.03	-0.12	0.07	-0.01	2 (0)	10 (2)
11	DHAHRAN	-0.10	0.29	0.25	0.32	0.23	0.39	0.16	0.24	0.28	0.13	0.14	0.10	11 (7)	1 (0)
12	AHSA	0.17	0.38	0.47	0.32	0.32	0.21	0.27	-0.18	0.33	0.21	0.35	0.17	11 (7)	1 (0)
14	MADINA	-0.22	-0.12	0.04	0.13	-0.35	-0.32	-0.38	-0.36	-0.35	-0.44	-0.39	-0.13	2 (0)	10 (7)
15	RIYADH (New)	-0.17	0.12	0.10	0.15	-0.06	0.09	-0.06	0.12	0.24	-0.05	0.15	0.08	8 (1)	4 (0)
16	RIYADH (Old)	-0.10	0.07	-0.01	0.22	0.14	0.12	-0.18	0.18	0.24	0.19	0.01	0.04	9 (0)	3 (0)
17	YENBO	0.13	0.35	0.25	0.22	0.43	0.10	-0.05	0.06	0.20	0.13	0.06	0.11	11 (4)	1 (0)
18	JEDDAH	0.51	0.41	0.47	0.39	0.41	0.43	0.33	0.40	0.48	0.51	0.38	0.41	12 (12)	0 (0)
19	MAKKAH	-0.02	-0.04	0.03	0.03	0.39	0.31	0.39	0.21	0.10	0.16	0.11	-0.10	9 (3)	3 (0)
20	TAIF	-0.04	0.05	0.00	-0.05	-0.03	-0.06	-0.18	-0.28	-0.26	-0.41	-0.21	-0.11	2 (0)	10 (3)
21	BAHA	0.18	0.55	0.32	0.31	0.36	0.41	0.29	0.36	0.38	0.31	0.23	0.13	12 (10)	0 (0)
22	W-DAWASIR	-0.38	-0.11	-0.10	0.11	0.11	0.19	0.00	0.10	-0.09	-0.12	-0.21	-0.18	4 (0)	7 (1)
23	BISHA	0.02	0.27	0.34	0.06	0.01	0.04	-0.14	0.35	0.08	0.17	0.32	0.10	11 (4)	1 (0)
24	ABHA	-0.03	0.21	0.14	0.14	0.18	0.34	0.15	0.34	0.29	0.21	0.16	0.13	11 (3)	1 (0)
25	KHAMIS MUSHAIT	0.00	0.46	0.35	0.35	0.25	0.05	0.00	0.21	0.28	0.31	0.19	0.11	11 (6)	1 (0)
26	NEJRAN	0.13	0.35	0.40	0.33	0.35	0.31	0.14	0.42	0.38	-0.12	0.22	0.15	12 (8)	0 (0)
27	SHARURRAH	-0.18	-0.05	-0.11	-0.04	-0.18	-0.32	-0.31	-0.22	-0.15	-0.41	-0.25	-0.09	0 (0)	12 (4)
28	SHARURRAH	-0.28	-0.26	-0.36	0.04	-0.25	-0.27	-0.36	-0.28	-0.37	-0.41	-0.16	-0.21	1 (0)	11 (9)
29	GIZAN	0.01	0.01	0.03	0.01	-0.04	0.04	0.10	-0.10	-0.07	-0.29	-0.03	0.00	6 (0)	5 (1)
	No. of stations Increasing	13 (2)	19 (11)	21 (15)	23 (12)	17 (13)	20 (11)	14 (8)	18 (11)	17 (11)	17 (8)	18 (6)	15 (1)		
	No of stations Decreasing	14 (5)	8 (1)	6 (2)	4 (0)	10 (2)	7 (4)	12 (4)	9 (5)	10 (5)	10 (4)	9 (2)	11 (1)		

Symbols Key:
↑ Increasing 99%   ↑ Increasing 95%   No trend   ↓ Decreasing 95%   ↓ Decreasing 99%

Fourteen stations have a positive Q and τ, for at least 10 months in a year therefore an uptrend in ETo namely; Turaif, Arar, Al jouf, Hafr Al-Baten, Hail, Gassim, Dhahran, Riyadh (old), Yenbo, Taif, W-dawaser, Bisha, Abha, and khamis Mushait. Another six stations showed a negative or zero Q during the whole year, therefore a downtrend in ET_o including; Tabuk, Wejh, Makkah, Nejran, Sharurrah, and Gizan. The other seven stations showed a mix of increasing and decreasing trends during the year and those are Rafha, Qaisumah, Ahsa, Madina, Riyadh (new), Jeddah, and Baha.

However, the up or down trends or downtrends in ET_o in the first mentioned group were not always significant as indicated by the upward arrows and summed in the last two columns of Tables 2 and 3. Only Yenbo had a confirmed significant trend at 95% level during the entire year. Other stations showed a significant up trends in ET_o for several months during the year including Hail and Taif , 10 months; Turaif and Arar, 9 months; Al jouf and Khamis Mushait, 8 months. The other stations among the uptrend group had significant uptrend in 3 months to 7 months in a year as shown in Table 2.

The number of stations with a decreasing trend is far less than those with increasing ET_o. Few stations showed a decreasing trend in ET_o for 9 months or higher including, Qaisumah, Tabuk, Wejh, Ahsa, Makkah, Nijran and Sharurrah. However, only Ahsa station had a significant decreasing trend for 7 months followed by Qaisumah and Sharurrah, 4 months, and Tabuk with only 3 months of declining in ET_o. The rest of stations, Wejh, Makkah, Nejran, Gizan, had a decreasing trend but this trend is not significant at 95% probability level.

The numbers of stations with increasing or decreasing trend are shown in the last two rows of Table 2 and 3. Figure 7, as well, illustrates the number of stations with significant/non-significant increasing/decreasing trend of ET throughout the studied areas. At least 15 stations or higher showed an increasing trend for the entire year except in January at which 14 stations showed a decreasing trend. March, April and June showed the highest number of stations with increasing ET_o. However, the significant increase in ET_o were confirmed for about 10 stations and for 9 months, February to October. During the months of October to January about 10 stations showed a decreasing trend but this decrease was significant for only 4 stations in September and October, one station in November and December and 3 stations in January.

Further inspection on the location of stations with increasing trends in ET_o revealed that most of these stations are located in the northern part of the Arabian peninsula north to the latitude line of 22 degrees. However, some other stations were located southern of this line at the southern west corner of Saudi Arabia. It seemed that stations located along the longitudinal line of 45 degrees showed an increasing trend. Actually the wind direction over the Arabian peninsula seemed to follow this line from south west to north in rainy seasons and from north to south west in the dry seasons.

To have an aerial graph for the regions with a decreasing or increasing trend in ET_o a contour map were plotted for each months and the results are shown in Figure 8. Certainly, regions with increasing trends are concentrated in the northern part of Saudi Arabia and extended to the south along the 45-degree longitudinal line. Significant and increasing regions, indicated by black and grey regions (P>95%) are prevail for most of the year except in January and July to some extent. In January most of SA areas have decreasing ET_o as shown in Figure 8 and also in Table 3 where 14 stations have a decreasing trends; although this decrease is not significant at 95% level except for 3 stations. The southeastern parts seemed to have decreasing trends most of the years, but also this trend is not significant at 95% except in July and October as indicated by the darker dotted regions.

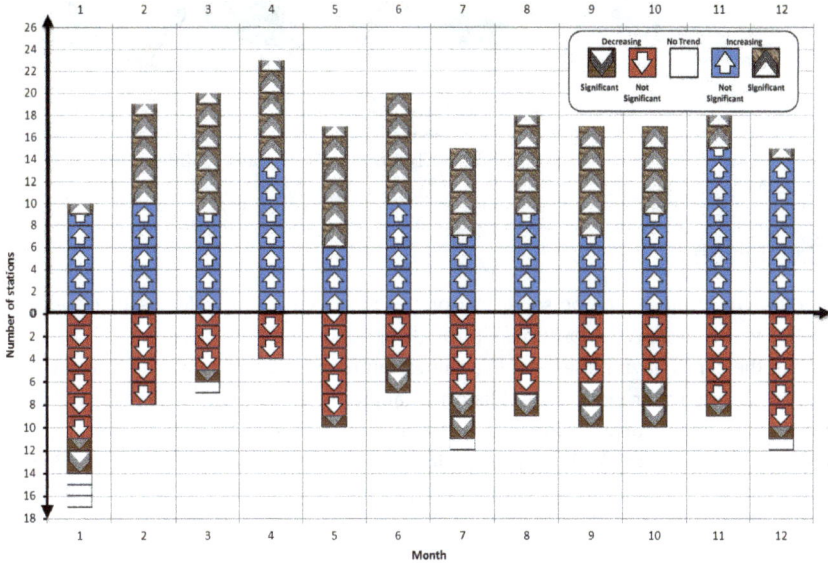

Figure 7. Number of stations with significant/non-significant increasing/decreasing trends of ET in the studied areas.

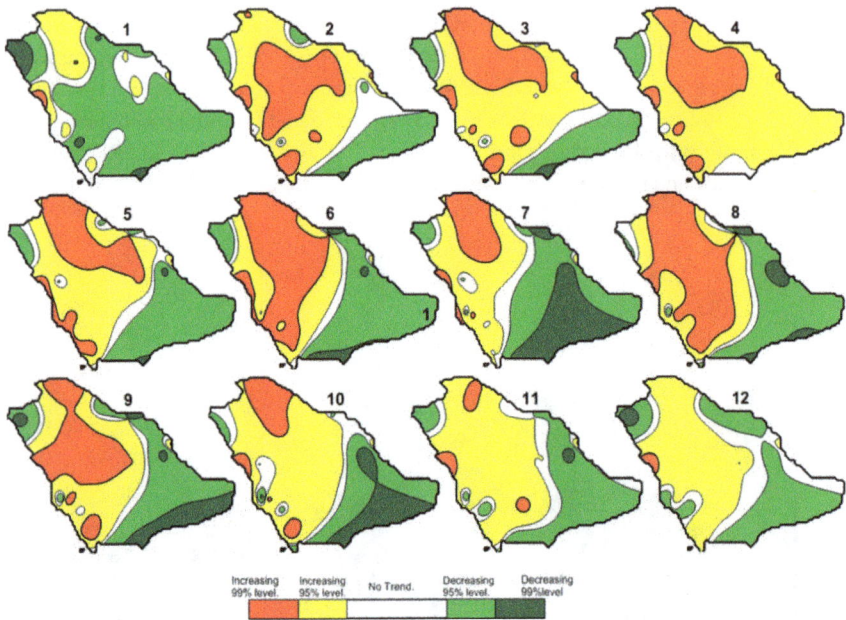

Figure 8. Contour plots showing the distribution ETo trends' direction and confidence level over the area of KSA.

4. Conclusion

Water scarcity problem can be solved by proper management of water usage. Most of the depleted water in KSA is consumed through agriculture. Identifying the ET_o trend and knowing the zones having the least ET_o values can help in determining the future plans of agricultural and water extensions. Historical analysis of daily ET_o in Saudi Arabia was carried out using Penman Monteith equation (FAO-56) for 29 meteorological stations distributed all over Saudi Arabia for the period 1980 to 2008. The long time average daily ET_o varied from about 5 mm/d in Jan to 15 mm/day in July which is one of the hottest months in the country. ET_o time series analysis using Mann-Kendall and Sen slope statistics revealed that ET_o has been increasing steadily during the study period. The average minimum and maximum daily ET_o increased steadily and ET_o average increased from about 9.6 to about 10.4 mm/day in 2008. Trend analysis revealed that about 14 of the weather stations showed a significant increasing trend in ET_o during the year for more than 7 months. Only 4 stations showed decreasing trends in three months, September, October and January.

Increasing ET_o trends prevail in the northern and south-west areas along the longitudinal line of 45 degrees while decreasing trends prevail in the north western spot along the red sea and south eastern parts along the Arabian Gulf. This demonstrates that ET_o fluctuation is increasing with time that can be considered a significant sign for climate variability in the

Arabian peninsula. This increase in ET_0 seemed to be mainly affected by the global warming or the increase in temperature in the Arabian peninsula which was confirmed by several studies mentioned in this paper. Analyses of longer historic data are needed to confirm these findings. This demonstrates that ET_0 fluctuation is increasing with time that can be considered a significant sign for climate change. Though, the findings of this research suggest the needs to consider ET_0 changes in the planning for agricultural and water resources projects. Thus to rank the areas with fixed and decreasing ET_0 trend as highly recommended zones for future agricultural projects, and to do the opposite with the increasing ETo trends' zones. Finally, if the low ranked zones are essential due to other circumstances, then the water management policy should consider the increment rate in ETo and its effect on water consumption.

Author details

Mohammad Elnesr* and Abdurrahman Alazba
King Saud University, Alamoudi Water Chair, Saudi Arabia

Acknowledgement

The authors wish to express their deep thanks and gratitude to "Shaikh Mohammad Bin Husain Alamoudi" for his kind financial support to the King Saud University, through the research chair "Alamoudi Chair for Water Researches" (AWC), where this paper is part of the AWC chair activities. Thanks should also be expressed to the Presidency of Meteorology and Environment in Riyadh, KSA, who kindly support this research my meteorological data.

5. References

Abderrahman W. A. ; Bader T. A. ; Asfahan Ullah Kahn ; Ajward M. H. (1991) Weather Modification Impact on Reference Evapotranspiration, Soil Salinity and Desertification in Arid Regions, A Case Study. J. Arid Environments. 20(3):277-286

Abderrahman, W. A. (2001) Water demand management in Saudi Arabia. Ch. 6 In: Water Management In Islam proc.. IDRC/UNU Press 2001 [Naser I. Faruqui, Asit K. Biswas, and Murad J. Bino (eds.)]. 170 pp. available online at: http://www.idrc.ca/en/ev-93954-201-1-DO_TOPIC.html

Abdreeahman, W.A., and Al-Harazin, I. M. (2003). The impacts of global climatic change on reference crop evapotranspiration, irrigation water demands, soil salinity, and desertification in Arabian Peninsula. 67:74p. In Proc. Intl. Conf. of "Desertification in the Third Millennium". Dubai, 12-15 February 2003. Ed. Alsharhan, A. S. Organized by Zayed International Prize for the Environment. 504 pp

---

* Corresponding Author

Abo-Ghobar, H. (2000). Estimation of reference evapotranspiration for southern region of Saudi Arabia. Irrig Sci 19: 81 - 86.

Abramowitz, M., Stegun, I. eds. (1972), Handbook of Mathematical Functions with Formulas, Graphs, and Mathematical Tables, New York: Dover Publications

Al-Ansari, N., E. Salameh and H. Al-Omari, (1999). Analysis of rainfall in the badia region, Jordan. Research Paper No. 1, Al-al-Bayt University, Jordan.

Albakry, A., Alsaleem, I., and ElBeishi, M. (2010). Geography of the Kingdom of Saudi Arabia and Some Other Countries. 3rd ED. ISBN: 9960-19-056-0, Pub. Saudi Ministry of Education.(www.moe.gov.sa) 164pp.

Allen, R. G., Pereira, L. S., Raes, D., Smith, M., (1998). Crop evapotranspiration. Guidelines for computing crop water requirements. FAO Irrigation and drainage paper 56. Food and Agriculture Organization of the United Nations, 300pp.

Allen, R.G., Walter, I.A., Elliott, R.L., Howell, T.A., Itenfisu, D., Jensen, M.E. and Snyder, R.L. (2005). The ASCE Standardized Reference Evapotranspiration Equation. Amer. Soc. of Civil Eng. Reston, Virginia. 192pp.

AQUASTAT. (2008). FAO's Information System on Water and Agriculture, Climate information tool. Website, http://www.fao.org/nr/water/aquastat/ countries/saudi_arabia/index.stm., accessed 25/2/2009.

Bandyopadhyay, A., A. Bhadra, N. S. Raghuwanshi, and R. Singh. (2009). Temporal trends in estimates of reference evapotranspiration over india. Journal of Hydrologic Engineering 14(5):508-515.

Burman, R.D., Jensen M.E. and Allen R.G. (1987). Thermodynamic factors in evapotranspiration. In: Proc. Irrig. and Drain. Spec. Conf., James L.G. and English M.J. (eds). ASCE, Portland, Ore., July. : 28-30.

Cohen, S. and Stanhill G. (1996). Contemporary climate change in Jordan Valley. J. Appl. Meteorol., 35: 1051-1058.

Dinpashoh, Y., (2006). Study of reference crop evapotranspiration in I.R. of Iran. Agric. Water Manage. 84: 123–129.

Doorenbos, J. and Pruitt, W. O. (1975). Guidelines for predicting crop water requirements, Irrigation and Drainage Paper 24, Food and Agriculture Organization of the United Nations, Rome, 179 p.

Elnesr, M., & Alazba, A. (2010). Spatio-Temporal Variability of Evapotranspiration over the Kingdom of Saudi Arabia. Applied Engineering in Agriculture, ASABE, 26(5), 833-842.

Elnesr, M., A. Alazba and M. Abu-Zreig, (2010a). Analysis of evapotranspiration variability and trends in the Arabian Peninsula. Am. J. Environ Sci., 6: 535-547. DOI: 10.3844/ajessp.2010.535.547
URL: http://www.thescipub.com/abstract/10.3844/ajessp.2010.535.547

Elnesr, M., M.M. Abu-Zreig and A.A. Alazba, (2010b). Temperature Trends and Distribution in the Arabian Peninsula. Am. J. Environ. Sci., 6: 191-203. DOI: 10.3844/ajessp.2010.191.203
URL: http://www.thescipub.com/abstract/10.3844/ajessp.2010.191.203

Fehrman r. L. (2007). Increasing Evapotranspiration Trends over the Mississippi River Basin. MSc. Thesis. faculty of the graduate school of Cornell Univ.

FU, G. , Stephen, C. P. , YU, J. (2009). A critical overview of pan evaporation trends over the last 50 years. Climatic change 97(1-2):193-214

Gao, G., D. Chen, C.-y. Xu, and E. Simelton. (2007). Trend of estimated actual evapotranspiration over china during 1960–2002. Journal of Geophysical Research 112(D11):D11120+.

Garbrecht, J., Van Liew, M., and Brown, G. O. (2004). Trends in precipitation, streamflow, and evapotranspiration in the great plains of the united states. Journal of Hydrologic Engineering 9:360-367.

Hakan A., Savaş K., Osman Ş. (2010). Trend Analysis of Hydrometeorological Parameters in Climate Regions of Turkey. Conference prerelease, BALWOIS 2010 – Ohrid, Republic of Macedonia –25, 29 May 2010 URL: www.balwois.com/balwois/administration/ full_paper/ffp-1457.pdf accessed 27/3/2010

Harrison, L. P. (1963). Fundamentals concepts and definitions relating to humidity. In Wexler, A (Editor) Humidity and moisture Vol 3, Reinhold Publishing Co., N.Y.

Hegerl, G.C., F. W. Zwiers, P. Braconnot, N.P. Gillett, Y. Luo, J.A. Marengo Orsini, N. Nicholls, J.E. Penner and P.A. Stott. (2007). Understanding and Attributing Climate Change. In: Climate Change 2007: The Physical Science Basis. Contribution of Working Group I to the Fourth Assessment Report of the Intergovernmental Panel on Climate Change [Solomon, et al. (eds.)]. Cambridge University Press, Cambridge, United Kingdom and New York, NY, USA. 690pp.

Hobbins, M. T., Ramı́rez, J. A., Brown, T. C. (2004). Trends in pan evaporation and actual evapotranspiration across the conterminous U.S.: Paradoxical or complementary? Geophysical Research Letters. 31(13): L13503+

Isaaks E. H., and Srivastava R. M. (1989). An Introduction to Applied Geostatistics. Oxford University Press. 198 Madison Av. NY: 279-330. 560pp.

Kendall M., and Gibbons J.D., (1990). Rank Correlation Methods. New York: Oxford University Press, fifth ed. 272pp.

Liu, B., M. Xu, M. Henderson, and W. Gong. (2004). A spatial analysis of pan evaporation trends in china, 1955–2000. Journal of Geophysical Research 109(D15):D15102+.

Milly, P.C.D., and Dunne K.A., (2001). Trends in evaporation and surface cooling in the Mississippi River basin. Geophys. Res. Lett., 28, 1219-1222.

MOMRA (2007). Ministry of Municipal and Rural Affairs map of Saudi districts. (official website) URL: http://www.momra.gov.sa/GeneralServ/mun/imap.htm, accessed Dec. 2009.

Murray F.W. (1967). On the computation of saturation vapor pressure. J. Appl. Meteor. 6: 203-204.

Mustafa M. A., Akabawi K. A. and Zoghet M. F. (1989). Estimation of Reference Crop Evapotranspiration for the Life Zones of Saudi Arabia. J. Arid Environ. 17: 293-300.

Myneni, R.B., C.D. Keeling, C.J. Tucker, G. Asrar, Nemani R.R., (1997). Increased plant growth in the northern high latitudes from 1981 to 1991. Nature, 386: 698-702.

NIMA (2003). National Imagery and Mapping Agency. The United States Government. No copyrights under Title 17 U.S.C.

Roderick, M.L. and Farquhar, G.D. (2004). Changes in Australian pan evaporation from 1970 to 2002. Intl. J. of Climatology 24: 1077-1090.

Schwartz, P. and Randall D. (2003). An abrupt climate change scenario and its implications for United States national security. Global Business Network, Emeryville, CA. Online source: www.edf.org/documents/3566_AbruptClimateChange.pdf. 22pp.

Sen, P.K., 1968. Estimates of the regression coefficient based on Kendall's τ. Journal of the American Statistical Association. 63:1379-1389 quoted from Brauner, S. (1997) Nonparametric Estimation of Slope: Sen's Method in Environmental Pollution. In The Environmental Sampling & Monitoring Primer Project. Ed. Gallagher D. Online: http://www.cee.vt.edu /ewr/environmental/teach/smprimer/Sen/Sen.html.

Shenbin, C., Yunfeng, L., and Thomas, A. (2006). Climatic Change on The Tibetan Plateau: Potential Evapotranspiration Trends From 1961–2000 Climatic Change 76: 291–319

Smadi M., (2006). Observed Abrupt Changes in Minimum and Maximum Temperatures in Jordan in the 20th Century. Am. J. Environ. Sci., 2 (3): 114-120.

Smadi, M.M. and Zghoul, A., (2006). A Sudden Change In Rainfall Characteristics In Amman, Jordan During The Mid 1950s. Am. J. of Environ. Sci. 2 (3): 84-91.

Song, Z. W., H. L. Zhang, R. L. Snyder, F. E. Anderson, and F. Chen. (2010). Distribution and trends in reference evapotranspiration in the north china plain. Journal of Irrigation and Drainage Engineering 136(4):240-247.

Surfer (2002). Surfer©, a surface mapping system. Windows™ Software. Golden Software Inc. URL: www.goldensoftware.com

Szilagyi J. (2001). Modeled Areal Evaporation Trends Over The Conterminous United States. J. Irrig. Drain. Eng. 127(4): 196-200.

Tebakari, T., J. Yoshitani, and C. Suvanpimol. (2005). Time-space trend analysis in pan evaporation over kingdom of Thailand. J. of Hydr. Eng. 10(3):205-215.

Tetens, O. 1930. Uber einige meteorologische begriffe. Z. Geophys. 6:297-309. quoted from FAO, 1990.

Thomas, A. (2000). Spatial And Temporal Characteristics Of Potential Evapotranspiration Trends Over China. Int. J. Climatol. 20: 381–396

Wynbrandt, J. (2004). A Brief History of Saudi Arabia. Facts on File, Inc. ISBN 0-8160-5203-4. 352pp

Inventory of GIS-Based Decision Support Systems Addressing Climate Change Impacts on Coastal Waters and Related Inland Watersheds

F. Iyalomhe, J. Rizzi, S. Torresan, V. Gallina, A. Critto and A. Marcomini

Additional information is available at the end of the chapter

1. Introduction

A Decision Support System (DSS) is a computer-based software that can assist decision makers in their decision process, supporting rather than replacing their judgment and, at length, improving effectiveness over efficiency [1]. Environmental DSS are models based tools that cope with environmental issues and support decision makers in the sustainable management of natural resources and in the definition of possible adaptation and mitigation measures [2]. DSS have been developed and used to address complex decision-based problems in varying fields of research. For instance, in environmental resource management, DSS are generally classified into two main categories: Spatial Decision Support Systems (SDSS) and Environmental Decision Supports Systems (EDSS) [3-5]. SDSS provide the necessary platform for decision makers to analyse geographical information in a flexible manner, while EDSS integrate the relevant environmental models, database and assessment tools – coupled within a Graphic User Interface (GUI) – for functionality within a Geographical Information System (GIS) [1,4-6]. In some detail, GIS is a set of computer tools that can capture, manipulate, process and display spatial or geo-referenced data [7] in which the enhancement of spatial data integration, analysis and visualization can be conducted [8-9]. These functionalities make GIS-tools useful for efficient development and effective implementation of DSS within the management process. For this purpose they are used either as data managers (i.e. as a spatial geo-database tool) or as an end in itself (i.e. media to communicate information to decision makers) [8].

At present the increasing trends of industrialisation, urbanisation and population growth has not only resulted in numerous environmental problems but has increased the complexity in terms of uncertainty and multiplicity of scales. Accordingly, there is a consensus on the consideration of several perspectives in order to tackle environmental problems, particularly, climate change related impacts in coastal zones which are characterised by the dynamics and interactions of socio-economic and biogeophysical phenomena. There is the need to develop and apply relevant tools and techniques capable of processing not only the numerical aspects of these problems but also knowledge from experts, to assure stakeholder participation which is essential in the decision making process [5] and to guarantee the overall effectiveness of assessment and management of coastal environments – including related inland watersheds (i.e. surface and groundwaters affected by, and affecting, coastal waters).

The scientific community projected that climate change would further exacerbate environmental problems due to natural and anthropogenic impacts – with specific emphasis in coastal areas [10]. This data, nevertheless, depends on global and regional policy measures especially in sectors such as energy, economy and agriculture which seem to be a major threat to global sustainable development. As a response to this, mitigation and adaptation measures are already identified through intense research activities, yet these may not limit the projected effects of climate change over the next few decades On one side there is the influence of socio-economic development and environmental response while on the other there is the significant uncertainty still associated with present climatic predictive models. Thus, model inputs need to take into account scenarios highly affected by present and future policy measures in order to further reduce uncertainty in their predictions and thereby guarantee robust adaptation strategies.

In addition, climate change effects have been linked to the increase in global average temperature according to the IPCC emission scenarios [11]. Resulting ocean thermal expansion is expected to generate significant impacts via sea level rise, seawater intrusion into coastal aquifers, enhanced coastal erosion and storm surge flooding, while increasing population in coastal cities, especially megacities on islands and deltas, further aggravates major impacts of climate change on marine coastal regions. The latter include transitional environments such as estuaries, lagoons, low lying lands, lakes, which are particularly vulnerable because of their geographical location and intensive socio-economic activities [12-13].

Accordingly, several environmental resource regulations have already included the need to assess and manage negative impacts derived from climate change through their implementation. For instance, the European Commission approved the Green and White papers [14-15], the Water Framework Directive (WFD) [16], which represent an integrated and sound approach for the protection and management of water-related resources in both inland and coastal zones and signed the protocol for Integrated Coastal Zone Management (ICZM) [17], useful in the promotion of the integrated management of coastal areas in relation to local, regional, national and international goals. Moreover, the principles of Integrated Water Resources Management (IWRM) aimed to address typical water quality and quantity concerns with the optimisation of water management and sustainability in collaboration with WFD policy declarations [18]. Likewise, relevant national legislations like

Shoreline Management Planning (SMP) in the United Kingdom [19], Hazard Emergency Management (HEM) in the United States [20] and Groundwater Resources Management (GRM) in Bangladesh and India [21] were ratified and further endorse the assessment and management of coastal communities in relation to climate change impacts.

Within this context, the development of innovative tools is needed to implement regulatory frameworks and the decision making process required to cope with climate related impacts and risks. To this end, DSS are advocated as one of the principal tools for the described purposes.

This work will attempt to examine GIS-based DSS resulting from an open literature survey. It will highlight major features and applicability of each DSS in order to help the reader in the selection of DSS tailored on his specific application needs.

2. Description of the examined Decision Support Systems (DSS)

The literature survey led to identify twenty DSS designed to support the decision making-process related to climate change and environmental issues in coastal environments – including inland watersheds. The identified DSS are listed in Table 1 with the indication of the developer, development years, and literature reference. In order to provide a description of major features and an evaluation of the applicability of the 20 examined DSS, the work adopted the sets of criteria reported in Table 2 and grouped them within three different categories: general technical criteria, specific technical criteria, and availability and applicability criteria. The general technical criteria underline relevant general features related to each DSS, which include: the target coastal regions and ecosystems domain; the regulatory frameworks and specific legislations supported by each DSS; the considered climate change impacts and related scenarios, as well as the objectives of the examined systems. The specific technical aspects include the main functionalities, analytical methodologies and inference engine (i.e.structural elements) of the systems. A final set of criteria concerned applicability, i.e. scale and study areas, flexibility, status and availability of the examined systems. Within the following sections the identified DSS, listed in Table 1, will be presented discussed according to these criteria.

Name	Developer	Year of Development	Reference Source
CLIME: Climate and Lake Impacts decision support system	Helsinki University of Technology, Finland	1998-2003	[22] http://clime.tkk.fi
CORAL: Coastal Management Decision Support Modelling for Coral Reef Ecosystem	Within a World Bank funded Project :LA3EU	1994-1995	[23]

Name	Developer	Year of Development	Reference Source
COSMO: Coastal zone Simulation MOdel	Coastal Zone Management Centre, Hague	1992	[24]
Coastal Simulator decision support system.	Tyndall Centre for Climate Change Research, UK.	2000-2009	[25]
CVAT: Community Vulnerability Assessment Tool	National Oceanic and Atmospheric Administration, US.	1999	[20] www.csc.noaa.gov/products/ nchaz/startup.htm
DESYCO: Decision Support SYstem for COastal climate change impact assessment	Euro-Mediterranean Centre for Climate Change, (CMCC) Italy.	2005-2010	[2]
DITTY: Information technology tool for the management of Southern European lagoons	Within the European region project: DITTY	2002- 2005	[26]
DIVA: Dynamic Interactive Vulnerability Assessment	Potsdam Institute for Climate Impact Research, Germany	2003-2004	[27] http://www.dinas-coast.net.
ELBE: Elbe river basin Decision Support System	Research Institute of Knowledge System- RIKS, Netherland	2000-2006	[28] www.riks.nl/projects/Elbe-DSS
GVT: Groundwater Vulnerability Tool	University of Thrace and Water Resource Management Authority, Greece.	2003-2004	[29]
IWRM: Integrated Water Resources Management Decision Support System	Institute of Water Modelling, Bangladesh	2002-2010	[21] www.iwmbd.org
KRIM decision support system	Within the KRIM Project in Germany.	2001-2004	[30] www.krim.uni-bremen.de
MODSIM decision support systems	Labadie of Colorado State University, US	1970	[31-32] www.modsim.engr.colostate.edu
RegIS-Regional Impact Simulator	Cranfield University, UK	2003-2010	[33] http://www.cranfield.ac.uk/s as/naturalresources/research/ projects/regis2.html

Name	Developer	Year of Development	Reference Source
RAMCO: Rapid Assessment Module Coastal Zone Management	Research Institute of Knowledge System- RIKS, Netherland	1996-1999	[34-35] http://www.riks.nl/projects/RAMCO
SimLUCIA: Simulator model for St LUCIA	Research Institute of Knowledge System- RIKS within the UNEP Project, Netherland	1988-1996	[36] http://www.riks.nl/projects/SimLUCIA
SimCLIM: Simulator model System for Climate Change Impacts and Adaptation	University of Waikato and CLIMsystem limited, New Zealand.	2005	[37] www.climsystems.com
STREAM: Spatial Tools for River Basins and Environment and Analysis of Management Options	Vrije Universiteit Amsterdam and Coastal Zone Management Centre, Hague	1999	[38] http://www.geo.vu.nl/users/ivmstream/
TaiWAP: Taiwan Water Resources Assessment Program to Climate Change	National Taiwan University, Taiwan	2008	[39]
WADBOS: decision support systems	Research Institute of Knowledge System- RIKS, Netherland	1996-2002	[40-41] www.riks.nl/projects/WADBOS

Table 1. List of existing DSS on coastal waters and related inland watersheds.

Categories	Criteria
General technical criteria	• Coping with regulatory framework. This indicates the particular legislation or policy, the DSS refers to and which phase of the decision-making process is supported at the National, Regional and Local level (e.g., EU WFD, ICZM, IWRM, SMP, GRM, and HEM). • Study/ field of application area. The coastal zones where this DSS has been applied and tested (e.g., coastal zone, lakes, river basin, lagoon, groundwater aquifer etc.) • Objective. It specifies the main aims of the DSS. • Climate change impacts. This refers to relevant impacts due to climate change on the system (e.g., sea-level rise, coastal flooding, erosion, water quality). • Climate Change Scenarios. The kind of scenarios considered by the DSS, which are relevant to the system analysis and connected to climate change (e.g., emission, sea level rise, climatic scenarios).

Categories	Criteria
Specific technical criteria	• Functionalities. These indicate relevant functionalities (key outcomes) of the system useful to the decision process: environmental status evaluation, scenarios import (climate change and socio-economic scenarios) and analysis, measure identification and/or evaluation, relevant pressure identification and indicators production. • Analytical methodologies. These indicate the methodologies included in the system such as risks analysis, scenarios construction and/or analysis, integrated vulnerability analysis, Multi-Criteria Decision Analysis (MCDA), socio-economic analysis, uncertainty analysis, ecosystem-based approach etc. • Structural elements. The three major components of the DSS: dataset (i.e., the typology of data), models (e.g., economic, ecological, hydrological and morphological), interface (i.e., addressing if it's user-friendly and desktop or web-based).
Availability and applicability	• Scale and area of application. This specifies the spatiality of the system (e.g., local, regional, national, supra-national and global) within the case study areas. • Flexibility. The characteristics of the system to be flexible, in terms of change of input parameters, additional modules or models and functionalities. It is also linked to the fact that it can be apply on different coastal regions or case study areas. • Status and Availability. This specifies if the system is under development or already developed and ready for use, and if it is restricted to the developer and case study areas only or the public can access it too and the website where information about the DSS can be found.

Table 2. List of criteria used for the description of existing DSS.

2.1. General technical criteria

As far the application domain, the considered DSS focus on coastal zones and related ecosystems (e.g. lagoons, groundwaters, river basins, estuaries, and lakes), specifically thirteen DSS are on coastal zones, seven concern coastal associated ecosystems and four focus on both (Table 3).

As far as regulatory frameworks (i.e. ICZM, WFD, IWRM) and national legislations are concerned, the examined DSS reflect the assessment and management aspects of the related decision making process. Within the coastal, marine and river basin environments, the assessment phase of these frameworks consists of the analysis of environmental, social, economic and regulatory conditions, while the management phase looks at the definition and implementation of management plans. Accordingly, support is provided by each DSS to the implementation of one or two frameworks in the assessment and/or management phase in relation to specific objectives and application domain. Specifically, the investigated DSS can provide the evaluation of ecosystem pressures, the assessment of climate change hazard, vulnerability and risks, the development and analysis of relevant policies, and the definition and evaluation of different management options. Eight out of the twenty examined DSS provide support for the ICZM implementation through an integrated assessment involving regional climatic, ecological and socio-economic aspects (Table 3, second column). With respect to the WFD (i.e. six DSS) and IWRM (i.e. seven DSS), the main focus is on the assessment of environmental or ecological status of coastal regions and related ecosystems and on the consideration of anthropogenic impacts and risks on coastal resources. These two groups of DSS consider also the river basins management via evaluation of adaptation options, which is essential for the management phase of the WFD and IWRM implementation. Particularly interesting are the approaches adopted by three DSS: CLIME, STREAM and COSMO. CLIME supports both the assessment and management phases of WFD through the analysis of present and future climate change impacts on ecosystems and the socio-economic influence on water quality of the European lakes. STREAM evaluates climate change and land use effects on the hydrology of a specific river basin, in order to support the management phase of IWRM and WFD via the identification of water resources management measures. Lastly, COSMO provides support for the ICZM through the identification and evaluation of feasible management strategies for climate change and anthropogenic impacts relevant for coastal areas. Moreover, RegIS, Coastal Simulator, CVAT and GVT specifically support the implementation of national legislations through the consideration of socio-economic and technological issues relevant for identifying suitable mitigation actions. To this purpose, these DSS promote the involvement of stakeholders through participatory processes.

The main objective of the examined DSS is the analysis of vulnerability, impacts and risks, and the identification and evaluation of related management options, in order to guarantee robust decisions required for sustainable management of coastal and inland water resources. Specifically, the objectives of the examined DSS are concerned with three major issues: (1) the assessment of vulnerability to natural hazards and climate change (four DSS: CVAT, GVT, SimLUCIA, TaiWAP); (2) the evaluation of present and potential climate change impacts and risks on coastal zones and linked ecosystems, in order to predict how coastal regions will respond to climate change (nine DSS); (3) the evaluation or analysis of management options for the optimal utilisation of coastal resources and ecosystems through the identification of feasible measures and adequate coordination of all relevant users/stakeholders (seven DSS: WADBOS, COSMO CORAL, DITTY, ELBE, MODSIM, RAMCO).

Name	Application domain	Regulatory framework of reference	Objective	Climate change impacts addressed	Climate change scenarios generating impacts
CLIME	• Lakes.	WFD for environmental assessment.	To explore the potential impacts of climate change on European lakes dynamics linked coast.	• Water quality.	• Emission scenarios. • Temperature scenarios.
CORAL	• Coral reef	IWRM and ICZM both for environmental assessment and management.	Sustainable management of coastal ecosystems in particular, coral reef.	• ND	• ND

Name	Application domain	Regulatory framework of reference	Objective	Climate change impacts addressed	Climate change scenarios generating impacts
COSMO	• Coastal zones.	ICZM for environmental management.	To evaluate coastal management options considering anthropic (human) forcing and climate change impacts.	• Sea-level rise.	• Sea-level rise scenarios.
Coastal Simulator	• Coastal zones.	National legislation for environmental assessment and management.	Effects of climate change /management decisions on the future dynamics of the coast.	• Storm surge flooding. • Coastal erosion.	• Emission scenarios. • Sea-level rise scenarios.
CVAT	• Coastal zones.	National legislation for environmental assessment and management.	To assess hazards, vulnerability and risks related to climate change and support hazard mitigation options.	• Storm surge flooding. • Coastal erosion. • Cyclone. • Typhoon. • Extreme events	• Past observations.
DESYCO	• Coastal zones. • Coastal Lagoons	ICZM for environmental assessment and management.	To assess risks and impacts related to climate change and support the definition of adaptation measures.	• Sea-level rise. • Relative sea-level rise • Storm surge flooding. • Coastal erosion. • Water quality	• Emission scenarios. • Sea level rise scenarios.
DITTY	• Coastal Lagoons.	IWRM and WFD for environmental management.	To achieve sustainable and rational utilization of resources in the southern European lagoons by taking into account major anthropogenic impacts.	• ND	• ND
DIVA	• Coastal zones.	ICZM for environmental assessment and management.	To explore the effects of climate change impacts on coastal regions.	• Sea-level rise. • Coastal erosion. • Storm surge flooding.	• Emission scenarios. • Sea-level rise scenarios.
ELBE	• River basin. • Catchment.	WFD for environmental management.	To improve the general status of the river basin usage and provide sustainable protection measure within coast.	• Precipitation and temperature variation.	• Emission scenarios.

Name	Application domain	Regulatory framework of reference	Objective	Climate change impacts addressed	Climate change scenarios generating impacts
GVT	• Coastal zones.	National legislation for environmental assessment.	To describe the vulnerability of groundwater resources to pollution in a particular coastal region.	• Groundwater quality. • Saltwater intrusion.	• Sea-level rise scenarios.
IWRM	• Coastal zones. • River basin	IWRM for environmental assessment and management.	To explore potential risks on coastal resources due to climate and water management policies.	• Sea-level rise. • Coastal erosion.	• Sea-level rise scenarios. • Emission scenarios.
KRIM	• Coastal zones.	ICZM for environmental assessment.	To determine how coastal systems reacts to climate change in order to develop modern coastal management strategies.	• Sea-level rise. • Extreme events. • Coastal erosion.	• Sea-level rise scenarios. • Extreme events scenarios.
MODSIM	• River basin.	IWRM for environmental management.	To improve coordination and management of water resources in a typical river basin.	• ND	• ND
RegIS	• Coastal zones.	SMP and Habitats regulation (UK) for environmental assessment and management.	To evaluate the impacts of climate change, and adaptation options.	• Coastal and river flooding. • Sea level rise	• Emission scenarios • Socio-economic scenarios • Sea level rise scenarios
RAMCO	• River basin. • Coastal zones.	WFD and ICZM for environmental assessment and management.	For effective and sustainable management of coastal resources at the regional and local scales.	• ND	• ND
SimLUCIA	• Coastal zones.	National legislation for environmental assessment.	To assess the vulnerability of low lying areas in the coastal zones and island to sea-level rise due to climate change.	• Sea-level rise. • Coastal erosion. • Storm surge flooding.	• Sea-level rise scenarios.
SimCLIM	• Coastal zones.	ICZM for environmental assessment and management.	To explore present and potential risks related to climate change and natural hazards (e.g. erosion, flood).	• Sea-level rise. • Coastal flooding. • Coastal erosion.	• Sea-level rise scenarios.

Name	Application domain	Regulatory framework of reference	Objective	Climate change impacts addressed	Climate change scenarios generating impacts
STREAM	• River basin. • Estuaries.	IWRM and WFD for environmental management.	To integrate the impacts of climate change and land-use on water resources management.	• Water quality variation. • Saltwater intrusion.	• Emission scenarios.
TaiWAP	• River basin.	IWRM for environmental assessment.	To assess vulnerability of water supply systems to impacts of climate change and water demand.	• Water quality variations.	• Emission scenarios.
WADBOS	• River basin. • Coastal zones.	WFD and ICZM for environmental assessment and management.	To support the design and analysis of policy measures in order to achieve an integrated and sustainable management.	• ND	• ND

Table 3. List of the examined DSS according to the general technical criteria (ND: Not Defined).

According to the climate change impacts considered by the examined DSS, the review highlights that fifteen out of the 20 DSS applications regard the assessment of climate change impacts and related risks (CC-DSS). These DSS consider climate change impacts relative to sea level rise, coastal erosion, storm surge flooding and water quality. In particular, DESYCO also consider relative sea level rise in coastal regions where there are records of land subsidence, whereas KRIM and CVAT assess impacts related to extreme events and natural hazards (e.g. typhoon, cyclone, etc.) respectively. Moreover, GVT is specifically devoted to groundwater quality variations.

The relevant climate change related scenarios considered by the examined DSS refer to emission of greenhouse gases, temperature increase, sea level rise and occurrence of extreme events. In addition, CVAT used previous observations as baseline scenarios for the assessment of natural hazards; while RegIS considered scenarios related to coastal and river flooding along with socio-economic scenarios in order to estimate their potential feedback on climate change impacts. Although most of these CC-DSS applications used sea level rise scenarios, only DIVA used global sea level rise scenarios to estimate related impacts like coastal erosion and storm surge flooding. KRIM is the only DSS considering extreme events scenarios in its analysis to support the development of robust coastal management strategies.

2.2. Specific technical criteria

The criteria related to the specific technical aspects are reported in Table 4. As far as the functionalities are concerned (Table 4, first column), the ones implemented by DESYCO,

COSMO, SimCLIM, KRIM and RegIS include the identification and prioritisation of impacts, targets and areas at risk from climate change, sectorial evaluation of impacts or integrated assessment approach, and vulnerability evaluation and problem characterisation, in order to effectively differentiate and quantify impacts and risks at the regional scale. Moreover, they also support the definition and evaluation of management options through GIS-based spatial analysis. Other DSS, i.e. DIVA, SimCLIM and KRIM, implement scenarios import and generation, environmental status evaluation, impacts and vulnerability analysis and evaluation of adaptation strategies to adequately achieve a sustainable state of coastal resources and ecosystems.

Name	Functionalities	Analytical methodologies	Structural elements
CLIME	• Identification of pressure generated by climatic variables. • Environmental status evaluation. • Water quality evaluation related to climate change. • Socio-economic evaluation. • Spatial analysis (GIS).	• Scenarios construction and analysis. • Probabilistic Bayesian network. • Uncertainty analysis.	• Climatic, hydrological, chemical, geomorphological data. • Climate, ecological and hydrological models. • Web-based user interface
CORAL	• Evaluation of management strategies • Spatial analysis (GIS).	• Scenarios construction and analysis. • Cost-effectiveness analysis. • Ecosystem-based.	• Environmental, socio-economic, ecological, biological data. • Economic and ecological models. • Desktop user interface.
COSMO	• Problem characterization (e.g. water quality variation, coastal erosion etc.) • Impact evaluation of different development and protection plans. • Indicator production. • Spatial analysis (GIS).	• Scenarios construction and analysis. • MCDA. • Ecosystem-based	• Socio-economic, climatic, environmental, hydrological data. • Ecological, economic and hydrological models. • Desktop user friendly interface
Coastal Simulator	• Environmental status evaluation. • Management strategies identification and evaluation. • Indicator production. • Spatial analysis (GIS).	• Scenarios construction and analysis. • Uncertainty analysis. • Risk analysis. • Ecosystem-based.	• Climatic, socio-economic, environmental, hydrological, geomorphological data. • Ecological, morphological climatic and hydrological models. • Desktop user interface.
CVAT	• Environmental status evaluation. • Hazard identification. • Indicators production.	• Hazard analysis. • Critical facilities analysis. • Society analysis.	• Environmental and socio-economic data. • Hydrological model. • Desktop user friendly interface

Name	Functionalities	Analytical methodologies	Structural elements
	• Mitigation options identification and evaluation. • Spatial analysis (GIS).	• Economic analysis. • Environmental analysis. • Mitigation options analysis.	
DESYCO	• Prioritization of impacts, targets and areas at risk from climate change. • Impacts, vulnerability and risks identification. • Indicators production. • Adaptation options definition • Spatial analysis (GIS).	• Regional Risk Assessment methodology. • Scenarios construction and analysis. • MCDA. • Risk analysis.	• Climatic, biophysical, socio-economic, geomorphological, hydrological data. • Desktop automated user interface.
DITTY	• Management options evaluation • Indicator production. • Spatial analysis (GIS).	• Scenarios construction and analysis. • Uncertainty analysis. • MCDA. • Social cost and benefits analysis. • DPSIR.	• Morphological, social, hydrological, ecological data. • Hydrodynamics, biogeochemical, socio-economic models. • Desktop user interface.
DIVA	• Scenarios generation and analysis. • Environmental status evaluation. • Indicators production. • Adaptation options evaluation. • Spatial analysis (GIS).	• Scenarios construction and analysis. • Cost-benefit analysis. • Ecosystem-based.	• Climatic, socio-economic, geography, morphological data. • Economic, ecological, geomorphological, climate models. • Desktop graphical user interface.
ELBE	• Environmental status evaluation. • Protection measures identification. • End-user involvement. • Spatial analysis (GIS).	• Scenarios construction and analysis.	• Hydrological, ecological, socio-economic, morphological data. • Economic, • Hydrological, models. • Desktop complex user interface.
GVT	• Environmental status evaluation. • Indicators production • Spatial analysis (GIS). • Impact and vulnerability evaluation	• Risks analysis. • Fuzzy logic. • MCDA.	• Data (environmental, climatic, hydrological, socio-economic). Hydrological, socio-economic and DEM models. • Desktop user interface.
IWRM	• Environmental status evaluation. • Indicators production.	• Scenarios construction and analysis. • Risk analysis.	• Climatic, environmental, socio-economic, geomorphological data. • Hydrodynamic, climate, economic models.

Name	Functionalities	Analytical methodologies	Structural elements
	• Adaptation measures evaluation. • Information for non-technical users. • Spatial analysis (GIS).	• Cost-benefit analysis. • Socio-economic analysis.	• Desktop user interface.
KRIM	• Environmental status evaluation. • Adaptation measures evaluation. • Information for non-technical users. • Spatial analysis (GIS).	• Scenarios construction and analysis. • Impact and risk analysis. • Ecosystem-based.	• Climatic, socio-economic, ecological, environmental, hydrological data. • Economic, ecological, hydrodynamic, geomorphological models. • Desktop user interface.
MODSIM	• Environmental status evaluation. • Management measures evaluation. • Spatial analysis (GIS).	• Statistical analysis. • Analysis of policies.	• Administrative, hydrological, socio-economic, environmental data. • Socio-economic, hydrological models. • Web-based user interface.
RegIS	• Indicators production • Management measures evaluation. • Information for non-technical users. • sectoral evaluation • Spatial analysis (GIS).	• Scenarios construction and analysis. • Impact analysis. • DPSIR. • Integrated assessment.	• Climatic, socio-economic, geomorphological, hydrological data. • Climate and flood metal-models. • Desktop user interface.
RAMCO	• Environmental status evaluation. • Indicators generation. • Management measures evaluation. • Spatial analysis (GIS).	• Scenarios construction and analysis. • Cellular automata. • Ecosystem-based.	• Socio-economic, environmental, climatic data. • Biophysical, socio-economic and environmental models. • Web-based user interface.
SimLUCIA	• Indicators production. • Impact and vulnerability evaluation. • Management and land-use measures evaluation. • Spatial analysis (GIS).	• Cellular Automata. • Scenarios construction and analysis. • Socio-economic analysis. • Bayesian probabilistic networks. • Ecosystem-based.	• Climatic, environmental, socio-economic data. • Land use, social and economic, climate models. • Web-based user interface.
SimCLIM	• Environmental status evaluation. • Impact and vulnerability evaluation. • Adaptation strategies evaluation	• Scenario construction and analysis. • Statistical analysis. • Risk analysis.	• Climatic, hydrological, socio-economic data. • Climate, hydrological, economic models. • Desktop user interface.

Name	Functionalities	Analytical methodologies	Structural elements
	• Spatial analysis (GIS).	• Cost/benefit analysis. • Ecosystem-based.	
STREAM	• Environmental status evaluation. • Indicators production. • Management measures evaluation spatial analysis (GIS).	• Scenarios construction and analysis.	• Climatic, socio-economic, ecological, hydrological data. • Climate, hydrological models. • Web-based user interface.
TaiWAP	• Environmental status evaluation.- • Indicators production. • Spatial analysis (GIS).	• Scenarios construction and analysis. • Impact and vulnerability analysis.	• Climatic, socio-economic, hydrological data. • Climate, hydrological, water system dynamic models. • Desktop user interface.
WADBOS	• Management measures identification and evaluation. • Spatial analysis (GIS).	• Scenarios construction and analysis. • Sensitivity analysis. • MCDA.	• Socio-economic, hydrological, environmental, ecological data. • Socio-economic, ecological, landscape models. • Desktop user interface.

Table 4. List of the examined DSS according to the specific technical criteria.

In order to effectively support the assessment and management of groundwater resources, GVT and DESYCO estimate indicators in assessing impacts, vulnerability and risks to estimate groundwater quality and coastal environmental quality, respectively. Similarly, STREAM, ELBE, RAMCO and DITTY employ environmental status evaluation, protection measures identification, and spatial analysis to support the management aspects of coastal ecosystems. Moreover, CLIME and CORAL specifically support the assessment and management of lakes and coral reefs via the adoption of management strategies and the evaluation and identification of pressures from climatic variables.

In particular, five out of the 20 examined DSS (i.e. CVAT, GVT, Coastal Simulator, SimLUCIA and RegIS) consider hazards identification, impacts and vulnerability evaluation, mitigation/ management options identification and evaluation and sectoral evaluation to achieve a comprehensive and integrated analysis of coastal issues at the local or regional scale. Among all considered DSS, RegIS is the one most oriented to stakeholders.

The second column of table 4 shows the methodologies adopted by each DSS. Seventeen out of 20 examined DSS consider scenarios analysis to enable coastal managers, decision makers and stakeholders to anticipate and visualise coastal problems in the foreseeable future, and to better understand which future scenario is most suitable for consideration in the evaluation process. A useful methodology is represented by the Multi-Criteria Decision Analysis (MCDA) technique that is considered by five DSS (i.e. COSMO, DESYCO, DITTY, GVT and WADBOS) in order to compare, select and rank multiple alternatives that involve several attributes based on several different criteria. Moreover, DITTY and RegIS also consider the DPSIR approach as a causal framework to describe the interactions between the coastal system, society and ecosystems to carry out an integrated assessment with the aim to protect the coastal environment, guarantee its sustainable use, and conserve its biodiversity in accordance to the Convention on Biodiversity (2003). An ecosystemic assessment was developed nine DSS (i.e. CORAL, COSMO, Coastal simulator, DIVA, RegIS, KRIM, RAMCO, SimLUCIA, SimCLIM) to support the analysis of the studied region through the representation of relevant processes and their feedbacks. Furthermore KRIM, IWRM, COSMO, SimCLIM and Coastal Simulator employ the risk analysis approach for impacts and vulnerability evaluation and also for general environmental status evaluation. A more detailed approach to risk analysis, through the regional risk assessment methodology (RRA), was adopted by DESYCO, Coastal Simulator and RegIS with huge emphasis on the local or regional scales. Finally, CLIME and SimLUCIA consider the Bayesian probability network to highlight the causal relationship between ecosystems (e.g. lakes) and climate change effects.

With regard to the structure of examined DSS (Table 4, third column), most of them employ analytical models useful to highlight the basic features and natural processes of the examined territory, such as the landscape and ecological models used by the WADBOS, the environmental model employed by RAMCO, the geomorphological model used within KRIM and the flood meta-model which interface other models considered by the RegIS. Moreover, the majority of these DSS utilise numerical models necessary to simulate relevant circulation and geomorphological processes that may influence climate change and related risks. DSS like CLIME, DESYCO, CVAT and TaiWAP adopt models useful to represent specific climatic processes (e.g. hydrological cycle and fate of sediment). More importantly, ten (i.e. WADBOS, SimLUCIA, RAMCO, MODSIM, GVT, ELBE, DIVA, CORAL, DITTY AND SimCLIM) out of the twenty examined DSS consider relevant socioeconomic models outputs in their analysis to critically support the integrated assessment of coastal zones. Finally, the majority of these DSS consider integrated assessment models in order to emphasise the basic relationship among different categories of environmental processes such as physical, morphological, chemical, ecological and socio-economic – and to provide inclusive information about the environmental and socioeconomic processes.

As far as the software interfaces are concerned, very few of the examined DSS are applied through web-based interfaces, in spite of the fact that web-based facilities enhance easy access to information within a large network of users. Furthermore, all the reviewed DSS consider GIS tools as basic media to express their results or outputs in order to provide fast and intuitive results representation to non-experts (i.e. decision makers and stakeholders) and empower them for robust decisions. In addition to maps, the output produced by each DSS are also graphs, charts, and statistical tables.

2.3. Applicability criteria

Table 5 shows the implementation of the criteria concerning applicability to the examined DSS. Applicability include three aspects: scale/study areas, flexibility and status/availability

(Table 2). The spatial scales considered were five: global, supranational, national, regional, and local, in order of decreasing size. The study areas are those reported in the literature cited in Table 1. The flexibility derives from the capability of a given DSS to include new modules and models in its structure, thus new input parameters, and the suitability to be used for regionally different case studies. In order to visualize the estimation of the overall flexibility of a system, highly flexible/flexible/moderately-to-no flexible were indicated as +++/++/+. Status and availability refer to different extent of development (e.g. research prototype, commercial software) and public accessibility/last updated version, respectively.

Name	Scale and area of application	Flexibility	Status and availability last updated version (year)
CLIME	• Supra-National, National, Local. • (Northern, western and central part of Europe).	+++ Flexible in structural modification and study area.	Available to the public. Demo. 2010.
CORAL	• Regional, Local. • (Coastal areas of Curacao; Jamaica and Maldives).	+++ Flexible in study area.	Not available to the public. Prototype. 1995.
COSMO	• National, Local. • (Coast of Netherland).	++ Flexible in study area.	Commercial application. 1998.
Coastal Simulator	• National, Regional, Local. • (Coast of Norfolk in East Anglia, UK).	+	Available only to the Tyndall Research Centre. Prototype. 2009.
CVAT	• Regional, Local. • (New Hanover County, North Carolina).	++ Flexible in study area.	Available to public. Prototype. 2002.
DESYCO	• Regional, Local. • (North Adriatic Sea).	++ Flexible in study area.	Not available to the public. Prototype. 2010.
DITTY	• Supranational, National, Regional. • (Ria Formosa-Portugal; Mar Menor-Spain; Etang de Thau-France; Sacca di Goro-Italy, Gera-Greece).	+++ Flexible in study area.	Not available to the public. 2006.
DIVA	• Global, National.	+++ Flexible in study area.	Available to the public. 2009.
ELBE	• Local. • (Elbe river basin Germany).	+	Available to the public. 2003.
GVT	• Regional, Local. • (Eastern Macedonia and Northern Greece).	+	Not available to the public. 2006.
IWRM	• Regional, Local. • (Halti-Beel, Bangladesh)	++ Flexible in study area.	Not available to the public. Prototype. 2009.

Name	Scale and area of application	Flexibility	Status and availability last updated version (year)
KRIM	• Regional. • (German North sea Coast, Jade-Weser area in Germany).	+	Not available to the public. Prototype. 2003.
MODSIM	• National, Regional. • (San Diego Water County, Geum river basin- Korea).	++ Flexible in study area.	Available to the public online. 2006.
RegIS	• Regional, Local. • (North-West, East Anglia).	++ Flexible in study area.	Available online to stakeholders. Prototype. 2008.
RAMCO	• Regional, Local. • (South-West Sulawesi coastal zone).	++ Flexible in the used dataset and concepts.	Not available to the public. Prototype. 1999.
SimLUCIA	• Local • (St Lucia Island, West India)	+	Available online to the public. Demo. 1996.
SimCLIM	• National, Regional, Local. • (Rarotonga Island, Southeast Queensland).	++ Flexible in structural modification and study area.	Available to the public. Demo. 2009.
STREAM	• Regional, Local. • (Ganges/Brahmaputra river basin, Rhine river basin, Yangtze river basin and Amudarya river basin).	+++ Flexible in structural modification and study area.	Available online to the public. Demo. 1999.
TaiWAP	• Regional, Local. • (Touchien river basin).	+	Available to National Taiwan University. Prototype. 2008.
WADBOS	• Regional, Local. • (Dutch Wadden sea).	+	Available online to the public. Demo. 2002.

Table 5. List of the examined DSS according to the applicability criteria. (+++, highly flexible; ++, flexible; +: modertly to no-flexible).

As far as the scale of application is concerned, all the examined DSS, except DIVA, have been applied only at the local and regional scales because they were developed for a specific geographical context. Moreover, five out of the 20 examined DSS (i.e. CLIME, CORAL, DITTY, DIVA and STREAM) considered global, supranational, national, regional and local scales during their implementation.

Five of the reported DSS are highly flexible systems because they are used to address several impacts related to different case studies. Although DIVA can be applied to any coastal area around the world, it is sometimes not considered a highly flexible tool in terms of structural modification due to its inability to change its default integrated dataset. Finally, ELBE and WADBOS are identified as moderately-to-no flexible systems because their structure and functionalities were based on the specific needs of particular river basins.

The applicability of DSS reflects their ability to be implemented in several contexts (i.e. case study areas and structural modification), for example to include new models and functionalities ensuring common approaches to decision making and the production of comparable results [42].

Finally, concerning the availability and the status of the development, Table 5 shows that nine DSS are available to the public, three are available with a restricted access (i.e. only to stakeholders or to the developers), one is a commercial software (i.e. COSMO) and seven are not available to the public. Sometimes the restriction of the access is due to the fact that results require special skill for their interpretation, so the public can use them only with the support of the developer team. Among examined DSS, only 11 were developed/updated during the last 5 years, and 4 over the previous five years (for a total of 15 during the last 10 years) with the remaining five DSS showing the last version dating back to the '90s.

The overall content of Table 5, together with the main features of each DSS reported in Tables 3 and 4, allow the reader to undertake a screening evaluation of available DSS in relation to the specific impacts from climate change to be addressed.

3. Conclusions

This work should be regarded as a preliminary attempt to describe and evaluate the main features of available DSS for the assessment and management of climate change impacts on coastal area and related inland watersheds. A further and comprehensive evaluation should be based on comparative application in selected and relevant case studies, in order to evaluate the DSS technical performance, especially in relation to datasets availability, that often represents the real limiting factor. Moreover, sensitivity and uncertainty analyses will provide further evidence of the reliability of the investigated DSS.

This review highlighted the relevance of developing climate change impact assessment and management at the regional scale (i.e. subnational and local scale), according to the requirements of policy and regulatory frameworks and to the methodological and technical features of the described DSS. In fact, most of the available DSS show a regional to local applicability with a moderate to high flexibility. Indeed climate change impacts are very dependent on regional geographical features, climate and socio-economic conditions and regionally-specific information can assist coastal communities in planning adaptation measures to the effects of climate change.

Despite the current situation that shows available DSS mainly focusing on the analysis of specific individual climate change impacts and affected sectors (15 out of the 20 examined DSS), the further developments should aim at the adoption of ecosystem approaches considering the complex dynamics and interactions between coastal systems and other systems closely related to them (e.g. coastal aquifers, surface waters, river basins, estuaries), and at the adoption of multi-risk approaches in order to consider the interaction among different climate change impacts that affect the considered region.

Finally, it is important to remark the need to involve the end users and relevant stakeholders since the initial steps of the development process of these tools, in order to satisfy their actual requirements, especially in the perspective of providing useful climate services, and to avoid the quite often and frustrating situation where time and resource demanding DSS are not used beyond scientific testing exercises.

Author details

F. Iyalomhe and V. Gallina
University Ca' Foscari Venice, Department of Environmental Sciences, Informatics and Statistics, Venezia, Italy

J. Rizzi, A. Critto and A. Marcomini *
University Ca' Foscari Venice, Department of Environmental Sciences, Informatics and Statistics, Venezia, Italy
Euro Mediterranean Centre for Climate Change, CMCC, Lecce, Italy

S. Torresan
Euro Mediterranean Centre for Climate Change, CMCC, Lecce, Italy

Acknowledgement

The authors gratefully acknowledge the Euro-Mediterranean Centre for Climate Change (CMCC; Lecce, Italy), GEMINA project and the European Community's Seventh Framework Programme, Marie Curie action, Project EPSEI (under the grant agreement no.269327), for financial support.

4. References

[1] Janssen, R. (1992). Multiobjective decision support for environmental management. Kluwer Academic. (Dordrecht) Boston.
[2] Torresan, S., Zabeo, A., Rizzi, J., Critto, A., Pizzol, L., Giove, S. and Marcomini, A. (2010). Risks assessment and decision support tools for the integrated evaluation of climate change impacts on coastal zones. International Congress on Environmental Modelling and Software Modelling for Environmental Sake, Fifth Biennial Meeting, Ottawa, Canada.
[3] Matthies, M., Giupponi, C. and Ostendorf, B. (2007). Environmental Decision Support Systems; Current Issues, Methods and tools Environ model soft 22(2):123-28.

---

* Corresponding Author

[4] Uran, O. & Janssen, R. (2003). Why spatial decision support systems not used? Some experiences from the Netherlands. Comput environ urban, 27(5):511–526.

[5] Poch, M., Comas, J., Rodriguez-Roda, I., Sachez-Marrie, M. and Cortes, U. (2004). Designing and building real environmental decision support systems. Environ modell soft 19:857-873.

[6] Fabbri, K.P., (1998). A methodology for supporting decision making in integrated coastal zone management. Ocean cost manage, 39(1-2), pp.51–62.

[7] Environmental Systems Research Institute – ESRI, (1992). Arc Version 6.1.2, Redlands, California, USA.

[8] Nobre, A.M. & Ferreira, J.G. (2009). Integration of ecosystem-based tools to support coastal zone management. J costal res, 1676-1670.

[9] Matthies, M., Giupponi, C., Ostendorf, B. (2007). Environmental Decision Support Systems; Current Issues, Methods and tools. Environ modell soft 22(2):123-28.

[10] IPCC, Climate Change (2007): Impacts, Adaptation and Vulnerability, Summary for Policymakers, Contribution of Working Group II to the Fourth Assessment Report of the Intergovernmental Panel on Climate Change, Geneva, 2007.

[11] Nakicenovic, N., Alcamo, J., Davis, G., de Varies, B., Fenhann, J., Gaffin, S., Gregory, K., Grubler, A., Jung, T.Y. and Kram, T. (2000). Special report on emissions scenarios: a special report of Working Group III of the Intergovernmental Panel on Climate Change, Pacific Northwest National Laboratory, Richland, WA (US), Environmental Molecular Sciences Laboratory (US).

[12] Nicholls, R.J., Cazenave, A., (2010). Sea-level rise and its impact on coastal zones. Science, 328(5985):1517-1520.

[13] Jiang L., Hardee K.., (2010). How do Recent Population Trends Matter to Climate Change? Popul re policy rev, 30(2):287-312

[14] EC, COM(2007) 354, 29.06.2007. Green Paper – Adapting to climate change in Europe – options for EU action, Brussels.

[15] EC, COM(2009) 147, 01.04.2009. White Paper – Adapting to climate change: Towards a European framework for action, Brussels.

[16] EC (2000) Directive 2000/60/EC "Directive 2000/60/EC of the European Parliament and of the Council Establishing a Framework for the Community Action in the Field of Water Policy" Official Journal (OJ L 327) on 22 December 2000.

[17] EC (2002) Recommendation the European Parliament and of the Council of 30 May 2002 Concerning the Implementation of Integrated Coastal Zone Management in Europe, 2002/413/EC.

[18] EC (2003) Common Implementation Strategy for the Water Framework Directive (2000/60/CE). Guidance Document n. 11. Planning process. Office for Official Publications of the European Communities, Luxembourg.

[19] Thumerer, T., A. P. Jones, and D. Brown. (2000) "A GIS Based Coastal Management System for Climate Change Associated Flood Risk Assessment on the East Coast of England." Int j geogr inf sci 14(3):265–281.

[20] Flax, L.K., Jackson, R.W. & Stein, D.N. (2002). Community vulnerability assessment tool methodology. Natural Hazards Review, 3:163.

[21] Zaman, A. M., S. M. M. Rahman, and M. R. Khan. "Development of a DSS for Integrated Water Resources Management in Bangladesh" (2009), 18th World IMACS / MODSIM Congress, Cairns, Australia 13-17 July.

[22] Jolma, A., Kokkonen, T., Koivusalo, H., Laine, H., Tiits, K. (2010). Developing a Decision Support System for assessing the impact of climate change on lakes. The Impact of Climate Change on European Lakes, 411–435.

[23] Westmacott, S. (2001). Developing Decision Support Systems for Integrated Coastal Management in the Tropics: Is the ICM Decision-Making Environment Too Complex for the Development of a Useable and Useful DSS? J environ manage, 62(1): 55-74.

[24] Feenstra, J.F., Programme, U.N.E. & Milieuvraagst, V. (1998). Handbook on methods for climate change impact assessment and adaptation strategies, United Nations Environment Programme.

[25] Nicholls, R., Mokrech, M. & Hanson, S. (2009). An integrated coastal simulator for assessing the impacts of climate change. In IOP Conference Series: Earth and Environmental Science. 6 092020.

[26] Agnetis A., Basosi R., Caballero K., Casini M., Chesi G., Ciaschetti G., Detti P., Federici M., Focardi S., Franchi E., Garulli A., Mocenni C., Paoletti S., Pranzo M., Tiribocchi A., Torsello L., Vercelli A., Verdesca D., VicinoA. (2006). Development of a Decision Support System for the management of Southern European lagoons. Centre for Complex Systems Studies University of Siena, Siena, Italy. Technical Report TR2006-1.

[27] Hinkel, J. & Klein, R.J.T. (2009). Integrating knowledge to assess coastal vulnerability to sea-level rise: The development of the DIVA tool. Global environ chang, 19(3):384–395.

[28] BfG, (2003). Plot phase for the design and development of a decision support system (DSS) for river basin management with the example of the Elbe (Interim Report 2002-2003 (in German). Bundesanstalt fur Gewässer-kunde (German Federal Institute of Hydrology), Koblenz. Germany.

[29] Gemitzi, A. et al., 2006. Assessment of groundwater vulnerability to pollution: a combination of GIS, fuzzy logic and decision making techniques. Environ geol, 49(5):653–673.

[30] Schirmer, M., B. Schuchardt, B. Hahn, S. Bakkenist & D. Kraft (2003): KRIM: Climate change risk construct and coastal defence. DEKLM German climate research programme. Proceedings, 269-273

[31] Salewicz, K. A. and M, Nakayama. (2004): Development of a web-based decision support system (DSS) for managing large international rivers. Global environ chang, 14:25-37

[32] Labadie, J.W. (2006). MODSIM: decision support system for integrated river basin management. Summit on Environmental Modeling and Software, the International Environmental Modeling and Software Society, Burlington, VT USA.

[33] Holman, I.P., Rounsevell, M.D.A., Berry, P.M., Nicholls, R.J. (2008). Development and application of participatory integrated assessment software to support local/regional impacts and adaptation assessment. Journal of Earth and Environmental Science. DOI: 10. 1007/s 10584-008-9452-7.

[34] de Kok, J.L., Engelen, G., White, R. and Wind, H. G. (2001). Modelling land-use change in a decision-support system for coastal-zone management. Environ model assess, 6(2):123–132.

[35] Uljee, I, Engelen, G. And White, R. (1996) Rapid Assessment Module for Coastal Zone Management (RAMCO). Demo Guide Version 1.0, Work document CZM-C 96.08, RIKS (Research Institute for Knowledge Systems).

[36] Engelen, G., White, R., Uljee, I. and Wargnies, S. (1995). Vulnerability Assessment of Low-lying Coastal Areas and Small Islands to Climate change and sea-level rise. Report to the United Nations Environment Programme, Caribbean Regional Co-ordinating Unit, Kingston, Jamaica (905000/9379). Maastrict: Research Institute Knowledge Systems (RIKS).

[37] Warrick, R. A. (2009): Using SimCLIM for modelling the impacts of climate extremes in a changing climate: a preliminary case study of household water harvesting in Southeast Queensland. 18th World IMACS/MODSIM Congress, Cairns, Australia 13-17 July 2009.

[38] Aerts, J., Kriek, M. & Schepel, M. (1999). STREAM (Spatial tools for river basins and environment and analysis of management options): set up and requirements'. Phys chem earth, Part B: Hydrology, Oceans and Atmosphere, 24(6):591–595.

[39] Liu, M.T., Tung, C.P., Ke, K.Y., Chuang, L.H, Lin, C.Y. (2009): Application and development of decision support systems for assessing water shortage and allocation with climate change. Paddy Water Environ, 7:301-311. DOI 10.1007/s/0333-009-0177-7.

[40] van Buuren, J.T., Engelen, G. & van de Ven, K. (2002). The DSS WadBOS and EU policies implementation.

[41] Engelen G. (2000). The WADBOS Policy Support System: Information Technology to Bridge Knowledge and Choice. Technical paper prepared for the National Institute for Coastal and Marine Management/RIKZ. The Hague, the Netherlands.

[42] Agostini, P., Suter, G.W.II, Gottardo, S., Giubilato, E. (2009). Indicators and endpoints for risk-based decision process with decision support systems. In Marcomini, A., Suter, G.W.II, Critto, A. (Eds). Decision Support Systems for Risk Based Management of Contaminated Sites. Springer Verlag, New York.

Permissions

The contributors of this book come from diverse backgrounds, making this book a truly international effort. This book will bring forth new frontiers with its revolutionizing research information and detailed analysis of the nascent developments around the world.

We would like to thank Prof. (Dr.) Bharat Raj Singh, for lending his expertise to make the book truly unique. He has played a crucial role in the development of this book. Without his invaluable contribution this book wouldn't have been possible. He has made vital efforts to compile up to date information on the varied aspects of this subject to make this book a valuable addition to the collection of many professionals and students.

This book was conceptualized with the vision of imparting up-to-date information and advanced data in this field. To ensure the same, a matchless editorial board was set up. Every individual on the board went through rigorous rounds of assessment to prove their worth. After which they invested a large part of their time researching and compiling the most relevant data for our readers. Conferences and sessions were held from time to time between the editorial board and the contributing authors to present the data in the most comprehensible form. The editorial team has worked tirelessly to provide valuable and valid information to help people across the globe.

Every chapter published in this book has been scrutinized by our experts. Their significance has been extensively debated. The topics covered herein carry significant findings which will fuel the growth of the discipline. They may even be implemented as practical applications or may be referred to as a beginning point for another development. Chapters in this book were first published by InTech; hereby published with permission under the Creative Commons Attribution License or equivalent.

The editorial board has been involved in producing this book since its inception. They have spent rigorous hours researching and exploring the diverse topics which have resulted in the successful publishing of this book. They have passed on their knowledge of decades through this book. To expedite this challenging task, the publisher supported the team at every step. A small team of assistant editors was also appointed to further simplify the editing procedure and attain best results for the readers.

Our editorial team has been hand-picked from every corner of the world. Their multi-ethnicity adds dynamic inputs to the discussions which result in innovative outcomes. These outcomes are then further discussed with the researchers and contributors who give their valuable feedback and opinion regarding the same. The feedback is then collaborated with the researches and they are edited in a comprehensive manner to aid the understanding of the subject.

Apart from the editorial board, the designing team has also invested a significant amount of their time in understanding the subject and creating the most relevant covers. They scrutinized every image to scout for the most suitable representation of the subject and create an appropriate cover for the book.

The publishing team has been involved in this book since its early stages. They were actively engaged in every process, be it collecting the data, connecting with the contributors or procuring relevant information. The team has been an ardent support to the editorial, designing and production team. Their endless efforts to recruit the best for this project, has resulted in the accomplishment of this book. They are a veteran in the field of academics and their pool of knowledge is as vast as their experience in printing. Their expertise and guidance has proved useful at every step. Their uncompromising quality standards have made this book an exceptional effort. Their encouragement from time to time has been an inspiration for everyone.

The publisher and the editorial board hope that this book will prove to be a valuable piece of knowledge for researchers, students, practitioners and scholars across the globe.

List of Contributors

Bharat Raj Singh
School of Management Sciences, Technical Campus, Lucknow, Uttar Pradesh, India

Onkar Singh
Harcourt Butler Technological Institute, Kanpur, Uttar Pradesh, India

Xinhua Liu
National Meteorological Center (NMC), Beijing, China
Key Laboratory for Semi-Arid Climate Change of the Ministry of Education, Lanzhou University, Lanzhou, China

Juddy. N. Okpara and Aondover. A.Tarhule
Department of Geography and Environmental Sustainability, University of Oklahoma, Norman, USA

Muthiah Perumal
Department of Hydrology, Indian Institute of Technology, Roorkee, India

François Girard
Direction de la Recherche Forestière, Ministère des Ressources Naturelles et de la Faune, Québec, Canada

Olivier Taugourdeau
UMR AMAP – Université Montpellier 2, Montpellier, France

Yves Caraglio and Sylvie-Annabel Sabatier
UMR AMAP - CIRAD, Montpellier, France

Maxime Cailleret
INRA - URFM, Avignon, France

Samira Ouarmim
Université du Québec en Abitibi-Témiscamingue, Rouyn-Norenda, Canada

Ali Thabeet
Aleppo University Faculty of Agronomy, Aleppo, Syria

Béla Nováky
Department of Hydraulic and Water Resources Engineering, Budapest University of Technology and Economics, Hungary

Gábor Bálint
VITUKI Environmental Protection and Water Management Research Institute, Budapest, Hungary

Jesús Efren Ospina-Noreña, Carlos Gay García, Ana Elisa Peña del Valle and Matt Hare
Programa de Investigación en Cambio Climático (PINCC) [Climate Change Research Program] of the Universidad Nacional Autónoma de México (UNAM) [National Autonomous University of Mexico]

Anthony A. Duah
CSIR Water Research Institute, Accra, Ghana

Yongxin Xu
Department of Earth Science, University of the Western Cape, Bellville, South Africa

Renhe Zhang, Bingyi Wu, Jinping Han and Zhiyan Zuo
Chinese Academy of Meteorological Sciences, Beijing, China

E. González-Sosa and N.M. Ramos-Salinas
Hydraulics Laboratory, Engineering Faculty, Universidad Autónoma de Querétaro, Ciudad Universitaria, Cerro de las Campanas, Querétaro State, México

C.A. Mastachi-Loza and R. Becerril-Piña
Centro Interamericano de Recursos del Agua (CIRA), Engineering Faculty, Universidad Autónoma del Estado de México. Atlacomulco-Toluca Street, México State, México

Mohammad Elnesr and Abdurrahman Alazba
King Saud University, Alamoudi Water Chair, Saudi Arabia

F. Iyalomhe and V. Gallina
University Ca' Foscari Venice, Department of Environmental Sciences, Informatics and Statistics, Venezia, Italy

J. Rizzi, A. Critto and A. Marcomini
University Ca' Foscari Venice, Department of Environmental Sciences, Informatics and Statistics, Venezia, Italy
Euro Mediterranean Centre for Climate Change, CMCC, Lecce, Italy

S. Torresan
Euro Mediterranean Centre for Climate Change, CMCC, Lecce, Italy